Advanced Structured Materials

Volume 45

More information about this series at http://www.springer.com/series/8611

Holm Altenbach · Gennadi I. Mikhasev
Editors

Shell and Membrane Theories in Mechanics and Biology

From Macro- to Nanoscale Structures

Springer

Editors
Holm Altenbach
Lehrstuhl für Technische Mechanik
Otto-von-Guericke-Universität Magdeburg
Magdeburg
Germany

Gennadi I. Mikhasev
Department of Bio- and Nanomechanics
Belarusian State University
Minsk
Belarus

ISSN 1869-8433 ISSN 1869-8441 (electronic)
ISBN 978-3-319-02534-6 ISBN 978-3-319-02535-3 (eBook)
DOI 10.1007/978-3-319-02535-3

Library of Congress Control Number: 2014949368

Springer Cham Heidelberg New York Dordrecht London

Printed on acid-free paper

Springer is part of Springer Science+Business Media (www.springer.com)

Preface

The first international conference *Shell and Membrane Theories in Mechanics and Biology: From Macro- to Nanoscale Structures* (SMT in MB—2013) has taken place at the Belarusian State University (Minsk, Belarus) from 16 to 20 September 2013. This book is a collection of papers presented on the conference or close to the topics of the conference. It contains 17 revised and extended research articles written by experienced researchers participating in the conference. The book will offer the state-of-the-art in mechanical, materials, and civil engineering, ranging from composite materials up to characterization of nanostructures. Examples taken from novel trends in biomechanics. Well-known international experts present their research on materials modeling and evaluation up to recent printing and visualization for advanced analyzes and evaluation. The conference was organized by Belarusian State University (BSU, Belarus) together with the Joint Institute of Mechanical Engineering of the National Academy of Sciences of Belarus under support of the Belarusian Republican Foundation for Fundamental Research.

The main purpose of the conference was to bring together mechanical engineers, mathematicians, physicists, and other specialists carrying out researches in such areas like theories of shells, plates, and membranes in order to discuss important results and new ideas support and promote interdisciplinary research activity in these broad but allied fields. The new trends in application in mechanical, civil, and aerospace engineering, as well as in new branches like medicine and biology demand the improvements of the theoretical foundations of these theories. The forum was addressed to the broad spectrum of scientists from different fields of mechanics, biomechanics, mathematics, physics, medicine, biology, etc. Although all papers related to the shell-like structures, membranes and films or to adjacent fields in mechanics, physics, etc., were welcomed, the focus was paid on presentations which relate to new theories, approaches, methods, and applications describing nonclassical effects (e.g., the small-size effects at the micro- and nanoscale level, etc.). Applied studies in industry, medicine, biology, and nanotechnology were also presented.

The editors wish to thank all the authors for their participation and cooperation, which made this volume possible. Finally, we would like to thank the team of Springer-Verlag, especially Dr. Christoph Baumann, for their excellent cooperation during the preparation of this volume

Magdeburg, June 2014 Holm Altenbach
Minsk Gennadi I. Mikhasev

Contents

Contributors

Lenser A. Aghalovyan Institute of Mechanics of NAS of Armenia, Yerevan, Armenia

Holm Altenbach Lehrstuhl für Technische Mechanik, Institut für Mechanik, Fakultät für Maschinenbau, Otto-von-Guericke-Universität Magdeburg, Magdeburg, Germany

Jan Awrejcewicz Department of Automation, Biomechanics and Mechatronics, Lodz University of Technology, Łódź, Poland

Svetlana M. Bauer Department of Theoretical and Applied Mechanics, St. Petersburg State University, St. Petersburg, Russia

Igor Berinskii Department of Theoretical and Applied Mechanics, St. Petersburg State Polytechnical University, St. Petersburg, Russia; Laboratory for Discrete Models in Mechanics, Institute for Problems in Mechanical Engineering, St. Petersburg, Russia

Uroš Bohinc Slovenian National Building and Civil Engineering Institute, Ljubljana, Slovenia

Marina G. Botogova Belarusian State University, Minsk, Belarus

Boštjan Brank Faculty of Civil and Geodetic Engineering, University of Ljubljana, Ljubljana, Slovenia

Andris Chate Institute of Materials and Structures, Riga Technical University, Riga, Latvia

Sergei A. Chizhik Heat and Mass Transfer Institute of National Academy of Sciences of Belarus, Minsk, Belarus

Elizaveta S. Drozd Heat and Mass Transfer Institute of National Academy of Sciences of Belarus, Minsk, Belarus

Victor A. Eremeyev Lehrstuhl für Technische Mechanik, Institut für Mechanik, Fakultät für Maschinenbau, Otto-von-Guericke-Universität Magdeburg, Magdeburg, Germany; Southern Scientific Center of Russian Academy of Science and Southern Federal University, Rostov-on-Don, Russia

Antonino Favata Department of Civil, Environmental and Mechanical Engineering, University of Trento, Trento, Italy

Sergei B. Filippov Department of Theoretical and Applied Mechanics, St. Petersburg State University, Peterhof, St. Petersburg, Russia

Ralf Frotscher Laboratory of Engineering Mechanics and FEM, Aachen University of Applied Sciences, Aachen, Germany; Biomechanics Lab, Institute of Bioengineering, Aachen University of Applied Sciences, Jülich, Germany

Sergej Gluhih Institute of Materials and Structures, Riga Technical University, Riga, Latvia

Matthias Goßmann Laboratory of Medical and Molecular Biology, Institute of Bioengineering, Aachen University of Applied Sciences, Jülich, Germany

Adnan Ibrahimbegović Ecole Normale Supérieure de Cachan, LMT, Cachan, France

Andrejs Kovalovs Institute of Materials and Structures, Riga Technical University, Riga, Latvia

Anton Krivtsov Department of Theoretical and Applied Mechanics, St. Petersburg State Polytechnical University, St. Petersburg, Russia; Laboratory for Discrete Models in Mechanics, Institute for Problems in Mechanical Engineering, St. Petersburg, Russia

Vadim A. Krysko Department of Mathematics and Modeling, Saratov State Technical University, Saratov, Russia

Gennady M. Kulikov Tambov State Technical University, Tambov, Russia

Gennadi I. Mikhasev Department of Bio- and Nanomechanics, Belarusian State University, Minsk, Belarus

Maria E. Mychko Heat and Mass Transfer Institute of National Academy of Sciences of Belarus, Minsk, Belarus

Svetlana V. Plotnikova Tambov State Technical University, Tambov, Russia

Paolo Podio-Guidugli Accademia Nazionale dei Lincei and Department of Mathematics, University of Rome Tor Vergata, Rome, Italy

Hans-Jürgen Raatschen Laboratory of Engineering Mechanics and FEM, Aachen University of Applied Sciences, Aachen, Germany

Yury A. Rossikhin Research Center of Wave Dynamics, Voronezh State University of Architecture and Civil Engineering, Voronezh, Russia

Denis N. Sheydakov South Scientific Center of Russian Academy of Sciences, Rostov-on-Don, Russia

Marina V. Shitikova Research Center of Wave Dynamics, Voronezh State University of Architecture and Civil Engineering, Voronezh, Russia

Irina Slavashevich Department of Bio- and Nanomechanics, Belarusian State University, Minsk, Belarus

Manfred Staat Biomechanics Lab, Institute of Bioengineering, Aachen University of Applied Sciences, Jülich, Germany

Ayşegül Temiz-Artmann Laboratory of Medical and Molecular Biology, Institute of Bioengineering, Aachen University of Applied Sciences, Jülich, Germany

Peter E. Tovstik Department of Theoretical and Applied Mechanics, St. Petersburg State University, Peterhof, St. Petersburg, Russia

Eva B. Voronkova Department of Theoretical and Applied Mechanics, St. Petersburg State University, St. Petersburg, Russia

Kirill Yurkevich Department of Bio- and Nanomechanics, Belarusian State University, Minsk, Belarus

On Some Classes of 3D Boundary-Value Problems of Statics and Dynamics of Plates and Shells

Lenser A. Aghalovyan

Abstract Classes of the three-dimensional (3D) boundary-value problems for plates and shells, which can be successfully solved by the asymptotic method, are considered. The first, second, and mixed boundary-value problems of the elasticity theory, as well as the nonclassical boundary-value problems for determination of the stress-strain state of the Earth lithospheric plates are studied. The solutions of the 3D dynamic problems for layered plates are applied to diminution of impact of negative seismic waves on buildings and constructions. Connected and disconnected problems of thermoelasticity, the 3D dynamic problems of electroelasticity for beforehand polarized piezoceramic plates and shells are also solved.

1 Introduction

Thin-walled deformable solids such as beams, plates, shells, etc. are characterized by smallness of one geometrical dimensions (thickness) with respect to the characteristic one (length, width, radius of curvature). This property of similar solids permits us to introduce the small geometrical parameter $\varepsilon = h/\ell$, where h is the half-thickness, and ℓ is the characteristic tangential dimension. Then, when passing to dimensionless coordinates and components of the displacement vector, the governing equations turn out to be singularly perturbed by the small parameter ε. The mathematical theory of equations and systems like these was developed recently and, as a rule, such equations, in which the small parameter is the coefficient of the principal operator, were considered. In problems for thin elastic bodies the small parameter being the coefficient of the principal part of the differential operator results in specific difficulties. For instance, the number of boundary functions becomes infinite but countable. To solve similar equations and systems the asymptotic method is established as an effective tool. [8, 9, 11, 14] published the first papers where the solutions of the plate and shell theory problems were constructed by applying the asymptotic method.

L.A. Aghalovyan (✉)
Institute of Mechanics of NAS of Armenia, Marshall Baghramian Ave. 24B, 0019 Yerevan, Armenia
e-mail: aghal@mechins.sci.am

H. Altenbach and G.I. Mikhasev (eds.), *Shell and Membrane Theories in Mechanics and Biology*, Advanced Structured Materials 45, DOI 10.1007/978-3-319-02535-3_1

In this case the solution (I) is the combination of the solution of the inner problem (I^{int}) and the boundary layer solution (I_{b})

$$I = I^{\text{int}} + I_{\text{b}}. \quad (1)$$

The solution of the inner problem is sought in the form

$$I^{\text{int}} = \varepsilon^{q_I + s} I^{(s)}, \quad s = \overline{0, N}. \quad (2)$$

Here $s = \overline{0, N}$ means summing up by integral values of repeating index s from zero to the number of approximations N, and q_I characterizes the insensitivity of the corresponding required dimension. The parameter q_I may be established after substituting Eq. (2) into the dimensionless governing equations and equating coefficients at the same power of the small parameter ε. One can get the noncontradictory system for determination of the decomposition coefficients $I^{(s)}$. The values q_I strictly depend on the type of the boundary conditions given on the facial surfaces of a plate or a shell. When using the asymptotic method, finding the noncontradictory values q_I is a non-trivial task for any physical problem. Some authors consider the process of seeking values q_I as work of art [7]. Nowadays, the asymptotic theories for both isotropic and anisotropic plates [14] and shells [10] have been developed [3]. Let us consider some classes of spatial static and dynamic problems successfully solved by the asymptotic method.

2 Asymptotic Solution of the First 3D Boundary-Value Problem

Let us consider the problem on the determination of the stress-strain states of thermoelastic anisotropic plates $D = \{(x, y, z) : 0 \le x \le a, 0 \le y \le b, |z| \le h, \ell = \min(a, b), h \ll \ell\}$ with anisotropy of general type (21 constants of elasticity). Passing to the dimensionless coordinates $\xi = x/\ell$, $\eta = y/\ell$, $\varsigma = z/h$ and displacements $U = u/\ell$, $V = v/\ell$, $W = w/\ell$, the equilibrium and constitutive equations of thermoelasticity (Duhamel-Neumann's equations) take the following form

$$\frac{\partial \sigma_{jx}}{\partial \xi} + \frac{\partial \sigma_{jy}}{\partial \eta} + \varepsilon^{-1} \frac{\partial \sigma_{jz}}{\partial \varsigma} + \ell F_j(\xi, \eta, \varsigma) = 0, \quad j = x, y, z,$$

$$\frac{\partial U}{\partial \xi} = a_{i1}\sigma_{xx} + a_{i2}\sigma_{yy} + a_{i3}\sigma_{zz} + a_{i4}\sigma_{yz} + a_{i5}\sigma_{xz} + a_{i6}\sigma_{xy} + \alpha_{ii}\theta,$$

$$(U, V; \xi, \eta; i = 1, 2),$$

$$\varepsilon^{-1} \frac{\partial W}{\partial \varsigma} = a_{13}\sigma_{xx} + a_{23}\sigma_{yy} + a_{33}\sigma_{zz} + a_{34}\sigma_{yz} + a_{35}\sigma_{xz} + a_{36}\sigma_{xy} + \alpha_{33}\theta, \quad (3)$$

$$\frac{\partial U}{\partial \eta} + \frac{\partial V}{\partial \xi} = a_{16}\sigma_{xx} + a_{26}\sigma_{yy} + a_{36}\sigma_{zz} + a_{46}\sigma_{yz} + a_{56}\sigma_{xz} + a_{66}\sigma_{xy} + \alpha_{12}\theta,$$

$$\varepsilon^{-1}\frac{\partial U}{\partial \varsigma}+\frac{\partial W}{\partial \xi}=a_{1j}\boldsymbol{\sigma}_{xx}+a_{2j}\boldsymbol{\sigma}_{yy}+a_{3j}\boldsymbol{\sigma}_{zz}+a_{4j}\boldsymbol{\sigma}_{yz}+a_{5j}\boldsymbol{\sigma}_{xz}+a_{6j}\boldsymbol{\sigma}_{xy}+\alpha_{k3}\theta,$$
$$(U,V;\xi,\eta;\ j=5,4;\ k=1,2),$$

where a_{ij} are the elastic parameters, α_{ij} are the thermoelastic expansion coefficients, F_j are the volume forces (weight, seismic load, etc.), $\theta = T - T_0$ is the change of the temperature field (T_0 is the reference temperature). The conditions of the first boundary-value problem of the elasticity theory are assumed to be given on the plate faces $\zeta = \pm 1$. Otherwise, the stresses $\boldsymbol{\sigma}_{13}, \boldsymbol{\sigma}_{23}$ and $\boldsymbol{\sigma}_{33}$ are given as functions of tangential coordinates ξ, η. In addition, the volume forces and the temperature field are assumed to have an influence comparable with the influence of the surface forces.

The solution of the singularly perturbed system (3) has the form of Eq. (1). For the inner problem, we get the noncontradictory system for determination of $I^{(s)}$, if [3]

$$q_{\boldsymbol{\sigma}_{xx},\boldsymbol{\sigma}_{xy},\boldsymbol{\sigma}_{yy}}=-2,\quad q_{\boldsymbol{\sigma}_{xz},\boldsymbol{\sigma}_{yz}}=-1,\quad q_{\boldsymbol{\sigma}_{zz}}=0,\quad q_{u,v}=-2,\quad q_{w}=-3. \tag{4}$$

The volume forces and the temperature field will be presented in the equations for the input approximation if

$$F_x=\varepsilon^{-2+s}F_x^{(s)},\quad (x,y),\qquad F_z=\varepsilon^{-1+s}F_z^{(s)},\qquad \theta=\varepsilon^{-2+s}\theta^{(s)}, \tag{5}$$

where $F_x^{(0)} = \varepsilon^2 \ell F_x$, (x, y), $F_z^{(0)} = \varepsilon \ell F_z$, $\theta^{(0)} = \varepsilon^2\theta$, $F_j^{(s)} = 0$, $\theta^{(s)} = 0$ at $s \neq 0$.

Substituting Eq. (2) into Eq. (3) and taking into account Eqs. (4), and (5), one gets the noncontradictory system for determination of $I^{(s)}$. This system produces the following equations

$$\begin{aligned}
W^{(s)}&=w^{(s)}(\xi,\eta)+w_*^{(s)}(\xi,\eta,\varsigma),\\
U^{(s)}&=-\zeta\frac{\partial w^{(s)}}{\partial \xi}+u^{(s)}(\xi,\eta)+u_*^{(s)}(\xi,\eta,\varsigma),\quad (u,v;\xi,\eta),\\
\boldsymbol{\sigma}_{xx}^{(s)}&=\zeta\tau_{x1}^{(s)}(\xi,\eta)+\tau_{x0}^{(s)}(\xi,\eta)+\boldsymbol{\sigma}_{xx*}^{(s)}(\xi,\eta,\varsigma),\quad (x,y),\\
\boldsymbol{\sigma}_{xy}^{(s)}&=\zeta\tau_{xy1}^{(s)}(\xi,\eta)+\tau_{xy0}^{(s)}(\xi,\eta)+\boldsymbol{\sigma}_{xy*}^{(s)}(\xi,\eta,\varsigma),\\
\boldsymbol{\sigma}_{xz}^{(s)}&=\frac{1}{2}\zeta^2\tau_{xz2}^{(s)}+\zeta\tau_{xz1}^{(s)}(\xi,\eta)+\tau_{xz0}^{(s)}(\xi,\eta)+\boldsymbol{\sigma}_{xz*}^{(s)}(\xi,\eta,\varsigma),\quad (x,y),\\
\boldsymbol{\sigma}_{zz}^{(s)}&=\frac{1}{6}\zeta^3\tau_{z3}^{(s)}+\frac{1}{2}\zeta^2\tau_{z2}^{(s)}+\zeta\tau_{z1}^{(s)}(\xi,\eta)+\tau_{z0}^{(s)}(\xi,\eta)+\boldsymbol{\sigma}_{zz*}^{(s)}(\xi,\eta,\varsigma).
\end{aligned} \tag{6}$$

In Eq. (6), $Q_*^{(s)}$ are well-known functions for any s if the previous approximations are determined, and the functions $\tau_{ij}^{(s)}(\xi, \eta)$ are expressed through $u^{(s)}, v^{(s)}, w^{(s)}$ by using the well-known formulae [3].

Satisfying the boundary conditions for the first boundary-value problem at $\zeta = \pm 1$, one obtains the equations

$$\ell_{11}u^{(s)} + \ell_{12}v^{(s)} = P_1^{(s)}, \quad \ell_{12}u^{(s)} + \ell_{22}v^{(s)} = P_2^{(s)}, \tag{7}$$

$$B_{11}\frac{\partial^4 w^{(s)}}{\partial\xi^4} + 4B_{16}\frac{\partial^4 w^{(s)}}{\partial\xi^3\partial\eta} + 2(B_{12} + 2B_{66})\frac{\partial^4 w^{(s)}}{\partial\xi^2\partial\eta^2} + 4B_{26}\frac{\partial^4 w^{(s)}}{\partial\xi\partial\eta^3} + B_{22}\frac{\partial^4 w^{(s)}}{\partial\eta^4} = q^{(s)} \tag{8}$$

with respect to the functions $u^{(s)}$, $v^{(s)}$, $w^{(s)}$. At $s = 0$, Eq. (7) written in dimensional coordinates coincide with the classical equations of the generalized plane problem, and Eq. (8) coincides with the classical plate bending equation which has one plane of the elastic symmetry. At $s > 0$, the right parts of these equations change only (besides the loading summands), the terms with the coefficients of elasticity of the mutual influence characterizing the general anisotropy being the part of these equations as well.

Note, that we have previously considered only the plates having one plane of the elastic symmetry. The asymptotic method permits to define the stress-strain state of plates possessing by the general anisotropy, it reduces the problem to well-known equations for plates with a plane of the elastic symmetry [3].

By using only the solution of the inner problem, it is impossible to satisfy the boundary conditions in each point at the lateral surface. This gap is eliminated by constructing the solution for the boundary layer (I_b). This solution decreases exponentially far from the lateral surface of the plate. For example, the solution of the boundary layer localized near the boundary $\xi = 0$ has the following form

$$I_b = \varepsilon^{\chi+s} I_b^{(s)}(\eta, \zeta)\exp(-\lambda\gamma), \quad \gamma = \xi/\varepsilon, \tag{9}$$

where $\chi = 0$ for u_b, v_b, w_b and $\chi = -1$ for $\boldsymbol{\sigma}_{ijb}$. In the general case, the determination of the boundary layer is reduced to the solution of an ordinary sixth order differential equation. For isotropic and orthotropic plates this equation may be split at $s = 0$ into two independent equations of forth and second order with the in-plane and out-of-plane boundary layers corresponding them. For the in-plane boundary layer λ_n are complex conjugate, and for the out-of-plane boundary layer λ_n are real, $\Re\lambda_n > 0$ characterizes the speed of decreasing on the boundary layer when removing from the lateral surface. The classical theory of plates and shells neglects the boundary layers, but the more precise theories by E. Reissner and S. Ambartsumyan take into account the out-of-plane boundary layer. Depending on the type of the spatial boundary conditions on the lateral surface, this or other boundary layer may be neglected in applied problems [3].

For the case of shells the solution structure (1) remains invariant, but in the inner problem there are not unique asymptotics for Eqs. (2), and (4). The asymptotics for description of the momentless and moment states of the shell are established, the optimal correlations and equations within the framework of common concepts and assumptions of the Kirchhoff-Love theory provide maximum accuracy for the determination of the stress-strain state of the shell [3].

3 Second and Mixed 3D Boundary-Value Problems for Anisotropic Plates and Shells

Classical and improved theories of plates and shells consider only the first boundary-value problem of elasticity theory. But there are applied problems, when on faces of plates and shells the conditions of the second or mixed problem of elasticity theory (nonclassical boundary-value problems of thin bodies) should be fulfilled. These problems are important in the calculations of the foundations of seismically safe constructions, when studying the interaction of rigid and pliable bodies, etc. The problem is to find the solution of Eq. (3) satisfying the boundary conditions of the second boundary problem

$$u(\xi, \eta, \pm 1) = u^{\pm}(\xi, \eta), \quad (u, v, w) \tag{10}$$

or mixed boundary problem

$$\begin{aligned} &\boldsymbol{\sigma}_{jz}(\xi, \eta, 1) = \boldsymbol{\sigma}^{+}_{j\eta}(\xi, \eta), \quad j = x, y, z, \\ &u(\xi, \eta, -1) = u^{-}(\xi, \eta), \quad (u, v, w). \end{aligned} \tag{11}$$

It may be directly proved that the asymptotics of Eqs. (2), (4) are not applicable for the solution of problems (3) and (10) or (3) and (11), which mean that the hypotheses of the Kirchhoff-Love theory for the solution of these problems are not applicable. The asymptotic method allows us to solve this problem effectively. There are rather simple asymptotics proposed in [1, 2]. Then the solution of the inner problem has the form of Eq. (2), but now

$$q_{\sigma_{ij}} = -1, q_{u,v,w} = 0. \tag{12}$$

Substituting Eqs. (2), and (12) into Eq. (3), one gets the system

$$\begin{aligned} &\frac{\partial \boldsymbol{\sigma}^{(s-1)}_{jx}}{\partial \xi} + \frac{\partial \boldsymbol{\sigma}^{(s-1)}_{jy}}{\partial \eta} + \frac{\partial \boldsymbol{\sigma}^{(s)}_{jz}}{\partial \varsigma} + F^{(s)}_{j} = 0, \quad j = x, y, z; \\ &F^{(0)}_{j} = \varepsilon^{2} \ell F_{j}, \quad F^{(s)}_{j} = 0, \quad s \neq 0, \\ &\frac{\partial U^{(s-1)}}{\partial \xi} = a_{i1}\boldsymbol{\sigma}^{(s)}_{xx} + a_{i2}\boldsymbol{\sigma}^{(s)}_{yy} + a_{i3}\boldsymbol{\sigma}^{(s)}_{zz} + a_{i4}\boldsymbol{\sigma}^{(s)}_{yz} + a_{i5}\boldsymbol{\sigma}^{(s)}_{xz} + a_{i6}\boldsymbol{\sigma}^{(s)}_{xy} + \alpha_{ii}\theta^{(s)}, \\ &(U, V; \xi, \eta; i = 1, 2), \end{aligned}$$

$$\frac{\partial W^{(s)}}{\partial \varsigma} = a_{13}\boldsymbol{\sigma}^{(s)}_{xx} + a_{23}\boldsymbol{\sigma}^{(s)}_{yy} + a_{33}\boldsymbol{\sigma}^{(s)}_{zz} + a_{34}\boldsymbol{\sigma}^{(s)}_{yz} + a_{35}\boldsymbol{\sigma}^{(s)}_{xz} + a_{36}\boldsymbol{\sigma}^{(s)}_{xy} + \alpha_{33}\theta^{(s)},$$

$$\frac{\partial U^{(s)}}{\partial \varsigma}+\frac{\partial W^{(s-1)}}{\partial \xi}=a_{1j}\sigma_{xx}^{(s)}+a_{2j}\sigma_{yy}^{(s)}+a_{3j}\sigma_{zz}^{(s)}+a_{4j}\sigma_{yz}^{(s)}$$
$$+a_{5j}\sigma_{xz}^{(s)}+a_{6j}\sigma_{xy}^{(s)}+\alpha_{k3}\theta^{(s)},\quad (U,V;\xi,\eta;\ j=5,4;\ k=1,2),$$
$$\frac{\partial U^{(s-1)}}{\partial \eta}+\frac{\partial V^{(s-1)}}{\partial \xi}=a_{16}\sigma_{xx}^{(s)}+a_{26}\sigma_{yy}^{(s)}+a_{36}\sigma_{zz}^{(s)}+a_{46}\sigma_{yz}^{(s)}$$
$$+a_{56}\sigma_{xz}^{(s)}+a_{66}\sigma_{xy}^{(s)}+\alpha_{12}\theta^{(s)},\quad (\theta^{(0)}=\varepsilon\theta,\theta^{(s)}=0,\ s\neq 0) \tag{13}$$

with respect to $I^{(s)}$. The system (13) should be integrated w.r.t. ζ. The solution will depend on six unknown functions $\sigma_{xz0}^{(s)}(\xi,\eta)$, $\sigma_{yz0}^{(s)}(\xi,\eta)$, $\sigma_{zz0}^{(s)}(\xi,\eta)$, $u^{(s)}(\xi,\eta)$, $v^{(s)}(\xi,\eta)$, $w^{(s)}(\xi,\eta)$, which can be determined if the six conditions (10) or (11) are satisfied. So, unlike the first boundary problem in the second and mixed boundary problems the solution of the inner problem is expressed by functions given at the faces of the plate. If the functions $u^{\pm}$, $v^{\pm}$, $w^{\pm}$, $\sigma_{jz}^{\pm}$ are algebraic polynomials in tangential coordinates ξ,η, the iteration process breaks and the mathematically exact solution of the inner problem (solution for the boundary layer) is obtained.

Asymptotics (2), (12) hold for layered anisotropic plates. Having the solved system (13) for each layer and satisfying the boundary conditions at the faces and conditions of the full contact between the layers (discontinuity of displacements and corresponding stress tensor components), the general asymptotic solution (becoming the mathematically exact one) will be obtained. As an illustration we present the solution for a two-layered rectangular anisotropic (21 elasticity parameters) plate $\Omega=\{(x,y,z):0\le x\le a, 0\le y\le b, -h_2\le z\le h_1, \ell=\min(a,b), \max(h_1,h_2)=h, h\ll\ell\}$ with the boundary conditions

$$\begin{aligned}&u(-h_2)=v(-h_2)=w(-h_2)=0,\\&\sigma_{jz}^{+}(h_1)=\sigma_{jz}^{+}=\text{const},\quad j=x,y,z.\end{aligned} \tag{14}$$

The iteration process breaks at the stage of the initial approximation, we having the solution

$$\sigma_{xx}^{(i)}=A_{13}^{(i)}\sigma_{zz}^{+}+A_{14}^{(i)}\sigma_{yz}^{+}+A_{15}^{(i)}\sigma_{xz}^{+},\ \sigma_{yy}^{(i)}=A_{23}^{(i)}\sigma_{zz}^{+}+A_{24}^{(i)}\sigma_{yz}^{+}+A_{25}^{(i)}\sigma_{xz}^{+},$$
$$\sigma_{xy}^{(i)}=A_{63}^{(i)}\sigma_{zz}^{+}+A_{64}^{(i)}\sigma_{yz}^{+}+A_{65}^{(i)}\sigma_{xz}^{+},\ \sigma_{xz}^{(i)}=\sigma_{xz}^{+},\ \sigma_{yz}^{(i)}=\sigma_{yz}^{+},\ \sigma_{zz}^{(i)}=\sigma_{zz}^{+},$$
$$u^{(i)}=h_2(D_{53}^{(i)}\sigma_{zz}^{+}+D_{54}^{(i)}\sigma_{yz}^{+}+D_{55}^{(i)}\sigma_{xy}^{+}),\quad (u,v,w;5,4,3),$$
$$D_{kj}^{(i)}=\zeta A_{kj}^{(i)}+A_{kj}^{(2)},\quad i=1,2;\quad \zeta=z/h,$$
$$A_{kl}=-a_{1l}B_{k1}-a_{2l}B_{k2}-a_{6l}B_{k6},\quad l,m=3,4,5,$$
$$A_{ml}=a_{m1}A_{1l}+a_{m2}A_{2l}+a_{m6}A_{6l}+a_{ml},\quad A_{ml}\neq A_{lm},$$
$$B_{ij}=(a_{ik}a_{jk}-a_{ij}a_{kk})/\Delta,\quad B_{kk}=(a_{ii}a_{jj}-a_{ij}^2)/\Delta,$$
$$\Delta=a_{11}a_{22}a_{66}+2a_{12}a_{26}a_{16}-a_{11}a_{26}^2-a_{22}a_{16}^2-a_{66}a_{12}^2 \tag{15}$$

with $i \neq j \neq k \neq i;\ i, j, k = 1, 2, 6,\ B_{ij} = B_{ji}$. The exact solution (15) permits us to answer, particularly, a very important question: how the modulus of the foundations can be determined if the Winkler-Zimmermann-Fuss model for the anisotropic and layered foundations is applicable? In this case from the solution (15) we write the relations between the displacements and stresses on the surface of the contact $z = 0$ between the layers, assuming that only the normal stresses σ_{zz}^{+} act (they get the index "c")

$$u^{(c)} = h_2 A_{53}^{(2)} \sigma_{zz}^{(c)}, \quad v^{(c)} = h_2 A_{43}^{(2)} \sigma_{zz}^{(c)}, \quad w^{(c)} = h_2 A_{33}^{(2)} \sigma_{zz}^{(c)}, \quad \sigma_{zz}^{(c)} = \sigma_{zz}^{+}. \tag{16}$$

From the solution (16) it follows that under the action of normal loading on the contact surface the tangential displacements, which will be related to greater values $A_{53}^{(2)}, A_{43}^{(2)}$, can be established. This means the inapplicability of the Winkler-Zimmermann-Fuss model to plates possessing general anisotropy. For orthotropic and isotropic foundations $A_{53}^{(2)} = A_{43}^{(2)} = 0$, $w^{(c)} = h_2 A_{33}^{(2)} \sigma_{zz}^{(c)}$ from which one gets

$$\sigma_{zz}^{(c)} = K w^{(c)}, \quad K = \frac{1}{h_2 A_{33}^{(2)}} = \frac{(1 - \nu_{12}^{(2)} \nu_{21}^{(2)}) E_z^{(2)}}{h_2 (1 - \nu_{12}^{(2)} \nu_{21}^{(2)} - \nu_{13}^{(2)} \nu_{31}^{(2)} - \nu_{23}^{(2)} \nu_{32}^{(2)} - 2\nu_{12}^{(2)} \nu 23^{(2)} \nu_{31}^{(2)})}. \tag{17}$$

For the n-layered orthotropic foundation, from the corresponding exact solution it follows

$$K = \frac{1}{\sum_{i=1}^{n} h_i A_{33}^{(i)}}. \tag{18}$$

Asymptotics (2), (12) may be successfully applied for the solution of the second and mixed spatial problems of shells.

4 3D Dynamic Problems of Plates and Shells

Asymptotics (2), (12) are effective for the solution of dynamic problems of plates and shells in the 3D formulation. In this case, if it is not applicable for the static first boundary-value problem according to (4), it is applicable for the dynamic problems. Let us present the following classes of forced vibrations of the anisotropic plate occupying the domain D:

(a) The plate is fixed on an absolutely rigid plane foundation

$$u(x, y, -h) = 0, \quad v(x, y, -h) = 0, \quad w(x, y, -h) = 0 \tag{19}$$

while at $z = h$ harmonically changing in time the stress tensor components are given by

$$\boldsymbol{\sigma}_{jz}(x, y, h) = \boldsymbol{\sigma}^{+}_{jz}(x, y)\exp(i\Omega t), \quad j = x, y, z \tag{20}$$

or displacements are the following

$$u(x, y, h) = u^{+}(x, y)\exp(i\Omega t), \quad (u, v, w); \tag{21}$$

(b) Oscillations of the displacement vector at the face surface $z = -h$ are defined by equations

$$u(x, y, -h) = u^{-}(x, y)\exp(i\Omega t), \quad (u, \text{v}, w) \tag{22}$$

while at $z = h$ one of the variants of the conditions

$$\boldsymbol{\sigma}_{jz}(x, y, h) = 0, \quad j = x, y, z; \quad u(x, y, h) = 0 \quad (u, v, w) \tag{23}$$

is given;

(c) The stress oscillations corresponding to the first boundary-value problem conditions

$$\boldsymbol{\sigma}_{jz}(x, y, \pm h) = \pm\boldsymbol{\sigma}^{\pm}_{jz}(x, y)\exp(i\Omega t), \quad j = x, y, z \tag{24}$$

are given, where $\boldsymbol{\sigma}^{\pm}_{jz}$, $u^{\pm}$, $v^{\pm}$, $w^{\pm}$ are known functions, Ω is the frequency of the external forced action.

We will consider steady-state vibrations. The solution of the dynamic spatial problem of the elasticity theory for an anisotropic body with 21 elastic moduli is sought in the following form:

$$\begin{aligned}&\boldsymbol{\sigma}_{\alpha\beta}(x, y, z, t) = \boldsymbol{\sigma}_{jk}(x, y, z)\exp(i\Omega t), \quad (\alpha, \beta = x, y, z\,; j, k = 1, 2, 3)\\&(u, v, w) = (u_x(x, y, z), u_y(x, y, z), u_z(x, y, z))\exp(i\Omega t).\end{aligned} \tag{25}$$

When passing to dimensionless coordinates ξ, η, ζ and displacements $U = u_x/\ell$, $V = u_y/\ell$, $W = u_z/\ell$, a new system turns out to be singularly perturbed by the small parameter ε. Its solution has the form of Eq. (1). The solution of the inner problem is sought in the form

$$\begin{aligned}&\boldsymbol{\sigma}^{\text{int}}_{jk} = \varepsilon^{-1+s}\boldsymbol{\sigma}^{(s)}_{jk}(\xi, \eta, \zeta), \quad s = \overline{0, N}, \quad j, k = 1, 2, 3,\\&(U^{\text{int}}, V^{\text{int}}, W^{int}) = \varepsilon^{s}(U^{(s)}, V^{(s)}, W^{(s)}).\end{aligned} \tag{26}$$

After substitution of Eq. (26) into the transformed Eq. (25) and passing to the dimensionless magnitudes, we get a new system w.r.t. coefficients $\boldsymbol{\sigma}^{(s)}_{jk}$, $U^{(s)}$, $V^{(s)}$, $W^{(s)}$. From this system it becomes possible to express $\boldsymbol{\sigma}^{(s)}_{jk}$ through $U^{(s)}$, $V^{(s)}$, $W^{(s)}$ which are determined from the equations [5]

$$
\begin{aligned}
&A_{55}\frac{\partial^2 U^{(s)}}{\partial\zeta^2}+A_{45}\frac{\partial^2 V^{(s)}}{\partial\zeta^2}+A_{35}\frac{\partial^2 w^{(s)}}{\partial\zeta^2}+\Delta\Omega_*^2U^{(s)}=R_u^{(s)}(\xi,\eta,\zeta),\\
&A_{54}\frac{\partial^2 U^{(s)}}{\partial\zeta^2}+A_{44}\frac{\partial^2 V^{(s)}}{\partial\zeta^2}+A_{34}\frac{\partial^2 w^{(s)}}{\partial\zeta^2}+\Delta\Omega_*^2V^{(s)}=R_v^{(s)}(\xi,\eta,\zeta),\\
&A_{53}\frac{\partial^2 U^{(s)}}{\partial\zeta^2}+A_{43}\frac{\partial^2 V^{(s)}}{\partial\zeta^2}+A_{33}\frac{\partial^2 w^{(s)}}{\partial\zeta^2}+\Delta\Omega_*^2W^{(s)}=R_w^{(s)}(\xi,\eta,\zeta),
\end{aligned}
\tag{27}
$$

where A_{ij} and Δ are expressed through a_{ij}, $\Omega_*^2=\rho h^2\Omega^2$, $R_u^{(0)}=R_v^{(0)}=R_w^{(0)}=0$, $R_u^{(s)}=R_v^{(s)}=R_w^{(s)}$ are known at $s>0$ functions determined through the data of the previous approximations. After solving the system of Eq. (27), the stresses are determined and all the variants of the boundary conditions (19–24) are satisfied.

For orthotropic plates $A_{45}=A_{35}=A_{54}=A_{34}=A_{53}=A_{43}=0$, and the system of Eq. (27) splits into three equations

$$
\begin{aligned}
&\frac{\partial^2 U^{(s)}}{\partial\zeta^2}+a_{55}\Omega_*^2U^{(s)}=R_u^{(s)}\quad (U,V;a_{55},a_{44};R_u,R_v),\\
&A_{11}\frac{\partial^2 W^{(s)}}{\partial\zeta^2}+\Omega_*^2W^{(s)}=R_w^{(s)},
\end{aligned}
\tag{28}
$$

which are independent at $s=0$. Here

$$
\begin{aligned}
&R_u^{(s)}=-\frac{\partial^2 W^{(s-1)}}{\partial\xi\partial\zeta}-a_{55}\left(\frac{\partial\boldsymbol{\sigma}_{11}^{(s-1)}}{\partial\xi}+\frac{\partial\boldsymbol{\sigma}_{12}^{(s-1)}}{\partial\eta}\right)\quad (u,v;\xi,\zeta;a_{55},a_{44};1,2)\\
&R_w^{(s)}=A_{23}\frac{\partial^2 U^{(s-1)}}{\partial\xi\partial\zeta}+A_{13}\frac{\partial^2 V^{(s-1)}}{\partial\eta\partial\zeta}-\left(\frac{\partial\boldsymbol{\sigma}_{13}^{(s-1)}}{\partial\xi}+\frac{\partial\boldsymbol{\sigma}_{23}^{(s-1)}}{\partial\eta}\right),\quad Q^{(m)}\equiv 0,\ m<0\\
&A_{11}=(a_{11}a_{22}-a_{12}^2)/\Delta,\quad \Delta=a_{11}a_{22}a_{33}+2a_{12}a_{23}a_{13}-a_{11}a_{23}^2-a_{22}a_{13}^2-a_{33}a_{12}^2.
\end{aligned}
\tag{29}
$$

The first two equations of system (28) describe the shear forced vibrations, and the third equation corresponds to the longitudinal vibrations.

The stresses are expressed through the displacements by the following equations:

$$
\begin{aligned}
&\boldsymbol{\sigma}_{11}^{(s)}=-A_{23}\frac{\partial W^{(s)}}{\partial\zeta}+A_{22}\frac{\partial U^{(s-1)}}{\partial\xi}-A_{12}\frac{\partial V^{(s-1)}}{\partial\eta},\\
&\boldsymbol{\sigma}_{22}^{(s)}=-A_{13}\frac{\partial W^{(s)}}{\partial\zeta}-A_{12}\frac{\partial U^{(s-1)}}{\partial\xi}+A_{33}\frac{\partial V^{(s-1)}}{\partial\eta},\\
&\boldsymbol{\sigma}_{33}^{(s)}=A_{11}\frac{\partial W^{(s)}}{\partial\zeta}-A_{23}\frac{\partial U^{(s-1)}}{\partial\xi}-A_{13}\frac{\partial V^{(s-1)}}{\partial\eta},
\end{aligned}
$$

$$\sigma_{12}^{(s)} = \frac{1}{a_{66}}\left(\frac{\partial U^{(s-1)}}{\partial \eta} + \frac{\partial V^{(s-1)}}{\partial \xi}\right), \quad \sigma_{13}^{(s)} = \frac{1}{a_{55}}\left(\frac{\partial U^{(s)}}{\partial \zeta} + \frac{\partial W^{(s-1)}}{\partial \xi}\right),$$

$$\sigma_{23}^{(s)} = \frac{1}{a_{44}}\left(\frac{\partial V^{(s)}}{\partial \zeta} + \frac{\partial W^{(s-1)}}{\partial \eta}\right), \quad Q^{(m)} \equiv 0, \quad m < 0,$$

$$A_{22} = \frac{(a_{22}a_{33} - a_{23}^2)}{\Delta}, \quad A_{33} = \frac{(a_{11}a_{33} - a_{13}^2)}{\Delta},$$

$$A_{12} = \frac{(a_{33}a_{12} - a_{13}a_{23})}{\Delta}, \quad A_{13} = \frac{(a_{11}a_{23} - a_{12}a_{13})}{\Delta},$$

$$A_{23} = \frac{(a_{22}a_{13} - a_{12}a_{23})}{\Delta}. \tag{30}$$

Here $U^{(s)} = U_0^{(s)} + u_\tau^{(s)}$, $\sum(U, V, W)$ is the solution of the system (28), where the first and the second terms are the solutions of the homogeneous and inhomogeneous equations (28) respectively. Determining the stresses by Eq. (30), it is easy to satisfy the boundary conditions (19–24). As application, let us introduce the solutions corresponding to the conditions (22) and the second group of Eq. (23) which simulate the seismic action on the structure foundations:

$$\begin{aligned}
U^{(s)} &= \frac{1}{\sin 2\Omega_* \sqrt{a_{55}}}((u^{-(s)} - u_\tau^{(s)}(\zeta = -1)) \sin \Omega_* \sqrt{a_{55}}(1 - \zeta) \\
&\quad - u_\tau^{(s)}(\zeta = 1) \sin \Omega_* \sqrt{a_{55}}(1 + \zeta)) + u_\tau^{(s)}(\xi, \eta, \zeta), \quad (U, V; a_{55}, a_{44}), \\
W^{(s)} &= \frac{1}{\sin \frac{2\Omega_*}{\sqrt{A_{11}}}}((w^{-(s)} - w_\tau^{(s)}(\zeta = -1)) \sin \frac{\Omega_*}{\sqrt{A_{11}}}(1 - \zeta) \\
&\quad - w_\tau^{(s)}(\zeta = 1) \sin \frac{\Omega_*}{\sqrt{A_{11}}}(1 + \zeta)) + w_\tau^{(s)}(\xi, \eta, \zeta), \\
u^{-(0)} &= u^-/\ell, \quad u^{-(s)} = 0, \quad s \neq 0, \quad (u, v, w).
\end{aligned} \tag{31}$$

The stresses are calculated by Eq. (30). The solution (31) will be bounded one if $\sin 2\Omega_* \sqrt{a_{55}} \neq 0$, $\sin 2\Omega_* \sqrt{a_{44}} \neq 0$, $\sin(2\Omega_*/\sqrt{A_{11}}) \neq 0$. If at least one of values of Ω vanishes, the solution corresponds to the resonance and coincide with principal values of the plate natural frequencies.

If the functions u^-, v^-, w^- are polynomials in the tangential coordinates, the iteration process breaks at the given step. As a result the mathematically exact solution of the inner problem is obtained. The described above approach may be used for the solution of problems on forced vibrations of layered plates as well as of shells.

The solution for the kth-layer is sought in the form of (25) with the index k. Then equations for the package of orthotropic plates analogous to (28) are obtained. Solving them, the boundary and contact conditions are satisfied. If the functions in the boundary conditions are polynomials in tangential coordinates, mathematically exact solutions are obtained. The analysis of the solution for a three-layered package shows an important result: if the middle layer is softer (rubber-like), then the lower layer displacements are harmonically changing w.r.t. the time. They influence weakly on the stress-strain state of the layer located upper the weak layer. This should be considered in applications of seismo-isolators in earthquake-safe buildings [4]. Asymptotics (2), (25) and (26) are applicable for the solution of the 3D dynamic problems for plates made of incompressible materials [6].

5 On Stress-Strain States of the Earth Lithosphere Plates and Possibilities of the Earthquakes Prediction

Modern science attributes the rise of strong earthquakes to tectonics of the Earth lithosphere plates ($\approx$95 % of the earthquakes, see [13]). It is well-known that the planet Earth ($R = 6{,}378$ km) is layered and inhomogeneous. It consists of earth crust, upper and lower mantles, outer and inner kernels. The distinctive feature of these layers is that essentially different velocities V_p and V_s of propagation of the longitudinal (primary) P and shear (secondary) S waves in the layers are observed. The thickness of the earth crust changes in the ranges of 20–70 km and 5–15 km below the continents and the oceans, respectively. In addition, the earth crust consist of different layers: sedimentary ($2 \leq V_\mathrm{p} \leq 5$ km/s, $h_1 = 10$–25 km), granite ($5.5 \leq V_\mathrm{p} \leq 6.0$ km/s, $h_2 = 30$–40 km), basalt ($6.5 \leq V_\mathrm{p} \leq 7.4$ km/s, $h_3 = 15$–20 km), where h_i is the thickness of the ith layers. In the limits of upper mantle at the depth of 100–250 km there is a layer (asthenosphere), where the velocity of transversal waves V_s is significantly reduced. The velocity V_p does not increase with the depth ($V_\mathrm{p} \approx$ const., $V_\mathrm{s} \approx 0$) which testifies the lower density of the material of this layer versus adjacent layers, i.e. there is a fluid-like incompressible medium permitting to withstand high pressures. The Earth crust and part of the upper mantle up to the boundary with asthenosphere compose the lithosphere. Lithosphere with the network of intersecting deep broken pieces is divided into several big blocks, which are called plates. The dimensions of the plates change from hundreds up to several thousand kilometer. The great lithosphere plates of the Earth are: Euroasian, Antarctic, Indo-Australian, Pacific Ocean, South-American, North-American, African, Anatolian, Arabian, etc.

The most part of earthquakes arises in zones the seismic and tectonic activities of which are ascribed to interactions of bordering on each other lithosphere plates (seismic zones). Two types of tectonic movements connected with earthquakes are selected: slow (age-old) and fast (jump-like) movements. Age-old movements (may last decades) are quasistatic ones, they resulting in accumulation of deformations in the lithosphere plates and in its separate parts. If they reach a critical value of the order 10^{-4} (by the data of the seismologist Rikitake of degree 4.5×10^{-5}), global destruction may be observed and the main part of the potential energy accumulated during years is separated in the form of the 3D longitudinal P and shear S waves, as well as surface waves.

Stated above facts point out the importance of determination of the stress-strain states of the Earth lithosphere plates and monitoring of their change in time. The presence of a dense network of modern seismostations and the satellite GPS systems

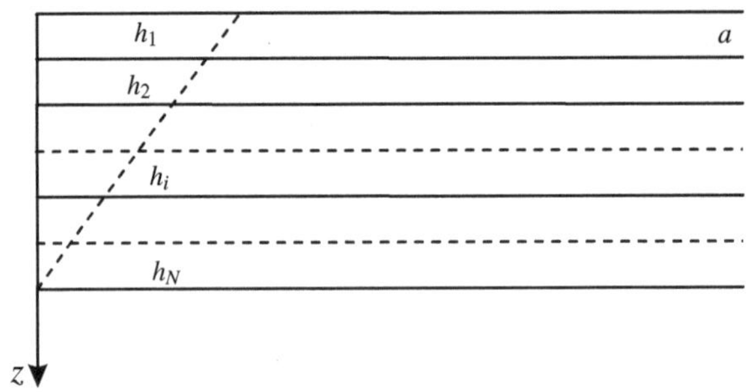

Fig. 1 Multilayered plate

permits us to follow the whole process in time. Here, by analogy with the exchange of meteodata, the collaboration of many countries located in the seismically dangerous zones is required. Later we will show that having the data of seismostations, satellite GPS systems and other measuring devices connected with the displacements of the surface of lithosphere plate, it is possible to determine its stress-strain state and to trace its change in time. The corresponding problem of the elasticity theory for a layered package is nonclassical as the boundary conditions (they are six) are given only on the face of the package: the corresponding stress tensor components are equal to zero (the surface is free), but the displacement vector components of the surface points are well-known as the data of seismostations, GPS and other measuring devices. It is assumed that the geological structure of the lithosphere plate is known, and the physical and mechanical characteristics as well as the thickness of the layers are estimated. In the general case, for determination of the stress-strain state of lithospheric plate under slow (age-old) tectonic movement the following problem is formulated: for a layered package of N orthotropic plates (isotropic plates are regarded as a special case) occupying the domain $D = \{(x, y, z) : 0 \leq x \leq a, 0 \leq y \leq b,\ 0 \leq z \leq h,\ h = \sum_{i=1}^{N} h_i,\ \min(a, b) = \ell,\ h \ll \ell\}$ (Fig. 1) it is required to find the solution of the three-dimensional equations of the elasticity theory.

The equations of equilibrium taking into account volume forces (layer weight) are the following:

$$\frac{\partial \sigma_{xx}^{(k)}}{\partial x} + \frac{\partial \sigma_{xy}^{(k)}}{\partial y} + \frac{\partial \sigma_{xz}^{(k)}}{\partial z} + F_x^{(k)} = 0, \quad (x, y, z), \quad k = 1, 2, \ldots, N. \tag{32}$$

The constitutive equations assuming elastic behavior and taking into account the influence of the temperature field $\theta = T(x, y, z) - T_0(x, y, z)$ by Duhamel-Neumann's equations are

$$\frac{\partial u_x^{(k)}}{\partial x} = a_{11}^{(k)} \sigma_{xx}^{(k)} + a_{12}^{(k)} \sigma_{yy}^{(k)} + a_{12}^{(k)} \sigma_{zz}^{(k)} + \alpha_{11}^{(k)} \theta^{(k)}, \quad (1, 2, 3; x, y, z)$$

$$\frac{\partial u_x^{(k)}}{\partial y} + \frac{\partial u_y^{(k)}}{\partial x} = a_{66}^{(k)} \sigma_{yz}^{(k)},$$

$$\frac{\partial u_x^{(k)}}{\partial z} + \frac{\partial u_z^{(k)}}{\partial x} = a_{55}^{(k)} \boldsymbol{\sigma}_{xz}^{(k)},$$
$$\frac{\partial u_y^{(k)}}{\partial z} + \frac{\partial u_z^{(k)}}{\partial y} = a_{44}^{(k)} \boldsymbol{\sigma}_{yz}^{(k)}. \tag{33}$$

The boundary conditions at $z = 0$ are assumed as follows:

$$\begin{aligned} &\boldsymbol{\sigma}_{jz}(x, y, 0, t) = 0, \quad j = x, y, z, \\ &u_j(x, y, 0, t) = u_j^+(x, y, t). \end{aligned} \tag{34}$$

The conditions of the perfect contact between the layers for an arbitrary kth layer is written in the form:

$$\begin{aligned} &\boldsymbol{\sigma}_{jz}^{(k)}(z = H_k) = \boldsymbol{\sigma}_{jz}^{(k+1)}(z = H_k), \quad H_k = \textstyle\sum_{i=1}^{k} h_i, \quad k = 1, 2, \ldots, N-1, \\ &u_j^{(k)}(z = H_k) = u_j^{(k+1)}(z = H_k), \quad j = x, y, z. \end{aligned} \tag{35}$$

In Eq. (34) the time t plays the role of a parameter which characterizes the moment $t = t_m$ when the measurements of points displacements of the package surface are realized. The boundary conditions on the lateral surface are not defined concretely since conditions (34), (35) are enough to determine the solutions of the inner problem. That is why these conditions correspond to the boundary layer solution exponentially decreasing far from the lateral surface. It may be built in a way described in [3], in the supplements they are usually neglected. In order to solve the stated boundary-value problem (32–35), we introduce again the dimensionless coordinates ξ, η, ζ and displacements $u = u_x/\ell, v = u_y/\ell, w = u_z/\ell$. As a result, we will obtain again a singularly perturbed system the solution of which is (1). The solution of the inner problem has the form

$$\begin{aligned} &\boldsymbol{\sigma}_{ij}^{(k)\text{int}} = \varepsilon^{-1+s} \boldsymbol{\sigma}_{ij}^{(k,s)}, \quad (i, j = x, y, z), \quad s = \overline{0, N}, \\ &u^{(k)\text{int}} = \varepsilon^s u^{(k,s)}, \quad (u, v, w), \quad s = \overline{0, S}, \quad k = 1, 2, \ldots, N. \end{aligned} \tag{36}$$

Substituting (36) into the transformed systems (32), (33), the recurrent system for determination of $\boldsymbol{\sigma}_{ij}^{(k,s)}, u^{(k,s)}, v^{(k,s)}, w^{(k,s)}$ will be obtained. Solving this system, one has

$$\boldsymbol{\sigma}_{jz}^{(k,s)} = \boldsymbol{\sigma}_{jz0}^{(k,s)}(\xi, \eta) + \boldsymbol{\sigma}_{jz*}^{(k,s)}(\xi, \eta, \varsigma), \quad j = x, y, z,$$

$$\boldsymbol{\sigma}_{xx}^{(k,s)} = -\frac{A_{23}^{(k)}}{A_{11}^{(k)}} \boldsymbol{\sigma}_{zz0}^{(k,s)} - \frac{\gamma_{11}^{(k)}}{A_{11}^{(k)}} \theta^{(k,s)} + \boldsymbol{\sigma}_{xx*}^{(k,s)}(\xi, \eta, \varsigma),$$

$$\boldsymbol{\sigma}_{yy}^{(k,s)} = -\frac{A_{13}^{(k)}}{A_{11}^{(k)}} \boldsymbol{\sigma}_{zz0}^{(s)} - \frac{\gamma_{22}^{(k)}}{A_{11}^{(k)}} \theta^{(k,s)} + \boldsymbol{\sigma}_{yy*}^{(s)}(\xi, \eta, \varsigma),$$

$$
\begin{aligned}
\sigma_{xy}^{(k,s)} &= \frac{1}{a_{66}^{(k)}}\left(\frac{\partial v^{(k,s-1)}}{\partial \xi} + \frac{\partial u^{(k,s-1)}}{\partial \eta}\right),\\
u^{(k,s)} &= a_{55}^{(k)}\varsigma\sigma_{xz0}^{(k,s)} + u_0^{(k,s)}(\xi,\eta) + u_*^{(k,s)}(\xi,\eta,\varsigma),\\
v^{(k,s)} &= a_{44}^{(k)}\varsigma\sigma_{yz0}^{(k,s)} + v_0^{(k,s)}(\xi,\eta) + \mathrm{v}_*^{(k,s)}(\xi,\eta,\varsigma),\\
w^{(k,s)} &= \frac{A_{33}^{(k)}}{A_{11}^{(k)}}\varsigma\sigma_{zz0}^{(k,s)} + \frac{\gamma_{33}^{(k)}}{A_{11}^{(k)}}\varsigma\theta^{(k,s)} + w_0^{(k,s)}(\xi,\eta) + w_*^{(k,s)}(\xi,\eta,\varsigma)\\
k &= 1,2,\ldots,N,
\end{aligned}
\tag{37}
$$

where

$$
\begin{aligned}
\sigma_{jz*}^{(k,s)} &= -\int_0^{\varsigma}\left(F_j^{(k,s)} + \frac{\partial\sigma_{jx}^{(k,s-1)}}{\partial\xi} + \frac{\partial\sigma_{jy}^{(k,s-1)}}{\partial\eta}\right)\mathrm{d}\varsigma, \quad j = x,y,z,\\
\sigma_{xx*}^{(k,s)} &= \frac{1}{A_{11}^{(k)}}\left(a_{22}^{(k)}\frac{\partial u^{(k,s-1)}}{\partial\xi} - a_{12}^{(k)}\frac{\partial v^{(k,s-1)}}{\partial\eta} - A_{23}^{(k)}\sigma_{zz*}^{(k,s)}\right),\\
\sigma_{yy*}^{(k,s)} &= \frac{1}{A_{11}^{(k)}}\left(a_{11}^{(k)}\frac{\partial \mathrm{v}^{(k,s-1)}}{\partial\eta} - a_{12}^{(k)}\frac{\partial u^{(k,s-1)}}{\partial\xi} - A_{13}^{(k)}\sigma_{zz*}^{(k,s)}\right),\\
u_*^{(k,s)} &= \int_0^{\varsigma}\left(a_{55}^{(k)}\sigma_{xz*}^{(k,s)} - \frac{\partial w^{(k,s-1)}}{\partial\xi}\right)\mathrm{d}\varsigma,\\
\mathrm{v}_*^{(k,s)} &= \int_0^{\varsigma}\left[a_{44}^{(k)}\sigma_{yz*}^{(k,s)} - \frac{\partial w^{(k,s-1)}}{\partial\eta}\right]\mathrm{d}\varsigma,\\
w_*^{(k,s)} &= \int_0^{\varsigma}\left(a_{13}^{(k)}\sigma_{xx*}^{(k,s)} + a_{23}^{(k)}\sigma_{yy*}^{(k,s)} + a_{33}^{(k)}\sigma_{zz*}^{(k,s)}\right)\mathrm{d}\varsigma,\\
\theta^{(k,0)} &= \varepsilon\theta^{(k)}, \quad \theta^{(k,s)} = 0 \ \text{ at } \ s \neq 0, \quad \sigma_{ij}^{(m)} \equiv 0,\\
u^{(k,m)} &= v^{(k,m)} = w^{(k,m)} \equiv 0 \ \text{ at } \ m < 0,\\
A_{11}^{(k)} &= a_{11}^{(k)}a_{22}^{(k)} - \left(a_{12}^{(k)}\right)^2, \quad A_{13}^{(k)} = a_{11}^{(k)}a_{23}^{(k)} - a_{12}^{(k)}a_{13}^{(k)},\\
A_{23}^{(k)} &= a_{22}^{(k)}a_{13}^{(k)} - a_{12}^{(k)}a_{23}^{(k)}, \quad A_{33}^{(k)} = a_{33}^{(k)}A_{11}^{(k)} - a_{13}^{(k)}A_{23}^{(k)} - a_{23}^{(k)}A_{13}^{(k)},\\
\gamma_{11}^{(k)} &= \alpha_{11}^{(k)}a_{22}^{(k)} - \alpha_{22}^{(k)}a_{12}^{(k)}, \quad \gamma_{22}^{(k)} = \alpha_{22}^{(k)}a_{11}^{(k)} - a_{11}^{(k)}a_{12}^{(k)},\\
\gamma_{33}^{(k)} &= \alpha_{33}^{(k)}A_{11}^{(k)} - a_{13}^{(k)}\gamma_{11}^{(k)} - a_{23}^{(k)}\gamma_{22}^{(k)},\\
& k = 1,2,\ldots,N, \quad Q^{(k,m)} \equiv 0 \ \text{ at } \ m < 0.
\end{aligned}
\tag{38}
$$

In the general case, the solution (36), (37) contains $6N$ unknown functions $\sigma_{jz0}^{(k,s)}$, $u_0^{(k,s)}$, $v_0^{(k,s)}$, $w_0^{(k,s)}$ which are uniquely determined from six boundary conditions

(34) and 6(N − 1) conditions of the contact (35). From Eq. (37) written for the first layer, taking into account conditions (34), one can determine the magnitudes for the first layer:

$$\begin{aligned}&\boldsymbol{\sigma}_{jz0}^{(1,s)}(\xi,\eta)=0,\quad \boldsymbol{\sigma}_{jz}^{(1,s)}(\xi,\eta,\zeta)=\boldsymbol{\sigma}_{jz*}^{(1,s)}(\xi,\eta,\zeta),\quad j=x,y,z,\\ &u_0^{(1,s)}(\xi,\eta)=u^{+(s)},\quad u^{+(0)}=u^{+}/\ell,\quad u^{+(s)}=0,\quad s\neq 0,\quad (u,v,w).\end{aligned}\tag{39}$$

Having data for the first layer and satisfying the contact conditions (35), the unknown magnitudes for the second layer are easily determined, then the data for the third layer are found, etc. For the magnitudes of the $(k+1)$ layer we have

$$\begin{aligned}&\boldsymbol{\sigma}_{jz}^{(k+1,s)}(\xi,\eta,\zeta)=\boldsymbol{\sigma}_{jz0}^{(k,s)}(\xi,\eta)+\boldsymbol{\sigma}_{jz*}^{(k,s)}(\xi,\eta,\zeta_k)\\ &\quad-\boldsymbol{\sigma}_{jz*}^{(k+1,s)}(\xi,\eta,\zeta_k)+\boldsymbol{\sigma}_{jz*}^{(k+1,s)}(\xi,\eta,\zeta),\\ &u^{(k+1,s)}(\xi,\eta,\zeta)=a_{55}^{(k+1)}(\zeta-\zeta_k)\boldsymbol{\sigma}_{xz0}^{(k+1,s)}+a_{55}^{(k)}\zeta_k\boldsymbol{\sigma}_{xz0}^{(k,s)}+u_0^{(k,s)}(\xi,\eta)\\ &\quad+u_*^{(k,s)}(\xi,\eta,\zeta_k)+u_*^{(k+1,s)}(\xi,\eta,\zeta)-u_*^{(k+1,s)}(\xi,\eta,\zeta_k),\\ &(u,v;x,y;a_{55},a_{44}),\\ &w^{(k+1,s)}(\xi,\eta,\zeta)=\frac{A_{33}^{(k+1)}}{A_{11}^{(k+1)}}(\zeta-\zeta_k)\boldsymbol{\sigma}_{zz0}^{(k+1,s)}+\frac{\gamma_{33}^{(k+1)}}{A_{11}^{(k+1)}}(\zeta-\zeta_k)\theta^{(k+1,s)}\\ &\quad+\frac{A_{33}^{(k)}}{A_{11}^{(k)}}\zeta_k\boldsymbol{\sigma}_{zz0}^{(k,s)}+\frac{\gamma_{33}^{(k)}}{A_{11}^{(k)}}\varsigma_k\theta^{(k,s)}+w_0^{(k,s)}(\xi,\eta)+w_*^{(k,s)}(\xi,\eta,\zeta)\\ &\quad+w_*^{(k+1,s)}(\xi,\eta,\zeta)-w_*^{(k+1,s)}(\xi,\eta,\zeta_k),\quad (k,1,2,\ldots,N-1).\end{aligned}\tag{40}$$

The other stresses are determined by Eq. (37). If the functions u_x^+, u_y^+, u_z^+ in the boundary conditions (34) are algebraic polynomials of tangential coordinates, the iteration process (36–40) of determination of all required magnitudes is broken at certain approximation. As a result, we obtain the mathematically exact solution of the inner problem (the solution for the spatial layers). It is testified by the data of seismostations, GPS and other systems. The displacements of the surface of lithosphere plate may be approximated by a polynomial and one gets the mathematically exact solution. At the given moment of time $t = t_*$ in n points of the surface of the lithosphere plate, the values of the displacement vector components $u_x^+(\xi_i, \eta_i, t_*)$, (x, y, z), $i = 1, 2, \ldots, n$ are known. The displacement of the surface points may be approximated by the Lagrange polynomial

$$u_x^+(\xi,\eta,t_*)=\sum_{i=1}^{n}u_x^+(\xi_i,\eta_i,t_*)\frac{\prod\limits_{j=1,j\neq i}^{n}(\xi-\xi_j)(\eta-\eta_j)}{\prod\limits_{j=1,j\neq i}^{n}(\xi_i-\xi_j)(\eta_i-\eta_j)},\quad (x,y,z)\tag{41}$$

and the mathematically exact solution corresponding to the time $t = t_*$ is obtained. Repeating the dimensions in time, we have a complete image about the stress-strain state of the lithosphere plate and its separate blocks, reveal the places and period of time, critical conditions may be created and the process may be dynamic-rapidly flowing (foreshock, earthquake, aftershock).

For the investigation of rapidly flowing processes it is necessary instead of Eq. (32) to use the equations of motion with new boundary conditions at $z = 0$:

$$\begin{aligned} &\sigma_{jz}(x, y, 0, t) = 0, \quad j = x, y, z, \\ &u_j(x, y, 0, t) = u_j^+(x, y)\exp(i\Omega t), \quad j = x, y, z, \end{aligned} \tag{42}$$

where Ω is the frequency of the surface vibrations of the lithosphere plate and fixed seismostations and GPS systems. The dynamic problem may be successfully solved by the asymptotic method. The solution is sought in the following form:

$$\begin{aligned} \sigma_{\alpha\beta}^{(k)}(x, y, z, t) &= \sigma_{jm}^{(k)}(x, y, z)\exp(i\Omega t), \\ u_\alpha^{(k)}(x, y, z, t) &= \bar{u}_\alpha^{(k)}(x, y, z)\exp(i\Omega t), \\ \alpha, \beta = x, y, z; &\quad j, m = 1, 2, 3; \quad k = 1, 2, \ldots, N. \end{aligned} \tag{43}$$

Then passing to the dimensionless coordinates ξ, η, ζ and displacements $U^{(k)} = \bar{u}_x^{(k)}/\ell$, $V^{(k)} = \bar{u}_y^{(k)}/\ell$, $W^{(k)} = \bar{u}_z^{(k)}/\ell$, the solution of the new system is realized as in the quasistatic problem.

6 Interaction of Plates and Shells with Different Physical Fields

The asymptotic method can be the base for studying the interaction of plates and shells with different physical fields. Let us turn our attention to connected dynamic problems of thermoelasticity for isotropic plates and problems of electroelasticity for beforehand polarized piezoceramic shell.

Two basic approaches in investigation of the influence of the temperature field on a deformable body have been formulated. The first approach is based on the temperature stress theory which does not take into account the connectedness of the strain and temperature fields. The classical equation of heat conductivity does not account terms characterizing the strain influence on the temperature field. Solving this equation the distribution of temperature in the body is determined. Thereafter, the equations of the elasticity theory taking into account the influence of the estimated temperature field by Duhamel-Neumann's equations are solved. The second approach is a subject of coupled thermoelasticity. Here the interaction of the strain and temperature fields are taken into account by means of calculation of the heat conductivity terms characterizing the influence of the change of volume strain rate on the temperature field. The asymptotic method permits us to find a general solution of the coupled problem and perform a qualitative analysis of the obtained results in the sense of applicability of this or another applied theory. We will show it on the example of the isotropic plate $D = \{x, y, z : |x| \le a, |y| \le b,\ |z| \le h, \min(a, b) = \ell,\ h \ll \ell\}$, when on its faces $z = \pm h$ the following temperature is given:

$$\theta(x, y, z = \pm h, t) = \theta^{\pm}(x, y) \sin \Omega t. \tag{44}$$

The displacements are defined

$$u_j(x, y, z = \pm h, t) = u_j^{\pm}(x, y) \sin \Omega t, \quad j = x, y, z. \tag{45}$$

It is required to find the solution of the system of the heat conductivity equation and the three-dimensional non-stationary equation of thermoelasticity

$$\begin{aligned} &\nabla^2\theta - \frac{1}{\chi}\frac{\partial\theta}{\partial t} - \eta_*\frac{\partial}{\partial t}\mathrm{div}\mathbf{u} = -\frac{W_*}{\lambda_*}, \quad \eta_* = \alpha_*\frac{E}{1-2\nu}\frac{T_0}{\lambda_*}, \\ &\frac{\partial\boldsymbol{\sigma}_{xx}}{\partial x} + \frac{\partial\boldsymbol{\sigma}_{xy}}{\partial y} + \frac{\partial\boldsymbol{\sigma}_{xz}}{\partial z} = \rho\ddot{u}_x, \quad (x, y, z). \end{aligned} \tag{46}$$

The constitutive equations of an isotropic elastic body (Duhamel-Neumann's equations) are the following:

$$\begin{aligned} &\frac{\partial u_x}{\partial x} = \frac{1}{2G(1+\nu)}(\boldsymbol{\sigma}_{xx} - \nu(\boldsymbol{\sigma}_{yy} + \boldsymbol{\sigma}_{zz})) + \alpha_*\theta, \\ &\frac{\partial u_x}{\partial z} + \frac{\partial u_z}{\partial x} = \frac{1}{G}\boldsymbol{\sigma}_{xz}, \quad (x, y, z). \end{aligned} \tag{47}$$

Satisfying the boundary conditions (44), (45), where E, G are Young's and shear moduli, ν is Poisson's ratio, ρ is the density, α_* is the thermal expansion coefficient, λ_* is the heat conductivity coefficient, $\chi = \frac{\lambda_*}{c_\varepsilon}$ is the thermal diffusivity, c_ε is the specific heat capacity under constant deformation, $\theta = T - T_0$ is the change of the temperature field, W_* is the specific density of the heat source, η_* is the coefficient of the strain field influence on the temperature field. The solution of the formulated problem will be sought in the following form:

$$\begin{aligned} &Q(x, y, z, t) = Q_1(x, y, z) \sin \Omega t + Q_2(x, y, z) \cos \Omega t, \\ &Q = \{u_j, \boldsymbol{\sigma}_{ij}, \theta\}, \quad i, j = x, y, z. \end{aligned} \tag{48}$$

When passing to dimensionless coordinates and displacements in (46), (47), one obtains the singularly perturbed system. The solution of the inner problem has the following form

$$Q_k(x, y, z) = \varepsilon^{\chi_Q + s} Q_k^{(s)}(\xi, \eta, \zeta), \quad s = \overline{0, S}, \quad k = 1, 2. \tag{49}$$

To determine $Q_k^{(s)}(\xi,\eta,\zeta)$ it is necessary to take into account the influence of the strain field on the temperature field. Then we get a noncontradictory system at $\chi_{u,v,w}=0$, $\chi_\sigma=-1$, $\chi_\theta=-1$. It is assumed that the contribution of the heat source W_* is of the same order as the order of the outer interactions contribution (44), (45). For the determination of desired values in general way the following equations are obtained

$$\frac{\partial^2 u_k^{(s)}}{\partial\zeta^2}+\gamma^2 u_k^{(s)}=R_{uk}^{(s)},\quad \gamma=\Omega h\sqrt{\frac{\rho}{G}},\quad Q_k^{(m)}=0\ \text{ at }\ m<0 \qquad (50)$$
$$(\xi,\eta;u_k,v_k),\quad k=1,2$$

and the system

$$\frac{\partial^2\theta_k^{(s)}}{\partial\zeta^2}+(3-2k)\left(q\theta_{3-k}^{(s)}+r\frac{\partial W_{3-k}^{(s)}}{\partial\zeta}\right)=R_{\theta k}^{(s)},\quad q=\frac{\Omega h^2}{\chi},\quad r=\Omega\eta_* h^2,$$
$$\frac{\partial^2 W_k^{(s)}}{\partial\zeta^2}+PW_k^{(s)}=\beta_*\frac{\partial\theta_k^{(s)}}{\partial\zeta}+R_{wk}^{(s)},\quad P=\gamma^2\frac{1-2\nu}{2(1-\nu)},\quad \beta_*=\alpha_*\frac{1+\nu}{1-\nu}, \qquad (51)$$

where $R_{uk}^{(0)}=0$, $R_{wk}^{(0)}=0$, $R_{\theta k}^{(0)}=(k-2)W^{(0)}$ and $R_{uk}^{(s)}, R_{wk}^{(s)}, R_{\theta k}^{(s)}$ are the known values estimated from the previous approximation. The stresses are expressed through displacements and $\theta_k^{(s)}$ by recurrent formulae. From (50) it follows that in the initial approximation the temperature field does not influence on the values of the tangential displacements. The system of Eqs. (50), and (51) contains parameters γ, P, β_*, q, r. Depending on the asymptotic orders of these parameters from system (50), (51) corresponding to this or another applied theory of thermoelasticity follow. Particularly, from the first Eq. (51) it follows that the connection of the thermoelastic problem depends on the order of the parameter $r=\Omega\eta_* h^2$, which takes place, for example, under high-frequency vibrations, and if r is unit order, then the problem should be solved as coupled problem. Equation (50) and the system (51) can be solved when all parameters are present. So, the asymptotic method permits us to solve the coupled temperature problem with the beforehand given exactness.

Consider now the three-dimensional problem solution of electro-elasticity for transversally isotropic piezoceramic shells with thickness of polarization. The middle surface of the shell of thickness $2h$ will be referred to the curvature lines α, β, and the rectilinear axis γ will be directed by outer normal to the middle surface. On the outer and inner surfaces of beforehand polarized by thickness of the shell the following conditions for the electric field potential are given:

$$\varphi^*(\alpha,\beta,\gamma=\pm h,t)=V_0^{\pm}(\alpha,\beta)\exp i\omega t. \qquad (52)$$

In addition, we have one of the combinations of the boundary conditions:

- first

$$\boldsymbol{\sigma}^*_{j\gamma}(\alpha,\beta,\gamma=\pm h,t)=\boldsymbol{\sigma}^{\pm}_{j\gamma}(\alpha,\beta)\exp i\omega t,\quad j=\alpha,\beta,\gamma, \tag{53}$$

- second

$$u^*_j(\alpha,\beta,\gamma=\pm h,t)=u^{\pm}_j(\alpha,\beta)\exp i\omega t,\quad j=\alpha,\beta,\gamma, \tag{54}$$

- mixed

$$\boldsymbol{\sigma}^*_{j\gamma}(\alpha,\beta,\gamma=h,t)=\boldsymbol{\sigma}^{+}_{j\gamma}(\alpha,\beta)\exp i\omega t, u^*_j(\alpha,\beta,\gamma=-h,t)=u^{-}_j(\alpha,\beta)\exp i\omega t. \tag{55}$$

In the case of the direct piezoelectric effect the functions $\boldsymbol{\sigma}^{\pm}_{j\gamma}$, $u^{\pm}_j$ are assumed to be known, and the functions $V^{\pm}_0$ should be determined. In the case of inverse piezoelectric effect, it is considered that $V^{\pm}$ are different from zero and $\boldsymbol{\sigma}^{\pm}_{j\gamma}$, $u^{\pm}_j$, $\boldsymbol{\sigma}^{+}_{j\gamma}$, u^{-}_j are equal to zero. The boundary conditions are not set on the end of the shell (we admit that the shell is long or closed).

It is required to find the solution of the following three-dimensional system of equations of electroelasticity of piezoceramics medium:

$$\begin{aligned}&\frac{\partial}{\partial\alpha}(H_2H_3\boldsymbol{\sigma}^*_{\alpha\alpha})+\frac{\partial}{\partial\beta}(H_1H_3\boldsymbol{\sigma}^*_{\alpha\beta})+\frac{\partial}{\partial\gamma}(H_1H_2\boldsymbol{\sigma}^*_{\alpha\gamma})-\boldsymbol{\sigma}^*_{\beta\beta}H_3\frac{\partial H_2}{\partial\alpha}-\boldsymbol{\sigma}^*_{\gamma\gamma}H_2\frac{\partial H_3}{\partial\alpha}\\&+\boldsymbol{\sigma}^*_{\alpha\beta}H_3\frac{\partial H_1}{\partial\beta}+\boldsymbol{\sigma}^*_{\alpha\gamma}H_2\frac{\partial H_1}{\partial\gamma}+P^*_\alpha H_1H_2H_3=\rho\frac{\partial u^*_\alpha}{\partial t^2}H_1H_2H_3,\quad(\alpha,\beta,\gamma;1,2,3).\end{aligned} \tag{56}$$

For these equations with the additional equations [12]

$$\begin{aligned}&\mathrm{div}\mathbf{D}^*=0,\quad \mathbf{E}^*=-\mathrm{grad}\varphi^*,\\
\boldsymbol{\sigma}^*_{\alpha\alpha}&=C^E_{11}\varepsilon^*_{\alpha\alpha}+C^E_{12}\varepsilon^*_{\beta\beta}+C^E_{13}\varepsilon^*_{\gamma\gamma}-e_{31}E^*_\gamma,\\
\boldsymbol{\sigma}^*_{\beta\beta}&=C^E_{12}\varepsilon^*_{\alpha\alpha}+C^E_{11}\varepsilon^*_{\beta\beta}+C^E_{13}\varepsilon^*_{\gamma\gamma}-e_{31}E^*_\gamma,\\
\boldsymbol{\sigma}^*_{\gamma\gamma}&=C^E_{13}(\varepsilon^*_{\alpha\alpha}+\varepsilon^*_{\beta\beta})+C^E_{33}\varepsilon^*_{\gamma\gamma}-e_{33}E^*_\gamma,\quad \boldsymbol{\sigma}^*_{\alpha\gamma}=C^E_{44}\varepsilon^*_{\alpha\gamma}-e_{15}E^*_\alpha,\\
\boldsymbol{\sigma}^*_{\beta\gamma}&=C^E_{44}\varepsilon^*_{\beta\gamma}-e_{15}E^*_\beta,\quad \boldsymbol{\sigma}^*_{\alpha\beta}=\frac{1}{2}(C^E_{11}-C^E_{12})\varepsilon^*_{\alpha\beta},\\
D^*_\alpha&=\varepsilon^{\sigma}_{11}E^*_\alpha+e_{15}\varepsilon^*_{\alpha\gamma},\quad D^*_\beta=\varepsilon^{\sigma}_{11}E^*_\beta+e_{15}\varepsilon^*_{\beta\gamma},\\
D^*_\gamma&=\varepsilon^{\sigma}_{33}E^*_\gamma+e_{31}(\varepsilon^*_{\alpha\alpha}+\varepsilon^{\sigma}_{\beta\beta})+e_{33}\varepsilon^*_{\gamma\gamma},\\
\varepsilon^*_{\alpha\alpha}&=\frac{1}{H_1}\frac{\partial u^*_\alpha}{\partial\alpha}+\frac{1}{H_1H_2}\frac{\partial H_1}{\partial\beta}u^*_\beta+\frac{1}{H_1H_3}\frac{\partial H_1}{\partial\gamma}u^*_\gamma,\\
\varepsilon^*_{\alpha\beta}&=\frac{H_1}{H_2}\frac{\partial}{\partial\beta}\left(\frac{u^*_\alpha}{H_1}\right)+\frac{H_2}{H_1}\frac{\partial}{\partial\alpha}\left(\frac{u^*_\beta}{H_2}\right),\quad(\alpha,\beta,\gamma;1,2,3)\end{aligned} \tag{57}$$

is required to satisfy one of the sets of boundary conditions (52), (53); (52), (54); (52), (55). In Eqs. (56), (57), $H_1 = A(1+\gamma/R_1)$, $H_2 = B(1+\gamma/R_2)$, $H_3 = 1$ are Lamé's coefficients, A, B are the coefficients of the first quadratic form, R_1, R_2 are the curvatures of the middle surface, $\boldsymbol{\sigma}_{ij}, \varepsilon_{ij}, u_j$ $(i, j = \alpha, \beta, \gamma)$ are the components of the stress tensor, strain tensor and displacement vector, φ is the potential of the electric field, E_j, D_j are the components of the vectors of the electric field strength and electric induction of piezoceramics, C_{ik}^E are the coefficients of piezoceramics elasticity under constant (zero) electric field, $\varepsilon_{11}^{\boldsymbol{\sigma}}, \varepsilon_{33}^{\boldsymbol{\sigma}}$ are the electric penetrability under constant (zero) strains, and e_{ik} are piezomoduli of the ceramics.

Taking into account the form of the boundary conditions (52)–(55), all the desired magnitudes are represented in the form of

$$Q^*(\alpha, \beta, \gamma, t) = Q(\alpha, \beta, \gamma) \exp i\omega t, \quad Q = \{\boldsymbol{\sigma}_{ij}, u_j, D_j, E_j, \varphi\}, \quad i, j = \alpha, \beta, \gamma. \tag{58}$$

The components of the symmetric tensor $\boldsymbol{\sigma}_{ij}$ may be substituted by the components of nonsymmetric tensor τ_{ij} as follows [3, 10]

$$\begin{gathered}
\tau_{\alpha\alpha} = \left(1+\frac{\gamma}{R_2}\right)\boldsymbol{\sigma}_{\alpha\alpha}, \quad \tau_{\beta\beta} = \left(1+\frac{\gamma}{R_1}\right)\boldsymbol{\sigma}_{\beta\beta}, \quad \tau_{\gamma\gamma} = \left(1+\frac{\gamma}{R_1}\right)\left(1+\frac{\gamma}{R_2}\right)\boldsymbol{\sigma}_{\gamma\gamma}, \\
\tau_{\beta\gamma} = \left(1+\frac{\gamma}{R_1}\right)\boldsymbol{\sigma}_{\beta\gamma}, \quad \tau_{\alpha\gamma} = \left(1+\frac{\gamma}{R_2}\right)\boldsymbol{\sigma}_{\alpha\gamma}, \quad \tau_{\alpha\beta} = \left(1+\frac{\gamma}{R_2}\right)\boldsymbol{\sigma}_{\alpha\beta}, \\
\left(1+\frac{\gamma}{R_1}\right)\tau_{\alpha\beta} = \left(1+\frac{\gamma}{R_2}\right)\tau_{\beta\alpha}.
\end{gathered} \tag{59}$$

After substitution of (58) into Eqs. (56), (57) we pass to dimensionless coordinates and displacements

$$\xi = \frac{\alpha}{R}, \ \eta = \frac{\beta}{R}, \ \zeta = \frac{\gamma}{h} = \varepsilon^{-1}\frac{\gamma}{R}, \ u = \frac{u_\alpha}{R}, \ v = \frac{u_\beta}{R}, \ w = \frac{u_\gamma}{R}, \ \varepsilon = \frac{h}{R}, \ h \ll R, \tag{60}$$

where R is the characteristic size of the middle surface of the shell. As a result we obtain a system which is singularly perturbed by a small parameter ε. The solution of the inner problem is sought in the form

$$Q^{\text{int}}(\alpha, \beta, \gamma) = \varepsilon^{\chi_Q + s} Q^{(s)}(\xi, \eta, \zeta), \quad s = \overline{0, S}. \tag{61}$$

Simultaneously it is assumed that we have established a dynamic process, which means that in the equations of the initial approximation following from the motion equations the inertial terms should be presented. It will hold, when $\rho h^2 \omega^2$ will be the magnitude of an order of one. The noncontradictory iteration process is obtained at $\chi_u = 0$ for all displacements, $\chi_\sigma = -1$ for all elastic stresses, $\chi_\varphi = \chi_E = \chi_D = -1$ for the characteristics of the electric field. As a result, the following recurrent differential equations for the determination of all desired magnitudes are obtained

$$\frac{\partial^2 u^{(s)}}{\partial \zeta^2} + \bar{\alpha}^2 u_k^{(s)} = R_u^{(s)}, \quad (u, v; \bar{\alpha}, \bar{\beta}), \quad \bar{\alpha}^2 = \bar{\beta}^2 = \rho h^2 \omega^2 / C_{11}^E,$$
$$C_{33}^E \frac{\partial^2 w^{(s)}}{\partial \zeta^2} + e_{33} \frac{\partial^2 \psi^{(s)}}{\partial \zeta^2} + \rho h^2 \omega^2 w^{(s)} = R_w^{*(s)}, \quad \psi = \varphi / h, \tag{62}$$
$$\frac{\partial \psi^{(s)}}{\partial \zeta} = \Omega^{(s)}(\zeta, \eta) + \frac{e_{33}}{\varepsilon_{33}^\sigma} \frac{\partial w^{(s)}}{\partial \zeta} + R_\psi^{(s)},$$

where $R_u^{(0)} = 0$, $R_w^{(0)} = 0$, $R_\psi^{(0)} = 0$, $R_u^{(s)}$, $R_w^{(s)}$, $R_\psi^{(s)}$ are known functions at $s > 0$, if the previous approximations are constructed. When having the functions $u^{(s)}, v^{(s)}, w^{(s)}, \varphi^{(s)}$ determined by the recurrent formulae from system (62), and taking into account Eqs. (58), (61), we can determine the remaining magnitudes satisfying the boundary conditions (52), (53); (52), (54); (52), (55) and, hereby, solve the problems formulated above. In the case of plates, when the functions in Eqs. (52)–(55) are polynomials, the iteration process cuts off and we get the mathematically exact solution of the inner problem.

As illustration, we present the solution of the problem of inverse piezoelectric effects, when (53) $V_0^\pm = \pm V_0 = \text{const}$ and $\boldsymbol{\sigma}_{j\gamma}^\pm(\alpha, \beta) = 0$ in Eq. (52). With an accuracy of the first step of iteration we have the following equations

$$\tau_{\alpha\alpha}^* = \left(e_{31} - \frac{\bar{\gamma} e_{33}}{\Delta} (C_{13}^E \varepsilon_{33}^\sigma + e_{31} e_{33}) \cos \bar{\gamma} \zeta + \frac{e_{31} e_{33}^2}{\Delta} \sin \bar{\gamma} \right) \frac{V_0}{h} \exp i\omega t$$
$$\bar{\gamma}^2 = \frac{\rho \omega^2 h^2}{\delta_{33}}, \quad (\alpha, \beta),$$
$$\tau_{\gamma\gamma}^* = \frac{e_{33}}{\Delta} \bar{\gamma} \delta_{33} \varepsilon_{33}^\sigma (\cos \bar{\gamma} - \cos \bar{\gamma} \zeta) \frac{V_0}{h} \exp i\omega t, \quad \delta_{33} = (C_{33}^E \varepsilon_{33}^\sigma + e_{33}^2) / \varepsilon_{33}^\sigma,$$
$$\tau_{\alpha\gamma}^* = 0, \quad u_\alpha^* = 0, \quad (\alpha, \beta), \quad \tau_{\alpha\beta}^* = 0,$$
$$u_\gamma^* = -\frac{\varepsilon_{33}^\sigma \varepsilon_{33}}{\Delta} V_0 \sin \bar{\gamma} \zeta \exp i\omega t,$$
$$\varphi^* = \left(\frac{\gamma}{h} - \frac{e_{33}^2}{\Delta} \left(\sin \bar{\gamma} \zeta - \frac{\gamma}{h} \sin \bar{\gamma} \right) \right) V_0 \exp i\omega t,$$
$$D_\gamma^* = -\varepsilon_{33}^\sigma \left(1 + \frac{e_{33}^2}{\Delta} \sin \bar{\gamma} \right) \frac{V_0}{h} \exp i\omega t, \quad D_\alpha^* = D_\beta^* = 0,$$
$$\Delta = \varepsilon_{33}^\sigma \delta_{33} \bar{\gamma} \cos \bar{\gamma} - e_{33}^2 \sin \bar{\gamma} \tag{63}$$

for the components of the displacement vector, the stress tensor and the potential of an electric field. Solution (63) is the mathematically exact one for a spatial piezoceramic layer of thickness $2h$.

By using the developed asymptotic method it is possible to consider other classes of problems on interaction of plates and shells with physical fields. But the main difficulties will be in the correct choice of asymptotically desired magnitudes.

7 Conclusions

Wide class of 3D problems for plates and shells, which may be effectively solved by using the asymptotic method, has been presented. Static and dynamic problems of elasticity theory, problems on the interaction of plates and shells with various physical fields (for example, with temperature and electroelastic fields) have been considered. Nonclassical problems for layered plates permitting to determine the stress-strain states of the Earth lithosphere plates by the data of the seismic stations and the GPS systems and to follow their change in time have been also solved. On the basis of the analysis of the dynamic problem solution for a layered orthotropic plate, the necessity of the application of seismo-isolators in a seismo-safe construction has been proved.

Acknowledgments The work was supported by the State Committee of Science MES RA in framework of the research project No. 13-2c009 SCS.

References

1. Aghalovyan, L.A.: On the structure of solution of one class of plane problems of elasticity theory of anisotropic body. In: Interuniversity Transactions: Mechanics, vol 2, pp 7–12.Yerevan State University Publishing House (1982)
2. Aghalovyan, L.A.: To the determination of stress-strain state of two-layered strip and validity of Winkler hypothesis (in Russ.). In: Works of XIIInd All-Union conference on theory of plates and shells, vol I, pp 13–18. Publishing House of the Tallinn Polytechnic Institute, Tallin (1983)
3. Aghalovyan, L.A.: Asymptotic Theory of Anisotropic Plates and Shells (in Russ.). Nauka, Moscow (1997)
4. Aghalovyan, L.A., Hovhannisyan, R.Z.: Theoretical proof of seismoisolators application necessity (in Russ.). In: Proceedings of International Workshop on Base Isolated High-Rise Building, pp 185–199. Gasprint, Yerevan (2008)
5. Aghalovyan, M.L.: On the character of forced vibrations of plates under general anisotropy (in Russ.). In: Works of the International conference: XVIII session of the International school on models of mechanics of solid medium, pp 14–18. Publishing House of Saratov State University, Saratov (2007)
6. Aghalovyan, M.L.: On forced vibrations of rectangular plates of incompressible material under mixed boundary conditions (in Russ.). In: Works of the II International conference "Actual problems of mechanics of solid medium", vol 1, pp 28–32. Pub. House of YGUAS-2010, Yerevan (2008)
7. Babich, V.M., Buldirev, V.S.: Art of asymptotics (in Russ.). Herald of Leningrad University **13**, 5–12 (1977)
8. Friedrichs, K.O., Dressler, R.F.: A boundary-layer theory for elastic plates. Comm. Pure. Appl. Math **14**(1), 1–33 (1961)

9. Gol'denveizer, A.: Derivation of an approximate theory of bending of a plate by the method of asymptotic integration of the equations of the theory of elasticity. J. Appl. Math. Mech **26**(4), 1000–1025 (1962)
10. Goldenveizer, A.L.: Theory of Thin Elastic Shells (in Russ.). Nauka, Moscow (1976)
11. Green, A.E.: On the linear theory of thin elastic shells. Proc Roy Soc Series A, Mathematical and Physical Sciences A **266**(1325), 143–160 (1962)
12. Grinchenko, V.T., Ulitko, A.F., Shulga, N.A.: Electroelasticity (in Russ.). Naukova Dumka, Kiev (1989)
13. Kasahara, K.: Earthquake Mechanics. Cambridge University Press, Cambridge (1981)
14. Vorovich, II.: Some mathematical questions of theory of plates and shells (in Russ.). In: Works of IInd All-Union conference on theoretical and applied mechanics, vol 3, pp 116–136. Nauka, Moscow (1966)

On the Theories of Plates and Shells at the Nanoscale

Holm Altenbach and Victor A. Eremeyev

Abstract During the last 50 years the nanotechnology is established as one of the advanced technologies manipulating matter on an atomic and molecular scale. New materials, devices or other structures possessing at least one dimension sized from 1–100 nm are developed. The question arises how structures composed of nanomaterials should be modeled. In addition, if it is necessary to perform a structural analysis what kind of theory should be used. Two approaches are suggested—theories which take into account quantum mechanical effects since they are important at the quantum-realm scale and theories which are based on the classical continuum mechanics adapted to nanoscale problems. Here the second approach will be discussed in detail. It will be shown that the classical continuum mechanics is enough for a sufficient description of the mechanical behavior of nanomaterials and-structures if surface stresses are taken into account. There are also other approaches in the literature, but they are note discussed in detail in this paper. The basic equations for nanosized plates and shells will be discussed. It is shown that for this class of objects with the help of suggested equations such effects like the size-dependent changes of the stiffness parameters can be described in a proper manner. In contrast to the results for the size-dependence of the Young's modulus (the Young's modulus increases when the specimen diameter becomes very thin) the plate stiffness parameters can increase or decrease when the plate thickness is in the range of several nm. Finally, the theory of plates with surface stresses will be compared with the theory of three-layered plates.

H. Altenbach (✉) · V.A. Eremeyev
Lehrstuhl für Technische Mechanik, Institut für Mechanik, Fakultät für Maschinenbau, Otto-von-Guericke-Universität Magdeburg, Universitätsplatz 2, 39106 Magdeburg, Germany
e-mail: holm.altenbach@ovgu.de

V.A. Eremeyev
Southern Scientific Center of Russian Academy of Science and Southern Federal University, Rostov on Don, Russia
e-mail: victor.eremeyev@ovgu.de; eremeyev.victor@gmail.com

H. Altenbach and G.I. Mikhasev (eds.), *Shell and Membrane Theories in Mechanics and Biology*, Advanced Structured Materials 45, DOI 10.1007/978-3-319-02535-3_2

1 Introduction and Historical Remarks

Structural mechanics is the computation of strains, deflections, and internal forces or stresses (stress resultants) within structures. The aim of the analysis is the safe design of new structures or the evaluation of existing structures. The starting point for any structural analysis are the following input data: loads (mechanical, thermal, electrical, etc.), the structure's geometry, support and contact conditions, and the material behavior information. Finally, the stresses, the strains and displacements are estimated for the whole structure. The classical analysis is limited by consideration of linear-elastic behavior and the description of the structure within the reference configuration. Advanced structural mechanics may include such effects like stability and non-linear behavior.

There are several approaches to the analysis. In the general case this results in solving a system of three-dimensional coupled partial differential equations. Analytical solution can be obtained only in exceptional cases. Thus during the last 50 years the finite element method was developed as a powerful tool for approximate solutions of three-dimensional problems. At the same time lower-dimensional theories became a further development. Since the pioneering works of Euler or Bernoulli we know that considering the different order of the dimensions of a structure one can describe the structural mechanics problems by simplified equations. Well-known examples are the theories of rods, beams, plates and shells. The first two are based on the assumption that the characteristic cross-section dimension is much smaller than the length, the last two assume that the thickness is much smaller in comparison with other dimensions. Below we focus our attention on plates and shells only.

Shell- and plate-like structures are used in civil and aerospace engineering as basic elements of constructions. Such structures are applied as a model of analysis in other branches, e.g. mechanical engineering, but also in new one like medical engineering. New applications are primarily related to new materials—instead of steel or concrete now one has to analyze sandwiches, laminates, foams, nano-films, biological membranes, etc. The new trends in applications demand improvements of the theoretical foundations of the plate and shell theory, since new effects must be taken into account. For example, in the case of small-size plate- or shell-like structures (for example, nano-tubes) the surface effect plays an increasing role in the mechanical analysis of these structural elements if the size decreases.

Let us make a brief overview on some important steps in the development of the theory of plates and shells. One of the first researcher in the field of plates and shells was Ernst Florens Friedrich Chladni (* November 30, 1756 in Wittenberg; † April 3, 1827 in Breslau). He was a physicist, astronomer and musician. His most important contribution included research on vibrating plates showing various modes of vibration (Chladni's nodal patterns). In this sense he was the founder of acoustics. Marie-Sophie Germain (* April 1, 1776 in Paris; † June 27, 1831 in Paris) payed attention on Chladni's experimental works. She took part in a contest organized by the French Academy of Sciences. The aim of the contest was to give "the mathematical theory of the vibration of an elastic surface and to compare the theory to experimental evidence". She submitted a paper on this topic, but failed. Joseph-Louis Lagrange (* January 25th, 1736 in Turin; † April 10th, 1813 in Paris) derived an equation based on Germain's paper that was correct under special assumptions. Later in 1816 Germain published her third paper on the above mentioned topic and she was awarded the *prix extraordinaire*. She derived the correct vibration equation

$$N^2 \left(\frac{\partial^4 z}{\partial x^4} + \frac{\partial^4 z}{\partial x^2 \partial y^2} + \frac{\partial^4 z}{\partial y^4} \right) + \frac{\partial^2 z}{\partial t^2} = 0 \quad (N = \text{const}),$$

but the experimental results were not very accurate predicted. In addition, she had trouble with the boundary conditions.

The next big step in the development of the plate theory was done by Gustav Robert Kirchhoff (* March 12, 1824 in Königsberg; † October 17, 1887 in Berlin). The Kirchhoff plate theory is an extension of the Euler-Bernoulli beam theory to thin plates and based on the assumption that a mid-surface plane can be used to represent the three-dimensional plate in two-dimensional form. He introduced following kinematic assumptions: straight lines normal to the mid-surface remain straight and normal after deformation and the thickness of the plate does not change during the deformation [75]. The governing equation of Kirchhoff's plate theory is

$$\nabla^2 \nabla^2 w = \frac{q}{D} \quad \text{with} \quad D = \frac{Eh^3}{12(1-\nu^2)}$$

Similar kinematical assumptions were introduced by Augustus Edward Hough Love (* April 17, 1863 in Weston-super-Mare; † June 5, 1940 in Oxford) in the case of shells.

The main disadvantage of Kirchhoff's plate theory is that the governing equation is of 4th order—but in some cases one has to satisfy three boundary conditions. In addition, the transverse flexibility is not presented in a satisfying manner. This was the staring point for several improvements of the Kirchhoff theory. Theodore von Kármán (* May 11, 1881 in Budapest; † May 6, 1963 in Aachen) presented at the beginning of the last century [38], which can be related to the reference configuration (in-plane behavior) and to the actual configuration (out-of-plane behavior) at the same time. Finally, he obtained the Föppl-von Kármán equations describing large deflections of thin plates. Another improvement was given by Eric Reissner (* January 5, 1913 in Aachen; † November 1, 1996 in La Jolla, CA). He has been accounted the stresses ignored in the Kirchhoff theory. In this sense Reissner's theory is named first order shear-deformation plate theory. A similar theory was presented by Raymond David Mindlin (* September 17, 1906 in New York; † November 22, 1987 in Hanover, New Hampshire) in 1951, but in contrast to Reissner he used Poisson's approach of dimension reduction applying power series.

Later in the former Soviet Union various plate and shell theories were suggested. Ilia Vekua (* April 23, 1907 in Sheshelety; † December 2, 1977 in Tbilisi) was a Georgian mathematician, which proposed a shell theory using generalized analytic functions [88]. Khamid Mushtari (* July 22, 1900 in Orenburg; † January 23, 1981 in Kazan) was a Tatarian scientist in the field of solid mechanics and mathematics, presenting a theory with large rotations which was successfully used in aerospace industries [60]. This theory was similar to Donnell's and Vlasovs approach and so the theory is usually named Mushtari-Donell-Vlasov theory [22, 93]. Sergei Ambarcumyan (* March 17, 1922 in Gumry) suggested a first-order shear deformable theory with a special distribution law for the transverse stresses [13].

Another approach was used by Paul Mansur Naghdi (* March 29, 1924 in Teheran; † July 9, 1994 in Berkeley, CA) which was based on the direct approach introduced by Euler and developed by the Cosserat brothers. A summary concerning this approach one can find in [61]. Further historical remarks concerning the theory of plates and shell can be found, for example, in [12].

Relatively recently the developments of nanotechnologies lead to the derivation of models of plates and shells which can be used at the nano-scale. At this level the surface phenomena are important. The classical theory of capillarity was establish by Young and Laplace and then was extended for the case of solids by Gibbs. Gurtin [32] and Steigmann [84] proposed models of surface stresses describing the surface elastic properties. The model by [32] found many applications for materials at the micro- and nanoscales, see the reviews [23, 36, 69, 90, 91]. In particular, the surface effects are used for explanation of deviation of the properties of nanosized specimens from the ones of bulk materials. The enhancements of the theory of plates and shells taking into account of surface stresses are discussed in [3, 9–11, 24] and reference therein.

After the work by [58, 59] the surface stresses can be modeled within the framework of the second-gradient theory of elasticity. Non-local and gradient type models of plates and shells are presented, for example, by [14, 18, 34, 45–48, 56, 62, 71–73, 80, 83, 87, 89]. Below only the first approach based on the introduction of equations for the bulk behavior and the surface behavior is used.

2 Materials and Structures Under Consideration

Material science classifies structural materials into three categories

- metals,
- ceramics, and
- polymers

It is difficult to give an exact assessment of the advantages and disadvantages of these three basic material classes, because each category covers whole groups of materials within which the range of properties is often as broad as the differences between the three material classes.

The following characteristic properties can be established:

- Mostly metallic materials are of medium to high density. They have good thermal stability and can be made corrosion-resistant by alloying. Metals have useful mechanical characteristics and it is moderately easy to shape and join. For this reason metals became the preferred structural engineering material.

- Ceramic materials have great thermal stability and are resistant to corrosion, abrasion and other forms of attack. They are very rigid but mostly brittle and can only be shaped with difficulty. In many cases they fail immediately beyond the elastic range.
- Polymer materials (plastics) are of low density, have good chemical resistance but lack thermal stability. They have poor mechanical properties, but are easily fabricated and joined. Their resistance to environmental degradation, e.g. the photomechanical effects of sunlight, is moderate.

At present classical structural materials are more and more substituted by advanced materials. An example of advanced materials are composite materials which classification is given in Fig. 1. Another class of advanced materials are sandwich materials with solid and hollow cores (Fig. 2). Similar to the classical sandwiches short fibre reinforced composites have with respect to the technology (here injection molding) a layered sandwich-type microstructure (Fig. 3). Recently another type of advanced materials is used as a material for lightweight structures–plastic or metallic foams. Figure 4 shows closed-cell foams, Fig. 5—open-cell foams.

Classical laminates and sandwiches are modeled as usual as materials with piecewise constant properties. Now we have also materials with changing properties–so called functionally graded materials (FGM). Examples of FGM are shown on Fig. 6.

With the development of nanotechnology a new class of advanced materials was established: nanomaterials. These are materials with single units of which are sized (in at least one dimension) between 1 and 1,000 nm but is usually 1–100 nm (the usual definition of nanoscale, [15]). Examples of nanomaterials are presented on Fig. 7.

The above mentioned materials have a common feature concerning modeling: in all these cases multi-scale and homogenized models are suggested. On Figs. 8 and 9

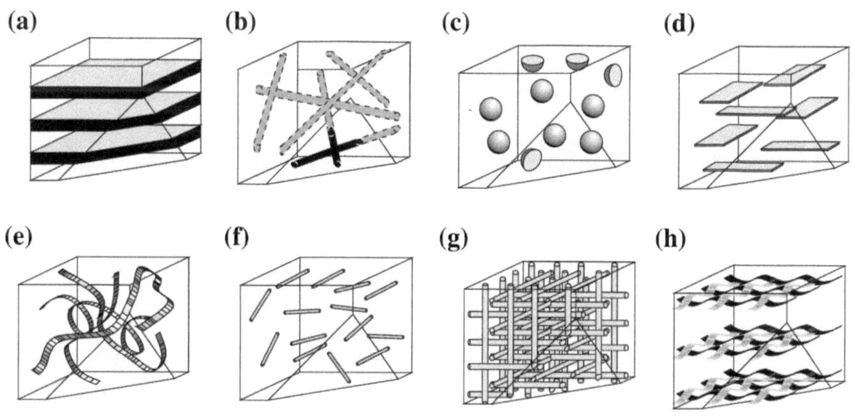

Fig. 1 Classification of composites: **a** laminate, **b** irregular reinforcement, **c** reinforcement with particles, **d** reinforcement with plate strapped particles, **e** random arrangement of continuous fibres, **f** irregular reinforcement with short fibres, **g** spatial reinforcement, **h** reinforcement with surface tissues (after [8])

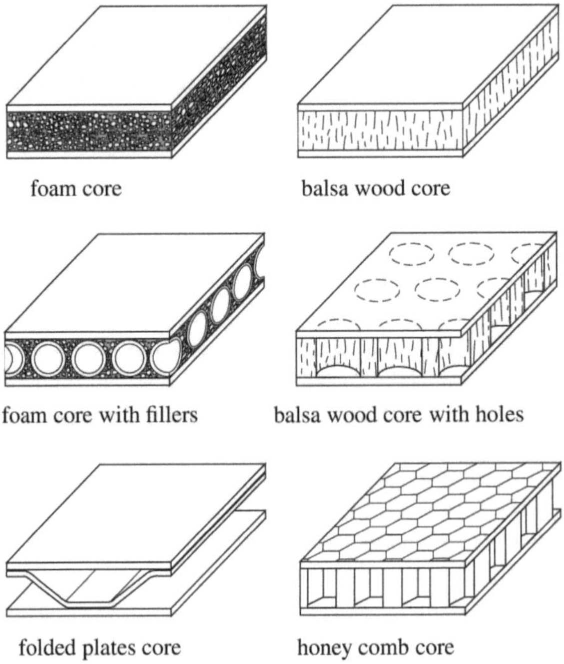

Fig. 2 Sandwich materials with solid and hollow cores [8]

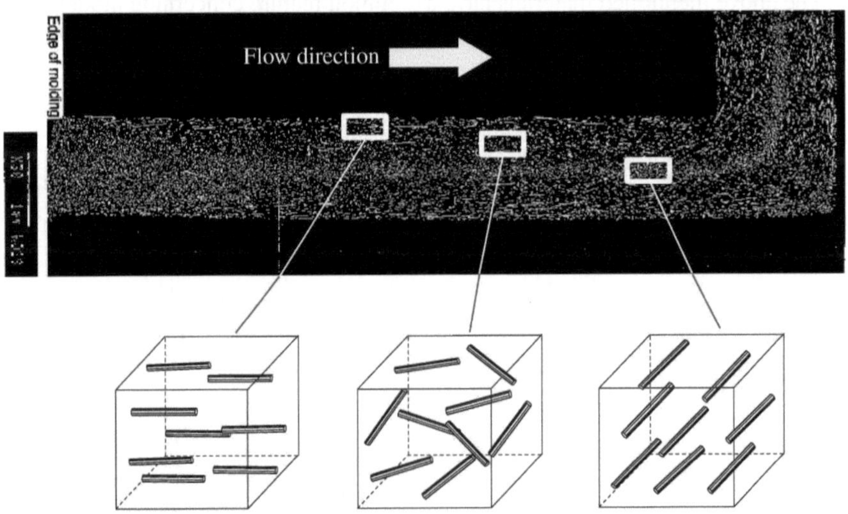

Fig. 3 Short fibre reinforced composite (after [81])

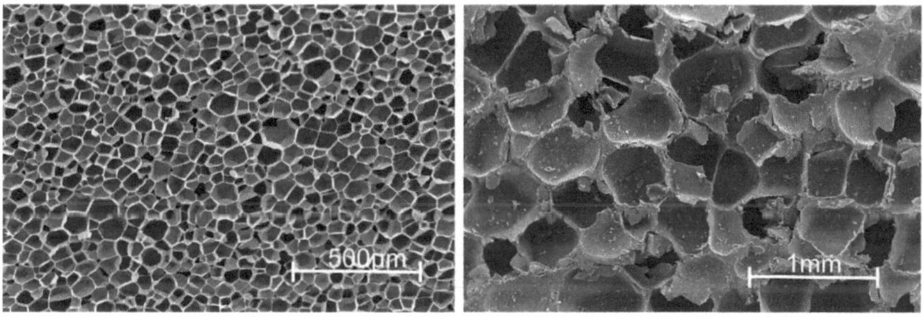

Fig. 4 Closed-cell polymeric foams with various densities after [43]

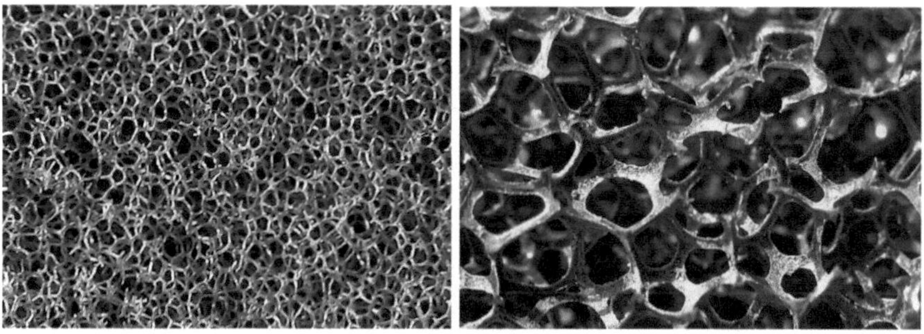

Fig. 5 Open-cell foams with various densities after [43]

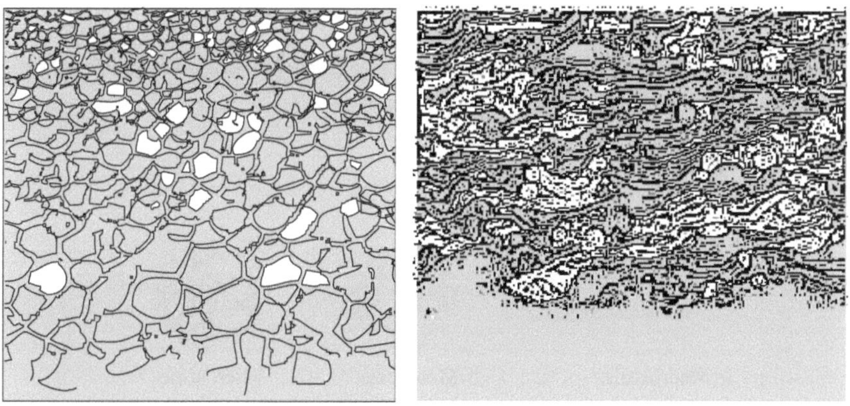

Fig. 6 Examples of FGMs with inhomogeneous microstructure: foam (*left*), thermal coating (*right*)

the model hierarchy in the case of laminated plates and open- and closed-cell foams is shown.

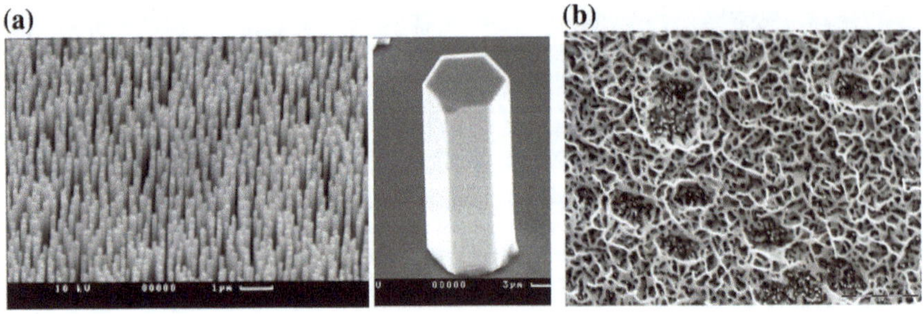

Fig. 7 Nanostructures: **a** ordered array of ZnO nanocrystals and single crystal, **b** ZnO nanofoam

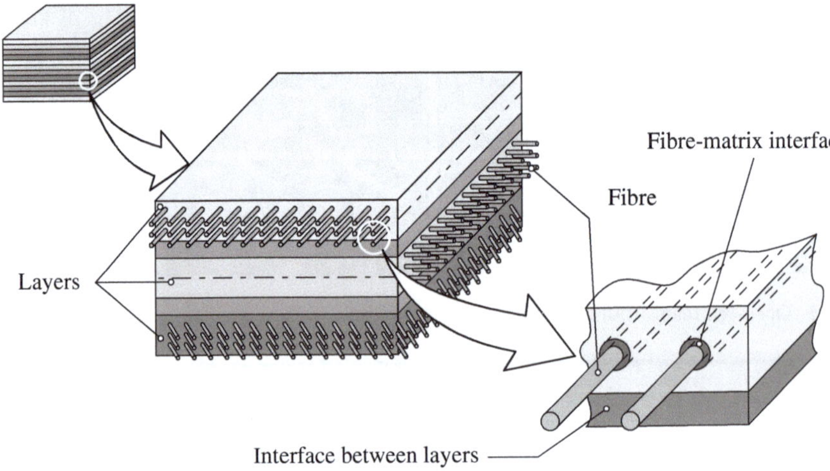

Fig. 8 Multi-scale modeling of laminated plates [8]

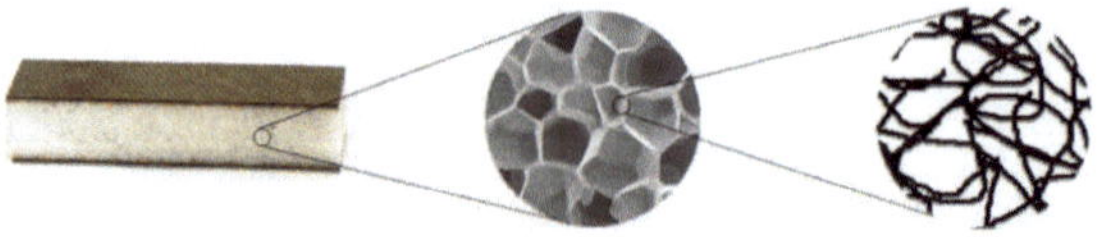

Fig. 9 Three scales for foam structures

3 Plate and Shell Theories: Fundamentals

The basic problem in civil and mechanical engineering is the analysis of the strength, the vibration behavior and the stability of structures with the help of structural models. The structural models can be classified by their

- spatial dimensions
- loadings
- kinematical and/or statical hypotheses

The starting point for any structural analysis is order of the dimensions in the three-dimensional space. We have to distinguish three basic models:

- The three spatial dimensions have the same order, no predominant direction for the dimensions exists. Typical examples of geometrical simple, compact structural elements in theory of elasticity are cube, prism, cylinder, sphere, etc.
- Two spatial dimensions have the same order, the third, which is related to the thickness is much smaller. Typical examples of surface structural elements in structural mechanics are: discs, plates, shells, folded structures, etc.
- Two spatial dimensions, which can be related to the cross-section, have the same order. The third dimension, which is related to the length of the structural element, has a much larger order in comparison with the cross-section dimensions. Typical examples in engineering mechanics are: rods, beams, torsion beam, etc.

Thin-walled structural elements (thin-walled light-weight profile structures) require an extension of the classical structural models. If the spatial dimensions are of significantly different order and the thickness of the profile is small in comparison to the other cross-section dimensions, and the cross-section dimensions are much smaller in comparison to the length of the structure one can introduce quasi-one-dimensional structural elements. Suitable theories are

- the thin-walled beam approach and
- the semi-membrane theory or generalized beam theory [93]

The further discussion are limited to structures subjected to the following definition.

Definition 2.1 (*Two-Dimensional Structure*) A two-dimensional load-barring structural element is a model for the analysis in engineering and structural mechanics with two geometrical dimensions, which are of the same order and which are significantly larger than the third (thickness) direction.

The mathematical consequence is that instead of a three-dimensional problem, which is presented by a system of partial differential equations, one can analyze a two-dimensional problem. The transition from the three-dimensional to the two-dimensional problem is not simple, but the solution effort decreases significantly. Within the definition the following model classes are included

- thin homogeneous plates,
- thin inhomogeneous plates (laminates, sandwiches),
- plates with structural anisotropy,
- moderately thick homogeneous plates,
- folded plates,
- membranes,
- biological membranes,
- nanotubes

Such models are applied in space and aircraft industries, automotive industries, shipbuilding industries, vehicle systems, civil engineering, medicine, etc. The developments in this field are widely discussed, for example, in [12, 30, 31, 61, 75, 77]. Recently several conferences were devoted to this topic: Euromech Colloquia 444 and 527 [4, 41], IUTAM Symposium on Relation of Shell, Plate, Beam and 3D Models [35], *Shell Structures Theory & Applications* conference [67] among others.

In the literature there are various formulation principles for two-dimensional theories. The corresponding equations can be deduced:

- starting from three-dimensional continuum mechanics equations or
- starting from two-dimensional equation describing the behavior or a deformable surface

If one starts from three-dimensional continuum mechanics equations there exist two possibilities to reduce the dimension

- the use of proper hypotheses or
- the use of mathematical approaches

All suggested methods have advantages and disadvantages, but hypotheses based theories are preferred by the engineers.

If the hypotheses based method is used we can introduce assumptions for the stress, strain or displacement states. For example, if we start with displacement approximations we can deduce the full set of governing taking and boundary conditions.

- Examples of displacement approximations for a plate (Fig. 10) are given as follows:
  - Kirchhoff [42]
    $u_\alpha(x_\beta, z) = u_\alpha^0(x_\beta) - z w_{,\alpha}(x_\beta), \quad w(x_\beta, z) = w(x_\beta)$
  - Mindlin [57]
    $u_\alpha(x_\beta, z) = u_\alpha^0(x_\beta) + z\varphi_\alpha(x_\beta), \quad w(x_\beta, z) = w(x_\beta)$
  - Levinson [50] and Reddy [74]
    $u_\alpha(x_\beta, z) = u_\alpha^0(x_\beta) - [w_{,\alpha}(x_\beta) + \varphi_\alpha(x_\beta)]\dfrac{4z^3}{3h^2}, \quad w(x_\beta, z) = w(x_\beta)$
  - Meenen and Altenbach [55]
    $u_\alpha(x_\beta, z) = u_\alpha^q(x_\beta)\phi^q(z) + w_{,\alpha}^q(x_\beta)\psi^q(z), \quad w(x_\beta, z) = w(x_\beta)^q\chi^q(z)$

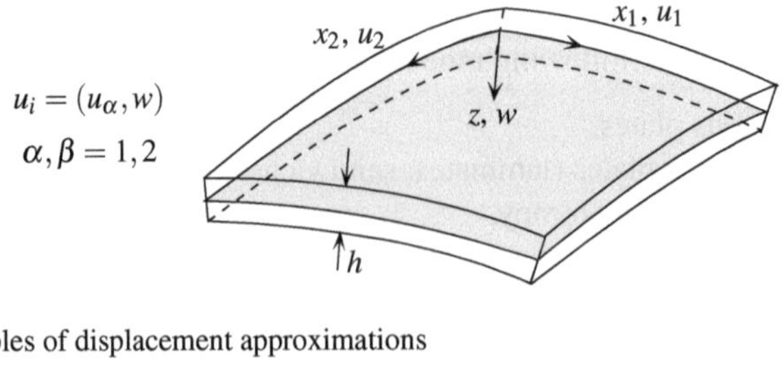

Fig. 10 Examples of displacement approximations

- Mathematical approaches are based on
  - Power series for the displacements, stresses and strains which were used by [17, 39, 40, 52, 55, 70] among others,
  - Special or trigonometric functions which were applied, for example, in [44, 86, 88],
  - Asymptotic integration which was used, for example, in [28]

Another approach is the so-called direct approach. The starting point is the à priori introduction of an two-dimensional deformable surface [2, 29, 61, 79, 94, 95]. This is a natural way of formulation the two-dimensional equations which was also mentioned indirectly by [82]: "...everyone, who is thinking about the foundations of Continuum Mechanics, will attend the world of images of the Cosserat brothers". Firstly the direct approach was discussed by Leonard Euler introducing the moment vector as independent quantity in the theory of beams. Later this approach was extended to three-dimensional problems and a first summary was given by [20, 21]. Later this approach was discussed and extended by [16, 26, 27, 37, 63, 65]. Special contributions to the two-dimensional direct approach were made by [7, 29, 61, 78] among others. In all these papers the independence of translations and rotations/forces and moments is assumed and which corresponds to the undergraduate courses in Statics and Rigid Body Dynamics.

In the adjacent part the direct approach to the theory of plates and shells will be considered.

Definition 2.2 (*Simple Shell*) A simple shell is a 2D-continuum in which the interaction between neighboring parts is due to forces and moments [95].

In addition let us make the following two assumptions:

1. The plate or shell (homogeneous or inhomogeneous in thickness direction) will be represented by a deformable surface.
2. Each material point of the surface is an infinitesimal body with 5 degrees of freedom (3 translations and 2 rotations).

The last one assumption can be easily extended to 6 degrees of freedom [5].

The kinematical model of a simple shell can presented as usual in continuum mechanics by comparison of the reference and actual configurations. In the reference configuration (undeformed state) we have

$$\{\mathbf{r}(q^1, q^2);\ \mathbf{d}_k(q^1, q^2)\},$$

where $\mathbf{r}(q^1, q^2)$ is the position vector, $\mathbf{d}_k(q^1, q^2)$ are orthonormal vectors. The actual configuration (deformed state) is given by

$$\{\mathbf{R}(q^1, q^2, t);\ \mathbf{D}_k(q^1, q^2, t)\},\ \mathbf{D}_k \cdot \mathbf{D}_m = \delta_{km},$$

where the capital letters have the same meaning as in the reference configuration. The motion of the directed surface can be presented by

$$\mathbf{R}(q,t), \quad \mathbf{P}(q,t) \equiv \mathbf{D}^k(q,t) \otimes \mathbf{d}_k(q)$$

$\mathbf{P}(q,t) \equiv \mathbf{P}(q^1,q^2,t)$ is a rotation tensor, $\det \mathbf{P} = +1$. The linear and angular velocities $\mathbf{v}(q,t), \boldsymbol{\omega}(q,t)$ are defined as

$$\mathbf{v} = \dot{\mathbf{R}}, \ \dot{\mathbf{P}} = \boldsymbol{\omega} \times \mathbf{P}, \ \mathbf{P}(q^1,q^2,0) = \mathbf{P}_0, \ \dot{f} \equiv \frac{\mathrm{d}f}{\mathrm{d}t}$$

and we obtain the local equations of motion:

- First Euler equation of motion

$$\widetilde{\nabla} \cdot \mathbf{T} + \rho \mathbf{F}_* = \rho(\mathbf{v} + \boldsymbol{\Theta}_1^{\mathrm{T}} \cdot \boldsymbol{\omega})^{\cdot}$$

- Second Euler equation of motion

$$\widetilde{\nabla} \cdot \mathbf{M} + \mathbf{T}_\times + \rho \mathbf{L} = \rho(\boldsymbol{\Theta}_1 \cdot \mathbf{v} + \boldsymbol{\Theta}_2 \cdot \boldsymbol{\omega})^{\cdot} + \rho \mathbf{v} \times \boldsymbol{\Theta}_1^{\mathrm{T}} \cdot \boldsymbol{\omega}$$

$\mathbf{T} = \mathbf{R}_\alpha \otimes \mathbf{T}^\alpha, \mathbf{M} = \mathbf{R}_\alpha \otimes \mathbf{M}^\alpha$ are the force and the moment tensor, respectively, $\mathbf{T}_\times \equiv \mathbf{R}_\alpha \times \mathbf{T}^\alpha$ is the vector invariant of the force tensor, $\mathbf{F}_*, \mathbf{L}$ are the mass density of external forces and moments, $\rho, \rho\boldsymbol{\Theta}_1, \rho\boldsymbol{\Theta}_2$ are the density, the first and the second tensor of inertia, respectively, $\widetilde{\nabla} \equiv \mathbf{R}^\alpha(q^1,q^2,t)\frac{\partial}{\partial q^\alpha}$ is the nabla operator.

Here and in the next following parts the direct tensor notation is used (see, for example, [49]).

The equation of the balance of energy can be formulated in the local form

$$\rho\dot{\mathscr{U}} = \mathbf{T}^{\mathrm{T}} \cdot\cdot \widetilde{\nabla}\mathbf{v} - \mathbf{T}_\times \cdot \boldsymbol{\omega} - \mathbf{M}^{\mathrm{T}} \cdot\cdot \widetilde{\nabla}\boldsymbol{\omega}$$

$\mathscr{U}$ is the mass density of the internal energy. Introducing the energetic tensors [54]

$$\mathbf{T}_{\mathrm{e}} = (\widetilde{\nabla}\mathbf{r})^{\mathrm{T}} \cdot \mathbf{T} \cdot \mathbf{P}, \ \mathbf{M}_{\mathrm{e}} = (\widetilde{\nabla}\mathbf{r})^{\mathrm{T}} \cdot \mathbf{M} \cdot \mathbf{P}$$

Another form of the balance of energy is given by

$$\rho\dot{\mathscr{U}} = \mathbf{T}_{\mathrm{e}}^{\mathrm{T}} \cdot\cdot \dot{\mathbf{E}} + \mathbf{M}_{\mathrm{e}}^{\mathrm{T}} \cdot\cdot \dot{\mathbf{F}}$$

$\mathbf{E}, \mathbf{F}$ are the first and the second deformation tensors

$$\mathbf{E} = \widetilde{\nabla}\mathbf{R} \cdot \mathbf{P} - \mathbf{a}, \ \mathbf{F} = (\boldsymbol{\Phi}_\alpha \cdot \mathbf{D}_k)\mathbf{r}^\alpha \otimes \mathbf{d}^k$$

with $\partial_\alpha \mathbf{P} = \boldsymbol{\Phi}_\alpha \times \mathbf{P} \Rightarrow 2\boldsymbol{\Phi}_\alpha = -[\partial_\alpha \mathbf{P} \cdot \mathbf{P}^{\mathrm{T}}]_\times$. $\mathbf{a}$ is the first two-dimensional metric tensor. Now we can define the reduced deformation tensors. The internal energy $\mathscr{U} = \mathscr{U}(\mathbf{E}, \mathbf{F})$ contains 12 scalar arguments. The number of arguments can be reduced due to some restrictions [95]:

- simple shells of constant thickness,

- non-polar materials,
- $\mathbf{L} \cdot \mathbf{D}_3 = \mathbf{0}, \quad \mathbf{M} \cdot \mathbf{D}_3 = \mathbf{0}$
- $\mathbf{M}_\mathrm{e}^\mathrm{T} \cdot\cdot [(\mathbf{F} - \mathbf{b} \cdot \mathbf{c}) \cdot \mathbf{c}] + \mathbf{T}_\mathrm{e}^\mathrm{T} \cdot\cdot [(\mathbf{E} + \mathbf{a}) \cdot \mathbf{c}] = \mathbf{0}$

the specific energy $\mathscr{U}$ must satisfy

$$\left(\frac{\partial \mathscr{U}}{\partial \mathbf{E}}\right)^\mathrm{T} \cdot\cdot [(\mathbf{E} + \mathbf{a}) \cdot \mathbf{c}] + \left(\frac{\partial \mathscr{U}}{\partial \mathbf{F}}\right)^\mathrm{T} \cdot\cdot [(\mathbf{F} - \mathbf{b} \cdot \mathbf{c}) \cdot \mathbf{c}] = 0, \ \frac{\partial \rho_0 \mathscr{U}}{\partial (\mathbf{F} \cdot \mathbf{n})} = \mathbf{0}$$

$\mathbf{b}$ is the second two-dimensional metric tensor and $\mathbf{c} = -\mathbf{n} \times \mathbf{a}$ is the discriminant tensor. The characteristic system of the first one equation is a system of 12th order

$$\frac{\mathrm{d}}{\mathrm{d}s}\mathbf{E} = (\mathbf{E} + \mathbf{a}) \cdot \mathbf{c}, \ \frac{\mathrm{d}}{\mathrm{d}s}\mathbf{F} = (\mathbf{F} - \mathbf{b} \cdot \mathbf{c}) \cdot \mathbf{c}$$

which have 11 independent integrals–the strain measures

$$\begin{aligned}
2\boldsymbol{\varepsilon} &= \left[(\mathbf{E} + \mathbf{a}) \cdot \mathbf{a} \cdot (\mathbf{E} + \mathbf{a})^\mathrm{T} - \mathbf{a}\right], \\
\boldsymbol{\Phi} &= (\mathbf{F} - \mathbf{b} \cdot \mathbf{c}) \cdot \mathbf{a} \cdot (\mathbf{E} + \mathbf{a})^\mathrm{T} + \mathbf{b} \cdot \mathbf{c} \cdot \boldsymbol{\varepsilon} + \mathbf{b} \cdot \mathbf{c}, \\
\boldsymbol{\gamma} &= \mathbf{E} \cdot \mathbf{n}, \\
\boldsymbol{\gamma}_* &= \mathbf{F} \cdot \mathbf{n}
\end{aligned}$$

The arbitrary function $\mathscr{U}(\boldsymbol{\varepsilon}, \boldsymbol{\Phi}, \boldsymbol{\gamma}, \boldsymbol{\gamma}_*)$ satisfies the first equation of the characteristic system. From the second equation follows that $\mathscr{U}$ does not depend on $\boldsymbol{\gamma}_*$. The tensors $\boldsymbol{\varepsilon}, \boldsymbol{\Phi}, \boldsymbol{\gamma}$ are called reduced deformation tensors. $\boldsymbol{\varepsilon}$ represents the plane tensile and shear strains, $\boldsymbol{\Phi}$ are the bending and torsional strains and $\boldsymbol{\gamma}$ are the transverse shear strains.

For a shell composed of a linear-elastic material with relatively small strains while the displacements and rotations can be relatively large the quadratic approximation for the strain energy can be introduced

$$\begin{aligned}
2\rho_0 \mathscr{U} &= 2\mathbf{T}_0 \cdot\cdot \boldsymbol{\varepsilon} + 2\mathbf{M}_0^\mathrm{T} \cdot\cdot \boldsymbol{\Phi} + 2\mathbf{N}_0 \cdot \boldsymbol{\gamma} + \boldsymbol{\varepsilon} \cdot\cdot {}^{(4)}\mathbf{C}_1 \cdot\cdot \boldsymbol{\varepsilon} + 2\boldsymbol{\varepsilon} \cdot\cdot {}^{(4)}\mathbf{C}_2 \cdot\cdot \boldsymbol{\Phi} \\
&\quad + 2\boldsymbol{\Phi} \cdot\cdot {}^{(4)}\mathbf{C}_3 \cdot\cdot \boldsymbol{\Phi} + \boldsymbol{\gamma} \cdot \boldsymbol{\Gamma} \cdot \boldsymbol{\gamma} + 2\boldsymbol{\gamma} \cdot \left({}^{(3)}\boldsymbol{\Gamma}_1 \cdot\cdot \boldsymbol{\varepsilon} + {}^{(3)}\boldsymbol{\Gamma}_2 \cdot\cdot \boldsymbol{\Phi}\right)
\end{aligned} \tag{1}$$

$\mathbf{T}_0, \mathbf{M}_0, \mathbf{N}_0, {}^{(4)}\mathbf{C}_1, {}^{(4)}\mathbf{C}_2, {}^{(4)}\mathbf{C}_3, {}^{(3)}\boldsymbol{\Gamma}_1, {}^{(3)}\boldsymbol{\Gamma}_2, \boldsymbol{\Gamma}$ are stiffness tensors of different rank. They express the effective elastic properties of the simple shell. The differences between various classes of simple shells are connected with different expressions of the stiffness tensors. The stiffness tensors do not depend on the deformations. Thus they may be found from tests based on the linear shell theory.

The constitutive equations can be obtained as the derivatives of the strain energy by the strains. In the simplest case ignoring the eigenstresses we get

- in-plane forces

$$\mathbf{T} \cdot \mathbf{a} = \rho_0 \frac{\partial \mathscr{U}}{\partial \boldsymbol{\varepsilon}} = {}^{(4)}\mathbf{C}_1 \cdot\cdot \boldsymbol{\varepsilon} + {}^{(4)}\mathbf{C}_2 \cdot\cdot \boldsymbol{\Phi} + \boldsymbol{\gamma} \cdot \boldsymbol{\Gamma}_1, \tag{2}$$

- transverse shear forces

$$\mathbf{T} \cdot \mathbf{n} = \rho_0 \frac{\partial \mathscr{U}}{\partial \boldsymbol{\gamma}} = \boldsymbol{\Gamma} \cdot \boldsymbol{\gamma} + \boldsymbol{\Gamma}_1 \cdot\cdot\, \boldsymbol{\varepsilon} + \boldsymbol{\Gamma}_2 \cdot\cdot\, \boldsymbol{\Phi}, \tag{3}$$

- moments

$$\mathbf{M}^{\mathrm{T}} = \rho_0 \frac{\partial \mathscr{U}}{\partial \boldsymbol{\kappa}} = \boldsymbol{\varepsilon} \cdot\cdot\, {}^{(4)}\mathbf{C}_2 + {}^{(4)}\mathbf{C}_3 \cdot\cdot\, \boldsymbol{\Phi} + \boldsymbol{\gamma} \cdot \boldsymbol{\Gamma}_2 \tag{4}$$

After the formulation of the governing equations one open question exists—the identification of the effective properties (stiffness, etc.). Various solutions of this problem are existing. To find the general structure of stiffness tensors the theory of symmetry must be applied. The classical theory of symmetry is not sufficient because it is valid for Euclidean tensors only. In the shell theory Euclidean and non-Euclidean tensors are involved. Details of the application of the theory of symmetry are presented in [95].

In the thickness direction of plates and shells we can assume homogeneous or inhomogeneous behavior. The second case is obtained if we have sandwiches and laminates (piecewise constant properties) or functionally graded materials (continuous distributed properties). Both particular cases can be modeled like a "microstructure". The question arises how the symmetries of the "microstructure" do affect the physical properties? The answer comes from the Curie-Neumann's principle [64, 66] in the physics of crystals: *The symmetry group of the reason belongs to the symmetry group of the consequence.*[1] The symmetry group of the reasons for the plates and shells is the intersection of

- symmetry of the material (fibre-reinforced material, rolled sheets, ...),
- symmetry of the surface shape (shell or plate),
- symmetry of the internal structure of the plate (for example, symmetry of the laminate stacking sequence with respect to the mid-surface, ...)

It is obvious that symmetry discussions for shells are more complex since the curvature has an influence.

Symmetries can be described in terms of the geometric operations which produce identical configurations. The set of symmetry operations and results of their combinations define a mathematical structure called a group. The symmetry operations which involve only rotations, reflection and inversion define the point group. The symmetries are described by orthogonal tensors

---

[1] Other formulations are:

- Any type of symmetry exhibited by the point group of a crystal is possessed by every physical property of the crystal.
- For a material element and for any of its physical properties, every material symmetry transformation of the material element is a physical symmetry transformation of the physical property.

- reflection ($\mathbf{n}$ is the unit normal to the mirror plane)

$$\mathbf{Q} = \mathbf{I} - 2\mathbf{n} \otimes \mathbf{n}, \ \det \mathbf{Q} = -1,$$

- rotation ($\mathbf{m}$ represents the axis and ψ is the angle of rotation)

$$\mathbf{Q}(\psi \mathbf{m}) = \mathbf{m} \otimes \mathbf{m} + \cos\psi(\mathbf{I} - \mathbf{m} \otimes \mathbf{m}) + \sin\psi \mathbf{m} \times \mathbf{I}, \ \det \mathbf{Q} = 1, \quad |\psi| < \pi,$$

- inversion

$$-\mathbf{I}$$

The identification procedure is related to solving boundary value problems or eigenvibration problems. Comparing similar two-dimensional and three-dimensional quantities like stresses and forces, etc. one can obtain the effective properties of the plate or shell. The following relations between 2D and 3D properties are assumed:

- forces and moments

$$\mathbf{T} = \langle \boldsymbol{\mu}^{-1} \cdot \boldsymbol{\sigma} \rangle, \quad \mathbf{M} = \langle \boldsymbol{\mu}^{-1} \cdot \boldsymbol{\sigma} \cdot \mathbf{c} z \rangle, \quad < (\ldots) > = \int_{-h/2}^{h/2} (\ldots) \mathrm{d}z,$$

  $\boldsymbol{\sigma}$ is the symmetric stress tensor of the classical theory of elasticity, $\boldsymbol{\mu}$ is the shifter tensor (see [61] among others) and $\mathbf{c}$ is the discriminant tensor.
- displacements and rotations

$$\rho_0(\mathbf{u} + \boldsymbol{\Theta}_1^{\mathrm{T}} \cdot \boldsymbol{\varphi}) = \langle \rho_0^* \mathbf{u}_* \rangle, \quad \rho_0(\boldsymbol{\Theta}_1 \cdot \mathbf{u} + \boldsymbol{\Theta}_2^{\mathrm{T}} \cdot \boldsymbol{\varphi}) = \langle \rho_0^* \mathbf{u}_* \cdot \mathbf{c} z \rangle$$

  $\mathbf{u}_*$ is the three-dimensional displacement vector.
- external force and moment

$$\rho_0 \mathbf{F}_* = \langle \rho_0^* \mathbf{F}^* \rangle + \mu^+ \boldsymbol{\sigma}_n^+ + \mu^- \boldsymbol{\sigma}_n^-,$$

$$\rho_0 \mathbf{L} = \mathbf{n} \times \langle \rho_0^* \mathbf{F}^* z \rangle + (h/2) \mathbf{n} \times (\mu^+ \boldsymbol{\sigma}_n^+ - \mu^- \boldsymbol{\sigma}_n^-)$$

  $\mu^{+(-)} = 1 - (+)hH + (h^2/4)G$, $\boldsymbol{\sigma}_n^{+(-)}$ are stress vectors on the upper and lower face surfaces of the plate or shell. H and G denotes the mean and Gaussian curvature, respectively.

Symmetry considerations (orthotropic material behavior, plane mid-surface) result in the following representation of the stiffness tensors

$$\begin{aligned}
\boldsymbol{A} &= A_{11}\mathbf{a}_1\mathbf{a}_1 + A_{12}(\mathbf{a}_1\mathbf{a}_2 + \mathbf{a}_2\mathbf{a}_1) + A_{22}\mathbf{a}_2\mathbf{a}_2 + A_{44}\mathbf{a}_4\mathbf{a}_4,\\
\boldsymbol{B} &= B_{13}\mathbf{a}_1\mathbf{a}_3 + B_{14}\mathbf{a}_1\mathbf{a}_4 + B_{23}\mathbf{a}_2\mathbf{a}_3 + \boldsymbol{B}_{24}\mathbf{a}_2\mathbf{a}_4 + B_{42}\mathbf{a}_4\mathbf{a}_2,\\
\boldsymbol{C} &= C_{22}\mathbf{a}_2\mathbf{a}_2 + C_{33}\mathbf{a}_3\mathbf{a}_3 + C_{34}(\mathbf{a}_3\mathbf{a}_4 + \mathbf{a}_4\mathbf{a}_3) + C_{44}\mathbf{a}_4\mathbf{a}_4,\\
\boldsymbol{\Gamma} &= \Gamma_1\mathbf{a}_1 + \Gamma_2\mathbf{a}_2, \qquad \boldsymbol{\Gamma}_1 = \mathbf{0}, \qquad \boldsymbol{\Gamma}_2 = \mathbf{0}
\end{aligned}$$

with $\mathbf{a}_1 = \mathbf{a} = \boldsymbol{e}_1\boldsymbol{e}_1 + \boldsymbol{e}_2\boldsymbol{e}_2, \mathbf{a}_2 = \boldsymbol{e}_1\boldsymbol{e}_1 - \boldsymbol{e}_2\boldsymbol{e}_2, \mathbf{a}_3 = \mathbf{c} = \boldsymbol{e}_1\boldsymbol{e}_2 - \boldsymbol{e}_2\boldsymbol{e}_1, \mathbf{a}_4 = \boldsymbol{e}_1\boldsymbol{e}_2 + \boldsymbol{e}_2\boldsymbol{e}_1$ $\boldsymbol{e}_1, \boldsymbol{e}_2$ are unit basic vectors. In addition, the orthogonality condition for $\mathbf{a}_i$ $(i = 1, 2, 3, 4)$ is fulfilled

$$\frac{1}{2}\mathbf{a}_i \cdot\cdot\, \mathbf{a}_j = \delta_{ij}, \quad \delta_{ij} = \begin{cases} 1, & i = j, \\ 0, & i \neq j \end{cases}$$

Let us assume for the elastic orthotropic law the following cases:

Case 1: Homogeneous plates–all properties are constant with respect to z.
Case 2: Inhomogeneous plates–all properties are functions of z.

The identification of the effective properties can be performed with the help of static boundary value problems (two-dimensional, three-dimensional) and the comparison of the forces and moments (in the sense of averaged stresses or stress resultants)

$$\boldsymbol{T} = <\boldsymbol{a}\cdot\boldsymbol{\sigma}>, \ \boldsymbol{M} = <\boldsymbol{a}\cdot\boldsymbol{\sigma} z\cdot\mathbf{c}>$$

Here the simplest case (plates) is considered. Details for homogeneous shells are given in [95], solutions for eigenvibrations are presented in [94].

- Problem 1: Tension and Bending
  The following two-dimensional kinematical field is given

$$\boldsymbol{u} = D_1 x_1\boldsymbol{e}_1 + D_2 x_2\boldsymbol{e}_2 - \frac{1}{2}\left(\frac{x_1^2}{R_1} + \frac{x_2^2}{R_2}\right)\boldsymbol{n}, \ \boldsymbol{\varphi} = -\frac{x_2}{R_2}\boldsymbol{e}_1 + \frac{x_1}{R_1}\boldsymbol{e}_2$$

  D_1, D_2, R_1 and R_2 are constants. Then the strains can be computed

$$\boldsymbol{\mu} = D_1\boldsymbol{e}_1\boldsymbol{e}_1 + D_2\boldsymbol{e}_2\boldsymbol{e}_2, \ \boldsymbol{\gamma} = 0, \ \boldsymbol{\kappa} = \frac{1}{R_1}\boldsymbol{e}_1\boldsymbol{e}_2 - \frac{1}{R_2}\boldsymbol{e}_2\boldsymbol{e}_1$$

  and we obtain the forces and moments from the constitutive equations.
  The three-dimensional strain tensor components are

$$\varepsilon_1 = D_1 + \frac{z}{R_1}, \ \varepsilon_2 = D_2 + \frac{z}{R_2}$$

  Stress tensor components (plane stress state is assumed) are

$$\sigma_1 = \frac{E_1}{1 - \nu_{12}\nu_{21}}(\varepsilon_1 + \nu_{21}\varepsilon_2), \ \sigma_2 = \frac{E_2}{1 - \nu_{12}\nu_{21}}(\varepsilon_2 + \nu_{12}\varepsilon_1)$$

Finally we have to compare both solutions

$$\begin{aligned}
A_{11} &= \frac{1}{4}\left\langle \frac{E_1 + E_2 + 2E_1\nu_{21}}{1-\nu_{12}\nu_{21}} \right\rangle, \; A_{12} = \frac{1}{4}\left\langle \frac{E_1 - E_2}{1-\nu_{12}\nu_{21}} \right\rangle,\\
A_{22} &= \frac{1}{4}\left\langle \frac{E_1 + E_2 - 2E_1\nu_{21}}{1-\nu_{12}\nu_{21}} \right\rangle,\\
B_{13} &= -\frac{1}{4}\left\langle \frac{E_1 + E_2 + 2E_1\nu_{21}}{1-\nu_{12}\nu_{21}} z \right\rangle, \; -B_{23} = B_{14} = \frac{1}{4}\left\langle \frac{E_1 - E_2}{1-\nu_{12}\nu_{21}} z \right\rangle,\\
B_{24} &= \frac{1}{4}\left\langle \frac{E_1 + E_2 - 2E_1\nu_{21}}{1-\nu_{12}\nu_{21}} z \right\rangle,\\
C_{33} &= \frac{1}{4}\left\langle \frac{E_1 + E_2 + 2E_1\nu_{21}}{1-\nu_{12}\nu_{21}} z^2 \right\rangle, \; C_{34} = -\frac{1}{4}\left\langle \frac{E_1 - E_2}{1-\nu_{12}\nu_{21}} z^2 \right\rangle,\\
C_{44} &= \frac{1}{4}\left\langle \frac{E_1 + E_2 - 2E_1\nu_{21}}{1-\nu_{12}\nu_{21}} z^2 \right\rangle
\end{aligned} \tag{5}$$

- Problem 2: Plane Shear

Let us assume the two-dimensional kinematical field

$$\boldsymbol{u} = S_2 x_2 \boldsymbol{e}_1 + S_1 x_1 \boldsymbol{e}_2 - S_2 x_1 x_2 \boldsymbol{n}, \; \boldsymbol{\varphi} = -S_2(x_1 \boldsymbol{e}_1 - x_2 \boldsymbol{e}_2)$$

S_1 and S_2 are constants. From the kinematical field the strains are computed. With the help of the constitutive equations the forces and moments are estimated.

The corresponding three-dimensional strain tensor component is

$$\gamma_{12} = u^*_{1,2} + u^*_{2,1} = S_1 + S_2 z$$

After calculation the stresses we can compare the results of the two-dimensional and three-dimensional solutions

$$A_{44} =< G_{12} >, \; B_{42} = - < G_{12} z >, \; C_{22} =< G_{12} z^2 > \tag{6}$$

- Problem 3: Torsion

Let us introduce a deformable strip ($|x_1| \leq l_1, |x_2| < \infty$) under constant torsion moment at the boundaries $x_1 = \pm l_1$. Then one gets the two-dimensional kinematical field

$$\boldsymbol{u} = u_2(x_1)\boldsymbol{e}_2, \quad \boldsymbol{\varphi} = -\varphi_1(x_1)\boldsymbol{e}_1$$

and the force and moment tensors

$$\boldsymbol{T} = (A_{44}u_{2,1} - B_{42}\varphi_{2,1})\mathbf{a}_4 + (\Gamma_1 - \Gamma_2)\varphi_2 \boldsymbol{e}_2 \boldsymbol{n}, \qquad \boldsymbol{M} = (B_{42}u_{2,1} - C_{22}\varphi_{2,1})\mathbf{a}_2$$

The dual three-dimensional problem (strip $|x_1| \leq l_1, |x_2| < \infty, |z| \leq h/2$) results in the displacements

$$u_1^* = w^* = 0, \quad u_2^* = u_2^*(x_1, z),$$

the stress tensor

$$\boldsymbol{\sigma} = G_{12}\frac{\partial u_2^*}{\partial x_1}\mathbf{a}_4 + G_{2n}\frac{\partial u_2^*}{\partial z}(\boldsymbol{e}_2\boldsymbol{n} + \boldsymbol{n}\boldsymbol{e}_2)$$

and equilibrium equation

$$G_{12}\frac{\partial^2 u_2^*}{\partial x_1^2} + \frac{\partial}{\partial z}\left(G_{2n}\frac{\partial u_2^*}{\partial z}\right) = 0$$

The solution with respect to the boundary conditions $\sigma_n = \tau_{1n} = \tau_{2n} = 0$ at the top and the bottom surfaces $|z| = h/2$ can be obtained by the following Fourier's ansatz: $u_2^*(x_1, z) = X(x_1)Z(z)$ which yields a Sturm-Liouville problem

$$\frac{\mathrm{d}}{\mathrm{d}z}\left(G_{2n}\frac{\mathrm{d}Z}{\mathrm{d}z}\right) + \lambda_*^2 G_{12} Z = 0, \quad \left.\frac{\mathrm{d}Z}{\mathrm{d}z}\right|_{|z|=\frac{h}{2}} = 0$$

and

$$\frac{\mathrm{d}^2 X}{\mathrm{d}x_1^2} - \lambda_*^2 X = 0$$

The lowest non-trivial positive solution λ_* one obtains from

$$X(x_1) = B\frac{\sinh \lambda_* x_1}{\lambda_* \cosh \lambda_* l_1} \text{ and } u_2^* = BZ(z)\frac{\sinh \lambda_* x_1}{\lambda_* \cosh \lambda_* l_1}$$

Finally, after comparison of the two-dimensional and the three-dimensional solutions one gets

$$\lambda = \lambda_* = \sqrt{\frac{(\Gamma_1 - \Gamma_2)A_{44}}{A_{44}C_{22} - B_{42}^2}}$$

T_{12} and M_{12} obtained by the two-dimensional and the three-dimensional approaches are in a full agreement. For the kinematical fields one gets

$$< G_{12}(u_2^* - u_2 - z\varphi_2)^2 >= \min(u_2, \varphi_2)$$

$$u_2 = \frac{M_{12}^* < G_{12}z >}{< G_{12} >< G_{12}z^2 > - < G_{12}z^2 >}\frac{\sinh \lambda_* x_1}{\lambda_* \cosh \lambda_* l_1},$$

$$\varphi_2 = -\frac{M_{12}^* < G_{12} >}{< G_{12} >< G_{12}z^2 > - < G_{12}z^2 >}\frac{\sinh \lambda_* x_1}{\lambda_* \cosh \lambda_* l_1}$$

In addition, one has to analyze the similar problem for the second direction in the two-dimensional case ($|x_1| < \infty$, $|x_2| \le l_2$) as in the three-dimensional case ($|x_1| < \infty$, $|x_2| \le l_2$, $|z| \le h/2$) with the constant torsion moment at the boundary $|x_2| \le l_2$. The following final results can be obtained

$$\frac{\mathrm{d}}{\mathrm{d}z}\left(G_{1n}\frac{\mathrm{d}Z^*}{\mathrm{d}z}\right) + \eta^2 G_{12} Z^* = 0, \quad \left.\frac{\mathrm{d}Z^*}{\mathrm{d}z}\right|_{|z|=\frac{h}{2}} = 0, \quad \eta = \sqrt{\frac{(\Gamma_1 + \Gamma_2)A_{44}}{A_{44}C_{22} - B_{42}^2}}$$

Finally, we get the expressions for the transverse shear stiffness tensor components

$$\Gamma_1 = \frac{1}{2}(\lambda^2 + \eta^2)\frac{A_{44}C_{22} - B_{42}^2}{A_{44}}, \qquad \Gamma_2 = \frac{1}{2}(\eta^2 - \lambda^2)\frac{A_{44}C_{22} - B_{42}^2}{A_{44}} \tag{7}$$

From the above mentioned stiffness values one gets the classical stiffness tensors for the isotropic homogeneous plate. The basic geometrical property is the thickness h, the plate is symmetrical with respect to the mid-plane which results in $\boldsymbol{B} \equiv \mathbf{0}$. Let us assume the following material data: the Young's modulus E and the shear modulus $G = E/2(1+\nu)$, ν is the Poisson's ratio. All material properties are constant. The in-plane and out-of-plane stiffness parameters can be computed by

$$A_{11} = \frac{Eh}{2(1-\nu)}, \quad A_{22} = \frac{Eh}{2(1+\nu)} = A_{44} = Gh,$$
$$C_{33} = \frac{Eh^3}{24(1-\nu)}, \quad C_{44} = \frac{Eh^3}{24(1+\nu)} = C_{22} = \frac{Gh^3}{12}$$

The classical plate (bending) stiffness [85] follows as

$$C_{33} + C_{44} = \frac{Eh^3}{12(1-\nu^2)},$$

the transverse shear stiffness can be estimated by

$$\Gamma = \lambda^2 C_{22} \quad \text{with} \quad \frac{\mathrm{d}^2 Z}{\mathrm{d}z^2} + \lambda^2 Z = 0, \quad \left.\frac{\mathrm{d}Z}{\mathrm{d}z}\right|_{|z|=\frac{h}{2}} = 0$$

$\cos\lambda z = 0$ yields the smallest eigenvalue $\lambda = \frac{\pi}{h}$

$$\Gamma = \frac{\pi^2}{h^2}\frac{Gh^3}{12} = \frac{\pi^2}{12}Gh$$

The value $\pi^2/12$ is the same like the shear correction of [57], Reissner's estimate $5/6$ [76] is close to this value.

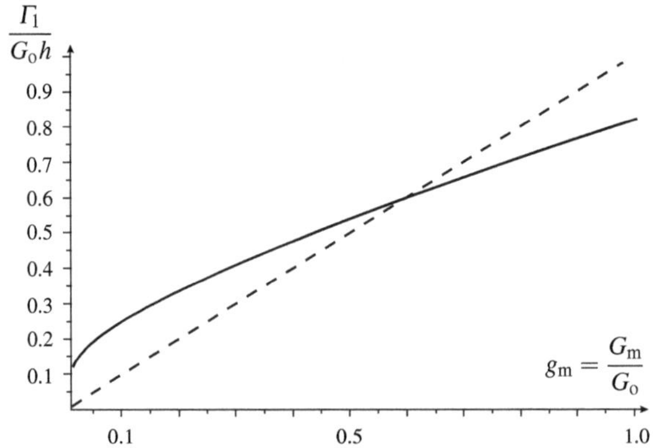

Fig. 11 Transverse shear stiffness of a FGM plate (Reissner's solution is presented by the *dashed* line, G_m, G_o shear modulus at the midplane, at outer faces, respectively)

The suggested approach can be applied in more general cases. As an example on Fig. 11 the transverse shear stiffness of a FGM plate is shown, for details see [1]. The following comments can be made

- Reissner's solution gives understated values of the transverse stiffness when the difference between elastic moduli is big enough.
- On the other hand, Reissner's solution gives overstated values when the elastic moduli do not differ.
- Reissner's formula gives us good coincidence with our results when the $g_m \sim 0.6$.

4 Nano-Sized Plates and Shells

The development of nanotechnologies extends the field of application of the classical or non-classical theories of plates and shells towards the new thin-walled structures. Nanomaterials have physical properties which are different from the bulk material. The classical linear elasticity can be extended to the nanoscale by taking into account the surface stresses. In particular, the surface stresses are responsible for the size-effect, that means the material properties of a specimen depend on its size. For example, Young's modulus of a cylindrical specimen increases significantly, when the cylinder diameter becomes very small. The surface stresses are the generalization of the scalar surface tension which is well-known phenomenon in the theory of capillarity.

The investigations of the surface phenomena were initiated by Laplace, Young and Gibbs. A summary of the investigations are given in the reviews of [68] or [23]. In [32, 68, 84] the surface stresses are considered. Recently two-dimensional theories of nanosized plates and shells were suggested. The theory of elasticity with surface stresses is applied to the modifications of the two-dimensional theories of nanosized plates in [33, 53] or [9]. Various theories of plates are formulated. The approaches can be classified, for example, by the starting point of the derivation. This can be the well-known three-dimensional continuum mechanics equations. In contrast, one can introduce à priory a two-dimensional deformable surface which is the basis for a more natural formulation of the two-dimensional governing equations. This so-called direct approach should be supplemented by the theoretical or experimental determination of the material parameters included in the constitutive equations.

4.1 Basic Equations of Linear Elasticity with Surface Stresses

Let V is a bounded domain in $I\!R^3$ with sufficiently regular boundary that a body occupies. Here we consider problems with mixed boundary conditions. Suppose Ω_1, a nonempty part of the boundary surface Ω of V, to be fixed: $\mathbf{u}|_{\Omega_1} = \mathbf{0}$ ($\mathbf{u}$ is the three-dimensional displacement vector). On the rest part $\Omega_2 = \Omega \backslash \Omega_1$ it is defined the stress vector $\mathbf{t}$ expressed through a given load $\boldsymbol{\phi}$ and $\mathbf{t}_S$ (the stress vector due the surface stresses) by the formula

$$\mathbf{t} = \boldsymbol{\phi} + \mathbf{t}_S,$$

where $\mathbf{t}_S$ is determined through the surface stress tensor $\boldsymbol{\tau}$, see [32, 68]. As a result, we have

$$\nabla \cdot \boldsymbol{\sigma} + \rho \mathbf{f} = \rho \ddot{\mathbf{u}}, \quad \mathbf{x} \in V, \tag{8}$$

$$\mathbf{u}|_{\Omega_1} = \mathbf{0}, \quad \mathbf{n} \cdot \boldsymbol{\sigma}|_{\Omega_2} = \mathbf{t}, \quad \mathbf{x} \in \Omega, \tag{9}$$

where $\boldsymbol{\sigma}$ is the stress tensor, ∇ the 3D gradient operator (3D nabla operator), ρ the body density, $\mathbf{f}$ the density of the volume forces, $\mathbf{n}$ the external unit normal to Ω, and the dot over a quantity denotes its partial derivative with respect to the time t. The surface stress vector is defined by

$$\mathbf{t}_S = \widetilde{\nabla} \cdot \boldsymbol{\tau}, \tag{10}$$

where $\boldsymbol{\tau}$ is the surface stress tensor on Ω and $\widetilde{\nabla}$ is the nabla operator on the surface Ω that relates with ∇ by the formula

$$\widetilde{\nabla} = \nabla - \mathbf{n}\frac{\partial}{\partial z},$$

and z is the coordinate along the normal to Ω.

Let us consider the special problem when the static conditions are given on the whole boundary

$$\mathbf{n} \cdot \boldsymbol{\sigma}|_{\Omega} = \mathbf{t}, \quad \mathbf{x} \in \Omega \tag{11}$$

For simplicity, we restrict ourselves to an isotropic material. The constitutive equation for the material is Hooke's law

$$\boldsymbol{\sigma} = 2\mu\mathbf{e} + \lambda\mathbf{I}\operatorname{tr}\mathbf{e} \quad \text{with} \quad \mathbf{e} = \mathbf{e}(\mathbf{u}) \equiv \frac{1}{2}\left[\nabla\mathbf{u} + (\nabla\mathbf{u})^{\mathrm{T}}\right] \tag{12}$$

For the surface stresses we assume the following constitutive equation:

$$\boldsymbol{\tau} = \tau_0\mathbf{a} + 2\mu_S\boldsymbol{\varepsilon} + \lambda_S\mathbf{a}\operatorname{tr}\boldsymbol{\varepsilon} \quad \text{with} \quad \boldsymbol{\varepsilon} = \boldsymbol{\varepsilon}(\mathbf{v}) \equiv \frac{1}{2}\left[\widetilde{\nabla}\mathbf{v}\cdot\mathbf{a} + \mathbf{a}\cdot(\widetilde{\nabla}\mathbf{v})^{\mathrm{T}}\right], \tag{13}$$

where $\mathbf{v}$ is the displacement of the film point $\mathbf{x}$ of Ω_2, τ_0 is the residual (initial) surface tension. Here $\mathbf{I}$ and $\mathbf{a} \equiv \mathbf{I} - \mathbf{n}\otimes\mathbf{n}$ are the three- and two-dimensional unit tensors, respectively, λ and μ are Lamé's coefficients of the bulk material whereas λ_S and μ_S are the elastic characteristics of the surface film Ω_2 (they are the surface analogues of Lamé's coefficients), $\mathbf{e}$ is the small strain tensor, and $\boldsymbol{\varepsilon}$ is the surface strain tensor. Following [32], we use the non-separation condition

$$\mathbf{u}|_{\Omega_2} = \mathbf{v}$$

This explicitly states that the displacements of the surface film Ω_2 coincide with the body displacements on the boundary. There are more general relations for the surface stresses that include residual stresses, anisotropy and other factors [6, 32, 69].

In equilibrium, the dynamic Eq. (8) changes to

$$\nabla\cdot\boldsymbol{\sigma} + \rho\mathbf{f} = \mathbf{0}. \tag{14}$$

Thus the equilibrium boundary value problem for an elastic body with surface stresses consists of Eq. (14) and the boundary conditions

$$\mathbf{u}|_{\Omega_1} = \mathbf{0}, \quad \left(\mathbf{n}\cdot\boldsymbol{\sigma} - \widetilde{\nabla}\cdot\boldsymbol{\tau}\right)\big|_{\Omega_2} = \boldsymbol{\phi}, \tag{15}$$

where $\boldsymbol{\sigma}$ and $\boldsymbol{\tau}$ satisfy relations (12) and (13), respectively. In Eq. (13) we set

$$\boldsymbol{\varepsilon} = \boldsymbol{\varepsilon}(\mathbf{u}) \equiv \frac{1}{2}\left[\widetilde{\nabla}\mathbf{u}\cdot\mathbf{a} + \mathbf{a}\cdot(\widetilde{\nabla}\mathbf{u})^{\mathrm{T}}\right]\big|_{\Omega_2}$$

If $\Omega_2 = \Omega$ then part Ω_1 is absent.

4.2 Transition to the Theory of Plates and Shells

Let us consider thin-walled 3D solid called also shell-like body (Fig. 12). We assume that the volume of the shell-like body V is bounded by two faces $\Omega_\pm$ and lateral surface Ω_ν. We also introduce the base (middle) surface ω. The radius-vector of

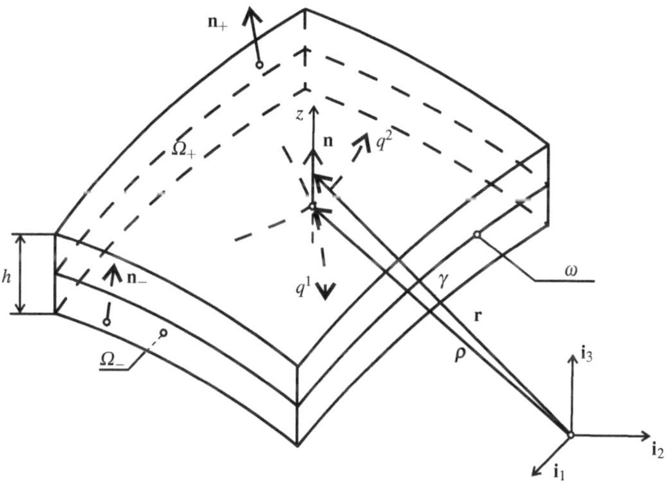

Fig. 12 Shell-like body

material points $\mathbf{r}$ is given as follows

$$\mathbf{r} = \boldsymbol{\rho}(q^1, q^2) + z\mathbf{n},$$

where $\boldsymbol{\rho}$ is the radius-vector of ω, $\mathbf{n}$ is the unit normal to ω, z is the transverse coordinate, $z \in [-h/2, h/2]$, h is the shell thickness, q^1, q^2 the coordinates on ω. Radius-vectors of $\Omega_\pm$ are given by $\mathbf{r}_\pm = \boldsymbol{\rho} \pm \mathbf{n}h/2$. For the sake of simplicity we assume that h is a constant.

Let as recall basic formulas used in the tensor calculus for description of tensor fields near ω, see [25, 49]. The basic and reciprocal bases on ω are given by

$$\boldsymbol{\rho}_\alpha = \frac{\partial \boldsymbol{\rho}}{\partial q^\alpha}, \quad \boldsymbol{\rho}_\alpha \cdot \boldsymbol{\rho}^\beta = \delta_\alpha^\beta, \quad \alpha, \beta = 1, 2,$$

where δ_α^β is the Kronecker symbol. The surface nabla-operator on ω is defined as

$$\widetilde{\nabla} = \boldsymbol{\rho}^\alpha \frac{\partial}{\partial q^\alpha}$$

We use q^1, q^2, z as coordinates in the neighborhood of ω. Then we have

$$\mathbf{r}_\alpha = \frac{\partial \mathbf{r}}{\partial q^\alpha} = \boldsymbol{\rho}_\alpha + z\frac{\partial \mathbf{n}}{\partial q^\alpha} = (\mathbf{a} - z\mathbf{b}) \cdot \boldsymbol{\rho}_\alpha, \quad \mathbf{r}_3 = \mathbf{r}^3 = \mathbf{n},$$
$$\mathbf{r}^\alpha = (\mathbf{a} - z\mathbf{b})^{-1} \cdot \boldsymbol{\rho}^\alpha, \quad \mathbf{r}_\alpha \cdot \mathbf{r}^\beta = \delta_\alpha^\beta, \quad \mathbf{b} = -\widetilde{\nabla}\mathbf{n},$$

where $\mathbf{b}$ is the curvature tensor of ω. Operators ∇ and $\widetilde{\nabla}^{\pm}$ are reduced to

$$\nabla = \mathbf{r}^{\alpha}\frac{\partial}{\partial q^{\alpha}} + \mathbf{n}\frac{\partial}{\partial z} = (\mathbf{a} - z\mathbf{b})^{-1}\cdot\boldsymbol{\rho}^{\alpha}\frac{\partial}{\partial q^{\alpha}} + \mathbf{n}\frac{\partial}{\partial z}$$
$$= (\mathbf{a} - z\mathbf{b})^{-1}\cdot\widetilde{\nabla} + \mathbf{n}\frac{\partial}{\partial z},$$

$$\widetilde{\nabla}^{\pm} = \left(\mathbf{a} \mp \frac{h}{2}\mathbf{b}\right)^{-1}\cdot\boldsymbol{\rho}^{\alpha}\frac{\partial}{\partial q^{\alpha}} = \left(\mathbf{a} \mp \frac{h}{2}\mathbf{b}\right)^{-1}\cdot\widetilde{\nabla}$$

Let us note that here the inverse tensor of $\mathbf{a} - z\mathbf{b}$ is defined according to the rule

$$(\mathbf{a} - z\mathbf{b})^{-1}\cdot(\mathbf{a} - z\mathbf{b}) = (\mathbf{a} - z\mathbf{b})\cdot(\mathbf{a} - z\mathbf{b})^{-1} = \mathbf{a}$$

We assume that on $\Omega_{\pm}$ the surface stresses act, so $\Omega_2 = \Omega_+ \bigcup \Omega_-$. As a results, this leads to the following boundary conditions on $\Omega_{\pm}$

$$\left(\mathbf{n}_{\pm}\cdot\boldsymbol{\sigma} \mp \widetilde{\nabla}^{\pm}\cdot\boldsymbol{\tau}_{S}^{\pm}\right)\big|_{\Omega_{\pm}} = \boldsymbol{\phi}_{\pm} \tag{16}$$

In (16) $\mathbf{n}_{\pm}$ are normals to $\Omega_{\pm}$ (see Fig. 12), $\boldsymbol{\tau}_{S}^{\pm}$ and $\boldsymbol{\phi}_{\pm}$ are surface stresses and loads on $\Omega_{\pm}$,

$$\boldsymbol{\tau}_{\pm} = \tau_0^{\pm}\mathbf{A} + \lambda_S^{\pm}\mathbf{a}\,\mathrm{tr}\,\boldsymbol{\varepsilon}_{\pm} + 2\mu_S^{\pm}\boldsymbol{\varepsilon}_{\pm}, \quad 2\boldsymbol{\varepsilon}_{\pm} = (\widetilde{\nabla}^{\pm}\mathbf{u}_S^{\pm})\cdot\mathbf{a} + \mathbf{a}\cdot(\widetilde{\nabla}^{\pm}\mathbf{u}_S^{\pm})^{\mathrm{T}}$$

Here $\mu_S^{\pm}$ $\lambda_S^{\pm}$ are the surface elastic moduli, and $\tau_0^{\pm}$ are residual surface stresses on $\Omega_{\pm}$. Let us note that the surface nabla operators $\widetilde{\nabla}^{\pm}$ are $\Omega_{\pm}$ differ from each other, in general.

For transition to the 2D equations of plates and shells we use the through-the-thickness integration procedure presented in, for example, [19, 49, 51]. Integrating (14) with respect to z and taking into account (16), we obtain

$$\widetilde{\nabla}\cdot\mathbf{T} + G_{+}\widetilde{\nabla}^{+}\cdot\boldsymbol{\tau}_{+} + G_{-}\widetilde{\nabla}^{-}\cdot\boldsymbol{\tau}_{-} + \mathbf{q} = \mathbf{0}, \tag{17}$$

where

$$\mathbf{T} = \langle(\mathbf{a} - z\mathbf{b})^{-1}\cdot\boldsymbol{\sigma}\rangle, \tag{18}$$

$$\mathbf{q} = G_{+}\boldsymbol{\varphi}_{+} - G_{-}\boldsymbol{\varphi}_{-} + \langle\mathbf{f}\rangle, \quad \langle(\ldots)\rangle = \int_{-h/2}^{h/2}(\ldots)G\,dz,$$

$$G = G(z) \equiv \det(\mathbf{a} - z\mathbf{b}), \quad G_{\pm} = G(\pm h/2).$$

$\mathbf{T}$ is the stress resultant tensor and $\mathbf{q}$ is the surface loads.

Cross-multiplying (14) by $z\mathbf{n}$ and integrating through the thickness using (16), we obtain the second equilibrium equation

$$\widetilde{\nabla} \cdot \mathbf{M} + \mathbf{T}_\times + \mathbf{m} + \frac{h}{2} G_+ \mathbf{n} \times \widetilde{\nabla}^+ \cdot \boldsymbol{\tau}_+ - \frac{h}{2} G_- \mathbf{n} \times \widetilde{\nabla}^- \cdot \boldsymbol{\tau}_- = \mathbf{0}, \qquad (19)$$

$$\mathbf{M} = -\langle (\mathbf{a} - z\mathbf{b})^{-1} \cdot z\boldsymbol{\sigma} \times \mathbf{n} \rangle, \qquad (20)$$

$$\mathbf{m} = \frac{h}{2} G_+ \mathbf{n} \times \boldsymbol{\varphi}_+ + \frac{h}{2} G_- \mathbf{n} \times \boldsymbol{\varphi}_- + \langle z\mathbf{n} \times \mathbf{f} \rangle.$$

Here $\mathbf{M}$ is the moment stress tensor and $\mathbf{m}$ is the surface moments, the subindex $\times$ stands for vectorial invariant of stress tensor [92], see [49]. In particular, for a diad $\mathbf{a} \otimes \mathbf{b}$ it is given by $(\mathbf{a} \otimes \mathbf{b})_\times = \mathbf{a} \times \mathbf{b}$.

Omitting algebraic manipulations and assuming $h\|\mathbf{b}\| \ll 1$ we reduce with accuracy of $O(h\|\mathbf{b}\|)$ Eqs. (18) and (20) to more simple expressions

$$\mathbf{T} = \langle \mathbf{a} \cdot \boldsymbol{\sigma} \rangle, \quad \mathbf{M} = -\langle \mathbf{a} \cdot z\boldsymbol{\sigma} \times \mathbf{n} \rangle$$

Thus, (17) and (19) transform to

$$\widetilde{\nabla} \cdot \mathbf{T} + \widetilde{\nabla} \cdot (\boldsymbol{\tau}_+ + \boldsymbol{\tau}_-) + \mathbf{q} = \mathbf{0}, \qquad (21)$$

$$\widetilde{\nabla} \cdot \mathbf{M} - \frac{h}{2} \widetilde{\nabla} \cdot \left[(\boldsymbol{\tau}_+ - \boldsymbol{\tau}_-) \times \mathbf{n} \right] + \mathbf{T}_\times + \left[(\boldsymbol{\tau}_+ - \boldsymbol{\tau}_-) \cdot \mathbf{b} \right]_\times + \mathbf{m} = \mathbf{0} \qquad (22)$$

Since $\mathbf{m} \cdot \mathbf{n} = 0$, from (22) it follows the so-called sixth equilibrium equation in the form

$$\mathbf{M} \cdot \cdot \mathbf{b} + \mathbf{T}_\times \cdot \mathbf{n} = 0$$

Equations (21) and (22) dictate the form of effective stress resultants tensors $\mathbf{T}^*$ and $\mathbf{M}^*$ as follows

$$\mathbf{T}^* = \mathbf{T} + \mathbf{T}_S, \quad \mathbf{M}^* = \mathbf{M} + \mathbf{M}_S, \qquad (23)$$

$$\mathbf{T}_S = \boldsymbol{\tau}_+ + \boldsymbol{\tau}_-, \quad \mathbf{M}_S = -\frac{h}{2} (\boldsymbol{\tau}_+ - \boldsymbol{\tau}_-) \times \mathbf{n}$$

For description of the shell deformations we use the standard approximation of the first order shear deformable plate and shell theory

$$\mathbf{u}(q^1, q^2, z) = \mathbf{w}(q^1, q^2) - z\boldsymbol{\vartheta}(q^1, q^2), \quad \mathbf{n} \cdot \boldsymbol{\vartheta} = 0. \qquad (24)$$

Within the theory it is assumed that the rotation vector $\boldsymbol{\vartheta}$ is kinematically independent on the vector of translations $\mathbf{w}$ of the middle surface. From (24) we derive the following formulae

$$\mathbf{u}_S^{\pm} = \mathbf{w} \mp \frac{h}{2}\boldsymbol{\vartheta}, \quad \boldsymbol{\varepsilon}_{\pm} = \boldsymbol{\varepsilon} \mp \frac{h}{2}\boldsymbol{\kappa}, \tag{25}$$

where

$$\boldsymbol{\varepsilon} = \frac{1}{2}\left(\widetilde{\nabla}\mathbf{w}\cdot\mathbf{a} + \mathbf{a}\cdot(\widetilde{\nabla}\mathbf{w})^{\mathrm{T}}\right), \quad \boldsymbol{\kappa} = \frac{1}{2}\left(\widetilde{\nabla}\boldsymbol{\vartheta}\cdot\mathbf{a} + \mathbf{a}\cdot(\widetilde{\nabla}\boldsymbol{\vartheta})^{\mathrm{T}}\right)$$

are 2D strain measures. Using (25), we obtain the following expressions of $\boldsymbol{\tau}_{\pm}$

$$\boldsymbol{\tau}_{\pm} = \tau_0^{\pm}\mathbf{a} + \lambda_S^{\pm}\mathbf{a}\operatorname{tr}\boldsymbol{\varepsilon} + 2\mu_S^{\pm}\boldsymbol{\varepsilon} \mp \frac{h}{2}\left(\lambda_S^{\pm}\mathbf{a}\operatorname{tr}\boldsymbol{\kappa} + 2\mu_S^{\pm}\boldsymbol{\kappa}\right)$$

Thus, we have the formulae

$$\begin{aligned}
\boldsymbol{\tau}_{+} + \boldsymbol{\tau}_{-} &= (\tau_0^{+} + \tau_0^{-})\mathbf{a} + (\lambda_S^{+} + \lambda_S^{-})\mathbf{a}\operatorname{tr}\boldsymbol{\varepsilon} + 2(\mu_S^{+} + \mu_S^{-})\boldsymbol{\varepsilon} \\
&\quad - \frac{h}{2}\left[(\lambda_S^{+} - \lambda_S^{-})\mathbf{a}\operatorname{tr}\boldsymbol{\kappa} + 2(\mu_S^{-} - \mu_S^{-})\boldsymbol{\kappa}\right], \\
\boldsymbol{\tau}_{+} - \boldsymbol{\tau}_{-} &= (\tau_0^{+} - \tau_0^{-})\mathbf{a} + (\lambda_S^{+} - \lambda_S^{-})\mathbf{a}\operatorname{tr}\boldsymbol{\varepsilon} + 2(\mu_S^{+} - \mu_S^{-})\boldsymbol{\varepsilon} \\
&\quad - \frac{h}{2}\left[(\lambda_S^{+} + \lambda_S^{-})\mathbf{a}\operatorname{tr}\boldsymbol{\kappa} + 2(\mu_S^{-} + \mu_S^{-})\boldsymbol{\kappa}\right]
\end{aligned}$$

In what follows we consider the same surface properties for both faces, that is $\tau_0^{+} = \tau_0^{-} = \tau_0$, $\mu_S^{+} = \mu_S^{-} = \mu_S$, $\lambda_S^{+} = \lambda_S^{-} = \lambda_S$. In this case the latter formulae can be simplified as follows

$$\boldsymbol{\tau}_{+} + \boldsymbol{\tau}_{-} = 2\tau_0\mathbf{a} + 2\lambda_S\mathbf{a}\operatorname{tr}\boldsymbol{\varepsilon} + 4\mu_S\boldsymbol{\varepsilon}, \qquad \boldsymbol{\tau}_{+} - \boldsymbol{\tau}_{-} = -h\left(\lambda_S\mathbf{a}\operatorname{tr}\boldsymbol{\kappa} + 2\mu_S\boldsymbol{\kappa}\right)$$

This leads to the following expressions for stress resultants:

$$\begin{aligned}
\mathbf{T}_S &= 2\tau_0\mathbf{a} + C_1^S\boldsymbol{\varepsilon} + C_2^S\mathbf{a}\operatorname{tr}\boldsymbol{\varepsilon}, \qquad \mathbf{M}_S = -\left[D_1^S\boldsymbol{\kappa} + D_2^S\mathbf{a}\operatorname{tr}\boldsymbol{\kappa}\right]\times\mathbf{n}, \\
C_1^S &= 4\mu_S, \quad C_2^S = 2\lambda_S, \quad D_1^S = h^2\mu_S, \quad D_2^S = h^2\lambda_S/2
\end{aligned} \tag{26}$$

Taking into account (23) from (26) it follows that the surface stresses do not influence on the transverse shear forces and the transverse shear stiffness. Indeed, $\mathbf{T}_S\cdot\mathbf{n} = \mathbf{0}$.

For $\mathbf{T}$ and $\mathbf{M}$ we assume the following constitutive relations

$$\begin{aligned}
\mathbf{T}\cdot\mathbf{a} - \frac{1}{2}(\mathbf{M}\cdot\cdot\mathbf{b})\mathbf{a}\times\mathbf{n} &= \frac{\partial W_S}{\partial\boldsymbol{\varepsilon}}, \quad \mathbf{T}\cdot\mathbf{n} = \frac{\partial W_S}{\partial\boldsymbol{\gamma}}, \quad \mathbf{M} = \frac{\partial W_S}{\partial\boldsymbol{\kappa}}, \\
2W_S &= \boldsymbol{\varepsilon}\cdot\cdot\mathbf{C}\cdot\cdot\boldsymbol{\varepsilon} + \boldsymbol{\kappa}\cdot\cdot\mathbf{D}\cdot\cdot\boldsymbol{\kappa} + \Gamma\boldsymbol{\gamma}\cdot\boldsymbol{\gamma}
\end{aligned}$$

Here W_S is the surface strain energy density, $\mathbf{C}$ and $\mathbf{D}$ are fourth-order tangential and bending stiffness tensors, respectively, $\boldsymbol{\gamma}$ is the vector of transverse shear deformations, $\boldsymbol{\gamma} = \widetilde{\nabla}(\mathbf{w}\cdot\mathbf{n}) - \boldsymbol{\vartheta}$, and Γ is the transverse shear stiffness. In the case of an isotopic shells the tangential and bending stiffness tensors take the form

$$\mathbf{C} = C_{11}\mathbf{a}_1\mathbf{a}_1 + C_{22}(\mathbf{a}_2\mathbf{a}_2 + \mathbf{a}_4\mathbf{a}_4), \quad \mathbf{D} = D_{22}(\mathbf{a}_2\mathbf{a}_2 + \mathbf{a}_4\mathbf{a}_4) + D_{33}\mathbf{a}_3\mathbf{a}_3 \quad (27)$$

Components C_{11}, C_{22}, D_{22}, D_{33} and Γ are given by

$$\begin{aligned} C_{11} &= \frac{Eh}{2(1-\nu)}, & C_{22} &= \frac{Eh}{2(1+\nu)}, \\ D_{22} &= \frac{Eh^3}{24(1+\nu)}, & D_{33} &= \frac{Eh^3}{24(1-\nu)}, \quad \Gamma = k\mu h, \\ E &= 2\mu(1+\nu), & \nu &= \frac{\lambda}{2(\lambda+\mu)}, \\ C \equiv C_{11} + C_{22} &= \frac{Eh}{1-\nu^2}, & D \equiv D_{11} + D_{22} &= \frac{Eh^3}{12(1-\nu^2)}, \end{aligned}$$

where C, D are tangential and bending stiffness of the shell, E and ν are the Young modulus and Poisson ratio of the shell material, and k is similar to the shear correction factor, see [31].

4.3 Theory of Plates

In the case of the plates the constitutive equations take more simple form. Indeed, here $\mathbf{b} = \mathbf{0}$, $\mathbf{n} = \mathbf{i}_3$. Effective stress resultants in the case of symmetric plates are given by [9, 24]

$$\begin{aligned} \mathbf{T}^* &= C_1\boldsymbol{\varepsilon} + C_2\mathbf{a}\operatorname{tr}\boldsymbol{\varepsilon} + \Gamma\boldsymbol{\gamma}\otimes\mathbf{i}_3, \quad \mathbf{M}^* = -\left[D_1\boldsymbol{\kappa} + D_2\mathbf{a}\operatorname{tr}\boldsymbol{\kappa}\right]\times\mathbf{i}_3, \\ C_1 &= C(1-\nu) + 4\mu^S, \quad C_2 = C\nu + 2\lambda^S, \\ D_1 &= D(1-\nu) + h^2\mu^S, \quad D_2 = D\nu + h^2\lambda^S/2 \end{aligned} \quad (28)$$

The effective tangential and bending stiffness are

$$C_{\text{eff}} \equiv C_1 + C_2 = C + 4\mu_S + 2\lambda_S, \quad D_{\text{eff}} \equiv D_1 + D_2 = D + h^2\mu_S + h^2\lambda_S/2$$

Constitutive relations (28) with the equilibrium Eqs. (21) and (22) lead to equilibrium equations for translations $\mathbf{w}$ and rotations $\boldsymbol{\vartheta}$. In particular, the equation for deflection $w = \mathbf{w}\cdot\mathbf{i}_3$ can be transformed to

$$D_{\text{eff}}\widetilde{\Delta}\widetilde{\Delta}w = \widetilde{\nabla}\cdot\mathbf{m} - \frac{D_{\text{eff}}}{\Gamma}\widetilde{\Delta}q_n + q_n, \qquad q_n = \mathbf{q}\cdot\mathbf{i}_3, \quad \widetilde{\Delta} = \widetilde{\nabla}\cdot\widetilde{\nabla}$$

Let us consider dependence of D_{eff}, C_1, C_2, D_1, D_2 on thickness h. The diagram D_{eff} vs. h is given in Fig. 13. Dimensionless parameters $\overline{C}_1 = C_1/C(1-\nu)$, $\overline{C}_2 = C_2/C\nu$, $\overline{D}_1 = D_1/D(1-\nu)$, $\overline{D}_2 = D_2/D\nu$ are shown in Fig. 14, [9, 24] for details. Here the material parameters take the values $\mu = 34.7$ GPa, $\nu = 0.3$, $\lambda_S = -3.48912$ N/m, $\mu_S = 6.2178$ N/m.

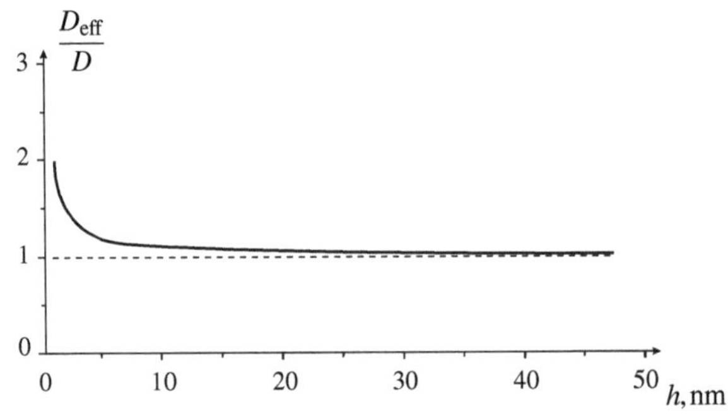

Fig. 13 Dependence of the bending stiffness on the thickness

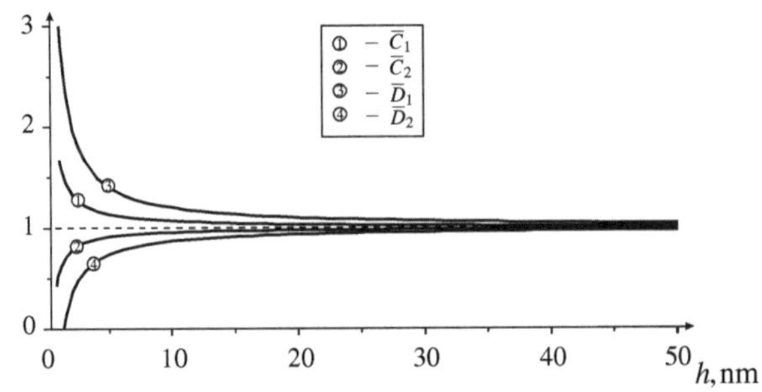

Fig. 14 Dependencies of the stiffness parameters on the thickness

4.4 Comparison with the Theory of Three-Layered Plates

Obviously, that there is similarity between the theory of plate with surface stresses and the theory of three-layered plates. Let us consider tension of a strip subjected by force P for two cases. The first is the tension of a strip with surface stresses while the second one is the tension of a sandwich plate, see Fig. 15. Let σ be stress in the bulk, τ be surface stress, and τ_{f} be the stress in the faces. Then we have elementary formulae

$$P = \sigma h + 2\tau, \quad P = \sigma(h - 2h_{\mathrm{f}}) + 2\tau_{\mathrm{f}} h_{\mathrm{f}},$$

where h and h_{f} are the thickness and thickness of surface layer, respectively. It is clear that

$$\tau = (\tau_{\mathrm{f}} - \sigma) h_{\mathrm{f}}$$

This gives one the possibility to interpret the surface stress τ as excess stress resultant for the surface layer of thickness h_{f}. As a result, the case of surface stresses can be obtained when $h_{\mathrm{f}} \to 0$: $\tau = \lim\limits_{h_{\mathrm{f}} \to 0} \tau_{\mathrm{f}} h_{\mathrm{f}}$.

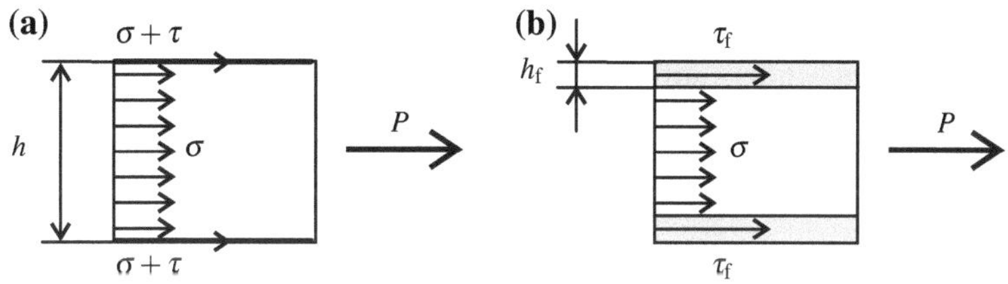

Fig. 15 Tension of **a** a strip with surface tension; **b** a three-layered strip

For detailed analysis we use the model of three-layered plates presented above. Let us consider the isotropic three-layered plate, see Fig. 15b. Here h_c is the thickness of the core, h_f the thickness of faces, and $h_c \gg h_f$. The thickness of the plate is $h = h_c + 2h_f$. The material properties are given by Young's moduli E_c, E_f and Poisson's ratio ν_c and ν_f of faces and core, respectively (or by shear moduli μ_c, μ_f).

Constitutive equations of three-layered plate are given by

$$\mathbf{T} = C_1 \boldsymbol{\varepsilon} + C_2 \mathbf{a} \operatorname{tr} \boldsymbol{\varepsilon} + \Gamma \boldsymbol{\gamma} \otimes \mathbf{i}_3, \quad \mathbf{M} = -\left[D_1 \boldsymbol{\kappa} + D_2 \mathbf{a} \operatorname{tr} \boldsymbol{\kappa}\right] \times \mathbf{i}_3, \tag{29}$$

where stiffness parameters can be computed by

$$\begin{gathered}
C_1 = 2C_{22}, \quad C_2 = C_{11} - C_{22}, \\
D_1 = 2D_{22}, \quad D_2 = D_{33} - D_{22}, \quad \Gamma = \ell^2 D_{22}, \\
C_{11} = \frac{1}{2}\left(\frac{2E_f h_f}{1-\nu_f} + \frac{E_c h_c}{1-\nu_c}\right), \quad C_{22} = \frac{1}{2}\left(\frac{2E_f h_f}{1+\nu_f} + \frac{E_c h_c}{1+\nu_c}\right), \\
D_{22} = \frac{1}{24}\left[\frac{E_f(h^3 - h_c^3)}{1+\nu_f} + \frac{E_c h_c^3}{1+\nu_c}\right], \quad D_{33} = \frac{1}{24}\left[\frac{E_f(h^3 - h_c^3)}{1-\nu_f} + \frac{E_c h_c^3}{1-\nu_c}\right],
\end{gathered}$$

where ℓ is the minimal non-zero root of the equation

$$\mu_0 \cos \ell \frac{h_f}{2} \cos \ell \frac{h_c}{2} - \sin \ell \frac{h_f}{2} \sin \ell \frac{h_c}{2} = 0, \quad \mu_0 = \mu_c / \mu_f$$

The bending stiffness of three-layered plate is

$$D_{\text{eff}} = D_{33} + D_{44} = \frac{1}{12}\left[\frac{E_f(h^3 - h_c^3)}{1-\nu_f^2} + \frac{E_c h_c^3}{1-\nu_c^2}\right]$$

Comparing (28) and (29), we can determine λ_S and μ_S within the parameters of the surface layers that is with parameters E_f, ν_f and h_f. Assuming $E_c = E$, $\nu_c = \nu$ and comparing tangential stiffness of three-layered plate and plate with surface stresses when $h_f \to 0$ with accuracy of $O(h_f^2)$, we obtain that

$$\mu_S \approx \frac{E_f h_f}{2(1+\nu_f)} \equiv \mu_f h_f, \quad \lambda_S \approx \frac{\nu_f E_f h_f}{1-\nu_f^2} \equiv \lambda_f h_f \frac{1-2\nu_f}{1-\nu_f}, \tag{30}$$

where λ_f is the Lamé modulus of the surface layer. The same formulas follow from comparison of the bending stiffness when $h_f \to 0$. Thus, we derive

$$\mu_S = \lim_{h_f \to 0} \mu_f h_f, \quad \lambda_S = \lim_{h_f \to 0} \lambda_f \frac{1 - 2\nu_f}{1 - \nu_f} h_f$$

The latter relations determine the surface elastic moduli μ_S and λ_S through the elastic moduli of the surface layer and its thickness. Formulas (30) are exact $h_f \to 0$. For finite values of h_f the accuracy is $O(h_f^2)$. A more general model of plates with surface stresses is presented by [11].

References

1. Altenbach, H., Eremeyev, V.: Direct approach-based analysis of plates composed of functionally graded materials. Arch. Appl. Mech. **78**, 775–794 (2008a)
2. Altenbach, H., Eremeyev, V.A.: Analysis of the viscoelastic behavior of plates made of functionally graded materials. ZAMM **88**(5), 332–341 (2008b)
3. Altenbach, H., Eremeyev, V.A.: On the shell theory on the nanoscale with surface stresses. Int. J. Eng. Sci. **49**(12), 1294–1301 (2011a)
4. Altenbach, H., Eremeyev, V.A. (eds.): Shell-like structures–non-classical theories and applications. Advanced Structured Materials, pp. 549–560. Springer, Berlin (2011b)
5. Altenbach, H., Eremeyev, V.A.: Cosserat-type shells. In: Altenbach, H., Eremeyev, V.A. (eds.) Generalized Continua from the Theory to Engineering Applications, pp. 131–178. Springer, Vienna (2013). CISM International Centre for Mechanical Sciences No. 541
6. Altenbach, H., Morozov, N.F. (eds.): Surface Effects in Solid Mechanics. Springer, Heidelberg (2013)
7. Altenbach, H., Zhilin, P.A.: A general theory of elastic simple shells (in Russ.). Uspekhi Mekhaniki **11**(4), 107–148 (1988)
8. Altenbach, H., Altenbach, J., Kissing, W.: Mechanics of Composite Structural Elements. Springer, Berlin (2004)
9. Altenbach, H., Eremeev, V.A., Morozov, N.F.: On equations of the linear theory of shells with surface stresses taken into account. Mech. Solids **45**(3), 331–342 (2010a)
10. Altenbach, H., Eremeyev, V.A., Morozov, N.F.: Surface viscoelasticity and effective properties of thin-walled structures at the nanoscale. Int. J. Eng. Sci. **59**, 83–89 (2012)
11. Altenbach, H., Eremeyev, V.A., Morozov, N.F.: Mechanical properties of materials considering surface effects. In: IUTAM Symposium on Surface Effects in the Mechanics of Nanomaterials and Heterostructures, pp. 105–115, Springer, Heidelberg (2013)
12. Altenbach, J., Altenbach, H., Eremeyev, V.A.: On generalized Cosserat-type theories of plates and shells—a short review and bibliography. Arch. Appl. Mech. **80**, 73–92 (2010b)
13. Ambarcumyan, S.A.: Theory of Anisotropic Plates: Strength, Stability, and Vibrations. Hemispher Publishing, Washington (1991)
14. Ashoori Movassagh, A., Mahmoodi, M.: A micro-scale modeling of Kirchhoff plate based on modified strain-gradient elasticity theory. Eur. J. Mech. A. Solids **40**, 50–59 (2013)
15. Buzea, C., Pacheco, I., Robbie, K.: Nanomaterials and nanoparticles: Sources and toxicity. Biointerphases **4**(2), MR17–MR71 (2007)
16. Capriz, G.: Continua with Microstructure. Springer, New York (1989)
17. Cauchy, A.L.: Sur l'équilibre et le mouvement d'une plaque solide. Exercises Mathématiques **3**, 328–355 (1828)
18. Challamel, N., Ameur, M.: Out-of-plane buckling of microstructured beams: Gradient elasticity approach. J. Eng. Mechanics **139**(8), 1036–1046 (2013)

19. Chróścielewski, J., Makowski, J., Pietraszkiewicz, W.: Statics and Dynamics of Multyfolded Shells. Wydawnictwo IPPT PAN, Warszawa, Nonlinear Theory and Finite Elelement Method (in Polish) (2004)
20. Cosserat, E., Cosserat, F.: Sur la théorie de l'élasticité. Ann Toulouse **10**, 1–116 (1886)
21. Cosserat, E., Cosserat, F.: Théorie des Corps Déformables. Herman, Paris (1909)
22. Donnell, L.H.: Beams, Plates, and Shells. McGraw-Hill, New York (1976)
23. Duan, H.L., Wang, J., Karihaloo, B.L.: Theory of elasticity at the nanoscale. Advances in Applied Mechanics, pp. 1–68. Elsevier, San Diego (2008)
24. Eremeyev, V.A., Altenbach, H., Morozov, N.F.: The influence of surface tension on the effective stiffness of nanosize plates. Doklady Phy. **54**(2), 98–100 (2009)
25. Eremeyev, V.A., Lebedev, L.P., Altenbach, H.: Foundations of Micropolar Mechanics. SpringerBriefs in Applied Sciences and Technologies. Springer, Heidelberg (2013).)
26. Eringen, A.C.: Microcontinuum Field Theory I: Foundations and Solids. Springer, New York (1999)
27. Eringen, A.C.: Microcontinuum Field Theory II: Fluent Media. Springer, New York (2001)
28. Goldenweiser, A.L.: Formulation of approximative theory of shells with the help of the asymptotic integration of the equations of the theory of elasticity (in Russ.). Prikl Mat i Mekh **26**(4), 668–686 (1962)
29. Günther, W.: Analoge Systeme von Schalengleichungen. Ing-Arch **30**, 160–188 (1961)
30. Grigolyuk, E.I., Kogan, A.F.: Present state of the theory of multilayered shells (in Russ.). Prikl Mekh **8**(6), 3–17 (1972)
31. Grigolyuk, E.I., Selezov, I.T.: Nonclassical theories of vibration of beams, plates and shells (in Russ.). In: Itogi nauki i tekhniki, Mekhanika tverdogo deformiruemogo tela, vol 5, VINITI, Moskva (1973)
32. Gurtin, M.E., Murdoch, A.I.: A continuum theory of elastic material surfaces. Arch. Ration. Mech. Anal. **57**(4), 291–323 (1975)
33. Huang, D.W.: Size-dependent response of ultra-thin films with surface effects. Int. J. Solids Struct. **45**(2), 568–579 (2008)
34. Ieşan, D.: Deformation of thin chiral plates in strain gradient elasticity. Euro. J. Mech. A. Solids **44**, 212–221 (2014)
35. Jaiani, G., Podio-Guidugli, P. (eds.): Relations of Shell, Plate, Beam, and 3D Models. Springer, Berlin (2008)
36. Javili, A., McBride, A., Steinmann, P.: Thermomechanics of solids with lower-dimensional energetics: On the importance of surface, interface, and curve structures at the nanoscale. A unifying review. Appl. Mech. Rev. *65*, 010,802-1-31 (2012)
37. Kafadar, C.B., Eringen, A.C.: Polar field theories. In: Eringen, A.C. (ed.) Continuum Physics. Academic Press, New York (1976)
38. von Kármán, T.: Festigkeitsprobleme im Maschinenbau. In: Encyk. d. Math. Wiss., vol. IV, pp. 311–385, Teubner, Leipzig (1910)
39. Kienzler, R.: Erweiterung der klassichen schalentheorie; der einfluß von dickenverzerrung und querschnittverwölbungen. Ingenieur-Archiv **52**, 311–322 (1982)
40. Kienzler, R.: On consistent plate theories. Arch. Appl. Mech. **72**, 229–247 (2002)
41. Kienzler, R., Altenbach, H., Ott, I. (eds.): Theories of Plates and Shells: Critical Review and New Applications. Springer, Berlin (2004)
42. Kirchhoff, G.R.: Über das Gleichgewicht und die Bewegung einer elastischen Scheibe. Crelles Journal für die reine und angewandte Mathematik **40**, 51–88 (1850)
43. Kraatz, A.: Anwendung der Invariantentheorie zur Berechnung des dreidimensionalen Versagens- und Kriechverhaltens von geschlossenzelligen Schaumstoffen unter Einbeziehung der Mikrostruktur. PhD thesis, Zentrum für Ingenieurwissenschaften, Halle (2007)
44. Krenk, S.: Theories for elastic plates via orthogonal polynomials. Trans. ASME J. Appl. Mech. **48**(4), 900–904 (1981)
45. Lazopoulos, K.: On the gradient strain elasticity theory of plates. Eur. J. Mech. A. Solids **23**(5), 843–852 (2004)

46. Lazopoulos, K.: On bending of strain gradient elastic micro-plates. Mech. Res. Commun. **36**(7), 777–783 (2009)
47. Lazopoulos, K., Lazopoulos, A.: Nonlinear strain gradient elastic thin shallow shells. Eur. J. Mech. A. Solids **30**(3), 286–292 (2011)
48. Lazopoulos, K., Alnefaie, K., Abu-Hamdeh, N., Aifantis, E.: The GRADELA plates and shells. In: Pietraszkiewicz, W., Górski, J. (eds.) Shell Structures: Theory and Applications, vol. 3, pp. 121–124. CRC Press, London (2014)
49. Lebedev, L.P., Cloud, M.J., Eremeyev, V.A.: Tensor Analysis with Applications in Mechanics. World Scientific, New York (2010)
50. Levinson, M.: An accurate, simple theory of the statics and dynamics of elastic plates. Mech. Res. Comm. **7**(6), 343–350 (1980)
51. Libai, A., Simmonds, J.G.: The Nonlinear Theory of Elastic Shells, 2nd edn. Cambridge University Press, Cambridge (1998)
52. Lo, K.H., Christensen, R.M., Wu, E.M.: A high-order theory of plate deformation. Part I: Homogeneous plates. Trans. ASME J. Appl. Mech. **44**(4), 663–668 (1977)
53. Lu, P., He, L.H., Lee, H.P., Lu, C.: Thin plate theory including surface effects. Int. J. Solids Struct. **43**(16), 4631–4647 (2006)
54. Lurie, A.I.: Theory of Elasticity. Foundations of Engineering Mechanics. Springer, Berlin (2005)
55. Meenen, J., Altenbach, H.: A consistent deduction of von kármán-type plate theories from threedimensional non-linear continuum mechanics. Acta Mech. **147**, 1–17 (2001)
56. Mikhasev, G.: On localized modes of free vibrations of single-walled carbon nanotubes embedded in nonhomogeneous elastic medium. ZAMM **94**(1–2), 130–141 (2014)
57. Mindlin, R.D.: Influence of rotatory inertia and shear on flexural motions of isotropic elastic plates. Trans. ASME J. App. Mech. **18**, 31–38 (1951)
58. Mindlin, R.D.: Second gradient of strain and surface-tension in linear elasticity. Int. J. Solids Struct. **1**(4), 417–438 (1965)
59. Mindlin, R.D., Eshel, N.N.: On first strain-gradient theories in linear elasticity. Int. J. Solids Struct. **4**(1), 109–124 (1968)
60. Mushtari, K., Galimov, K.: Nonlinear theory of thin elastic shells. NSF-NASA, Washington (1961)
61. Naghdi, P.M.: The theory of shells and plates. In: Flügge, S. (ed.) Handbuch der Physik, vol. VIa/2, pp. 425–640. Springer, Heidelberg (1972)
62. Nami, M., Janghorban, M.: Resonance behavior of FG rectangular micro/nano plate based on nonlocal elasticity theory and strain gradient theory with one gradient constant. Compos. Struct. **111**(1), 349–353 (2014)
63. Nowacki, W.: Theory of Asymmetric Elasticity. Pergamon, Oxford (1986)
64. Nye, J.: Physical Properties of Crystals. Oxford Science Publications, Oxford (2000)
65. Palmov, V.A.: Fundamental equations of the theory of asymmetric elasticity (in Russ.). Prikl. Mat. Mekh. **28**(3), 401–408 (1964)
66. Paufler, P.: Physikalische Kristallographie. Akademie, Berlin (1986)
67. Pietraszkiewicz, W., Gorski, J. (eds.): Shell Structures: Theory and Application, vol. 3. CRC Press, Boca Raton (2014)
68. Podstrigach, Y.S., Povstenko, Y.Z.: Introduction to mechanics of surface phenomena in deformable solids (in Russ.). Naukova Dumka, Kiev (1985)
69. Povstenko, Y.: Mathematical modeling of phenomena caused by surface stresses in solids. In: Altenbach, H., Morozov, N.F. (eds.) Surface Effects in Solid Mechanics, pp. 135–153. Springer, Heidlberg (2013)
70. Preußer, G.: Erweiterung der klassichen Schalentheorie; der Einfluß von Dickenverzerrung und Querschnittverwölbungen. Ingenieur-Archiv **54**, 51–61 (1981)
71. Ramezani, S.: A shear deformation micro-plate model based on the most general form of strain gradient elasticity. Int. J. Mech. Sci. **57**(1), 34–42 (2012)
72. Ramezani, S.: Nonlinear vibration analysis of micro-plates based on strain gradient elasticity theory. Nonlinear Dyn. **73**(3), 1399–1421 (2013)

73. Reddy, J., Srinivasa, A., Arbind, A., Khodabakhshi, P.: On gradient elasticity and discrete peridynamics with applications to beams and plates. Adv. Mater. Res. **745**, 145–154 (2013)
74. Reddy, J.N.: A simple higher-order theory for laminated composite plates. Trans. ASME J. Appl. Mech. **51**, 745–752 (1984)
75. Reddy, J.N.: Theory and Analysis of Elastic Plates and Shells. CRC Press, Boca Raton (2007)
76. Reissner, E.: On the theory of bending of elastic plates. J. Math. Phys. **23**, 184–194 (1944)
77. Reissner, E.: Reflections on the theory of elastic plates. Appl. Mech. Rev. **38**(11), 1453–1464 (1985)
78. Rothert, H., Zastrau, B.: Herleitung einer Direktortheorie für Kontinua mit lokalen Krümmungseigenschaften. ZAMM **61**, 567–581 (1981)
79. Rubin, M.B.: Cosserat Theories: Shells, Rods and Points, Solid Mechanics and Its Applications, vol. 79. Springer, Berlin (2000)
80. Sahmani, S., Ansari, R.: On the free vibration response of functionally graded higher-order shear deformable microplates based on the strain gradient elasticity theory. Compos. Struct. **95**, 430–442 (2013)
81. Saito, M., Kukula, S., Kataoka, Y., Miyata, T.: Practical use of statistically modified laminate model for injection moldings. Mater. Sci. Eng., A **285**(1–2), 280–287 (2000)
82. Schaefer, H.: Das Cosserat-Kontinuum. ZAMM **47**, 485–498 (2006)
83. Serpilli, M., Krasucki, F., Geymonat, G.: An asymptotic strain gradient Reissner-Mindlin plate model. Meccanica **48**(8), 2007–2018 (2013)
84. Steigmann, D.J., Ogden, R.W.: Elastic surface-substrate interactions. Proc. R. Soc. Lond. A **455**(1982), 437–474 (1999)
85. Timoshenko, S.P., Woinowsky-Krieger, S.: Theory of Plates and Shells. McGraw-Hill, New York (1985)
86. Touratier, M.: An effcient standard plate theory. Int. Eng. Sci. **29**(8), 901–916 (1991)
87. Tsiatas, G.: A new Kirchhoff plate model based on a modified couple stress theory. Int. J. Solids Struct. **46**(13), 2757–2764 (2009)
88. Vekua, I.N.: Shell Theory: General Methods of Construction. Pitman, Boston (1985)
89. Wang, B., Zhou, S., Zhao, J., Chen, X.: A size-dependent Kirchhoff micro-plate model based on strain gradient elasticity theory. Eur. J. Mech. A. Solids **30**(4), 517–524 (2011)
90. Wang, J., Duan, H.L., Huang, Z.P., Karihaloo, B.L.: A scaling law for properties of nano-structured materials. Proc. R. Soc. A **462**(2069), 1355–1363 (2006)
91. Wang, J., Huang, Z., Duan, H., Yu, S., Feng, X., Wang, G., Zhang, W., Wang, T.: Surface stress effect in mechanics of nanostructured materials. Acta Mech. Solida Sinica **24**, 52–82 (2011)
92. Wilson, E.B.: Vector Analysis. Yale University Press, New Haven (1901). (Founded upon the Lectures of J. W. Gibbs)
93. Wlassow, W.S.: Allgemeine Schalentheorie und ihre Anwendung in der Technik. Akademie, Berlin (1958)
94. Zhilin, P.A.: Mechanics of deformable directed surfaces. Int. J. Solids Struct. **12**, 635–648 (1976)
95. Zhilin, P.A.: Applied mechanics. Foundations of the theory of shells (in Russ.). St. Petersburg State Polytechnical University, St. Petersburg (2006)

Chaotic Vibrations of Conical and Spherical Shells and Their Control

Jan Awrejcewicz and Vadim A. Krysko

Abstract This chapter is aimed on investigation of non-linear dynamics of conical and spherical shells. The variational equations are derived, and then the problem is reduced to a set of non-linear ordinary differential and algebraic equations. Since further axially symmetric deformation of closed shallow rotational shells and circled plates subjected to uniformly distributed periodic load being normal to the middle plate/shell surface are studied, the polar co-ordinates are used and four types of boundary conditions are investigated. The obtained equations are solved numerically, and the results reliability and validity are discussed in either regular, bifurcation or chaotic regimes including constant and non-constant thickness of the mentioned structural members, taking into account an initial imperfection/deflection. The classical approaches (time histories and frequency power spectra) are used to monitor different transitions from periodic to chaotic vibrations. Novel non-linear dynamical phenomena exhibited by the studied plates/shells are detected and discussed versus the chosen control parameters. In particular, the so called vibration type charts (amplitude—frequency of excitation planes) versus the different shell slopes are reported, which are of a particular importance for direct engineering applications. Finally, it is demonstrated how one may control non-linear dynamics of the studied continuous systems by using their thickness, in order to avoid buckling and stability loss phenomena.

J. Awrejcewicz (✉)
Department of Automation, Biomechanics and Mechatronics,
Lodz University of Technology, 1/15 Stefanowski Str., 90-924 Łódź, Poland
e-mail: jan.awrejcewicz@p.lodz.pl

V.A. Krysko
Department of Mathematics and Modeling, Saratov State Technical University,
Politekhnicheskaya 77, 410054 Saratov, Russia
e-mail: tak@san.ru

H. Altenbach and G.I. Mikhasev (eds.), *Shell and Membrane Theories in Mechanics and Biology*, Advanced Structured Materials 45, DOI 10.1007/978-3-319-02535-3_3

1 Introduction

Han et al. [6] wrote in their paper that chaotic motions of plates and shells have not been studied yet. Though dynamics of plates/shells belongs to old issues, the today's engineering requires safe construction of light-weight, energy-saving and high-speed structures in both aviation and aerospace industries. These needs are associated with the occurrence of dangerous transient vibrations, buckling and large deformation phenomena exhibited by the plate/shell constructions, where the geometrical non-linearity and structural instability play a key role. In order to investigate large-amplitude non-linear vibrations of structural members including their bifurcation and chaotic dynamic phenomena, the recent theory of chaos and the qualitative theory of differential equations should be applied. It means that in order to predict the desired performances of the structural members and to guarantee their work in the required dynamical regimes, the associated bifurcation and chaotic phenomena should be investigated.

Feng and Sethna [5] derived the conditions for the Shilnikov-type homoclinic orbits while investigating chaotic dynamics of thin plates parametrically excited. The Melnikov method has been applied by Shu et al. [9] to predict chaos of a plate with large deflection. Chaotic dynamics of a parametrically excited rectangular thin plate via numerical simulations has been studied by Zhang [10].

This paper extends our previous studies of structural members like beams, plates and shells from the point of view of bifurcation and chaotic dynamical phenomena [1–4]. The chapter is organized in the following manner.

First, a mathematical model and algorithm of solutions is introduced. Then reliability of the obtained results is addressed. Next, regular and chaotic dynamics of spherical and conical shells with constant thickness is studied. Finally, a control of chaos is proposed. Concluding remarks finish the chapter.

2 Mathematical Model and Algorithm of Solutions

We consider a shallow elastic shell (Fig. 1), which can be treated as a plate with an initial imperfection, located in a closed 3D space in R^3 with the curvilinear system of coordinates α, β, γ Mikhlin [8]. We assume that Lamé parameters A, B and radii $R_1^{'}$, R_{12}, $R_2^{'}$ of the mean shell surface are continuous with their first derivatives regarding the functions α, β.

In the given co-ordinates the shell space Ω is defined as follows:

$$\Omega = \left\{\alpha, \beta, \gamma / (\alpha, \beta, \gamma) \in [0, a] \times [0, b] \times \left[-\frac{h}{2}, \frac{h}{2}\right]\right\}.$$

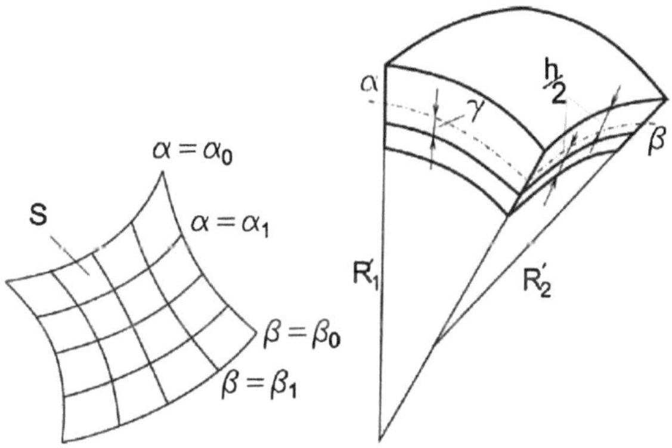

Fig. 1 Computational scheme of a shallow shell

PDEs governing non-linear dynamics of shells (Fig. 1) are obtained on the basis of the following hypotheses: shells are one-layer, made from an isotropic material (homogeneous and elastic), and the Kirchhoff-Love hypothesis holds.

The associated variational equation has the following form:

$$\delta \iint\limits_{\bar{S}} \left\{ \frac{D}{2} \left[(\Delta \bar{w})^2 - (1-\nu) L(\bar{w}, \bar{w}) \right] - \left[\Delta_k \bar{F} + L\left(\frac{1}{2}\bar{w} + \bar{w}_0, \bar{F} \right) \right] \bar{w} \right.$$

$$\left. - \frac{1}{2Eh} \left[\left(\Delta \bar{F} \right)^2 - (1+\nu) L\left(\bar{F}, \bar{F} \right) \right] \right\} ds - \iint\limits_{\bar{S}} \left[q - \frac{h}{\gamma g} (\ddot{\bar{w}} + \varepsilon \dot{\bar{w}}) \right] \delta \bar{w} d\bar{s} = 0, \tag{1}$$

where:

$$\Delta = \frac{1}{AB} \left(\frac{\partial}{\partial \bar{\alpha}} \frac{B}{A} \frac{\partial}{\partial \bar{\alpha}} + \frac{\partial}{\partial \bar{\beta}} \frac{A}{B} \frac{\partial}{\partial \bar{\beta}} \right),$$

$$\Delta_k = \frac{1}{AB} \left(\frac{\partial}{\partial \bar{\alpha}} \frac{1}{R'_1} \frac{B}{A} \frac{\partial}{\partial \bar{\alpha}} + \frac{\partial}{\partial \bar{\alpha}} \frac{1}{R_{12}} \frac{\partial}{\partial \bar{\beta}} + \frac{\partial}{\partial \bar{\beta}} \frac{1}{R_{12}} \frac{\partial}{\partial \bar{\alpha}} + \frac{\partial}{\partial \bar{\beta}} \frac{1}{R'_1} \frac{A}{B} \frac{\partial}{\partial \bar{\beta}} \right),$$

$$L(\bar{w}, \bar{F}) = \frac{\partial^2 \bar{w}}{\partial \bar{\alpha}^2} \cdot \frac{\partial^2 \bar{F}}{\partial \bar{\beta}^2} - 2 \frac{\partial^2 \bar{w}}{\partial \bar{\alpha} \partial \bar{\beta}} \cdot \frac{\partial \bar{F}^2}{\partial \bar{\alpha} \partial \bar{\beta}} + \frac{\partial^2 \bar{w}}{\partial \bar{\beta}^2} \cdot \frac{\partial^2 \bar{F}}{\partial \bar{\alpha}^2}.$$

In order to solve Eq. (1), in which the deflection function $\bar{w}$ and stress function $\bar{F}$ are independently variated, we cannot apply directly the Ritz procedure (the equation does not have the form of the functional variation equal to zero). In order to find the approximated value of element $\bar{w}$ and $\bar{F}$, we take the co-ordinate sequence $w_i(\alpha, \beta)$ and $\varphi_i(\alpha, \beta)$, satisfying the same requirements as Eq. (1).

In order to find the deflection w and stress function F, the systems of functions $\{\varphi_{ij}(x,y), \psi_{ij}(x,y)\}$, $i, j = 0, 1, 2, \ldots$, should satisfy the following five requirements:

1. $\varphi_{ij}(x,y) \in H_A$, $\psi_{ij}(x,y) \in H_A$, where H_A is the Hilbert space, which is called the energetic space;
2. $\forall i, j$ functions $\varphi_{ij}(x,y)$ and $\psi_{ij}(x,y)$ are linearly independent, continuous with their partial derivatives to the fourth order inclusive in space Ω;
3. $\varphi_{ij}(x,y)$ and $\psi_{ij}(x,y)$ satisfy the boundary conditions;
4. $\varphi_{ij}(x,y)$ and $\psi_{ij}(x,y)$ are compact in H_A;
5. $\varphi_{ij}(x,y)$ and $\psi_{ij}(x,y)$ should represent M first elements of the full system of the functions:

$$\bar{w} = \sum_{i=0}^{M_x} \bar{x}_i(\bar{t}) w_i(\bar{\alpha}, \bar{\beta}), \qquad \bar{F} = \sum_{i=0}^{M_x} \bar{y}_i(\bar{t}) \varphi_i(\bar{\alpha}, \bar{\beta}). \tag{2}$$

The approximating solutions have coefficients $\bar{x}_i(\bar{t})$ and $\bar{y}_i(\bar{t})$ which are the time-dependent functions. Substituting relations (2) into Eq. (1), carrying out the variational operation, and comparing to zero the terms standing by $\delta\bar{x}_i$, $\delta\bar{y}_i$, we get the following system of ODEs regarding $\bar{x}_i(\bar{t})$ and $\bar{y}_i(\bar{t})$:

$$\begin{gathered} K_{ik}(\ddot{x}_k + \varepsilon \dot{x}_k) + B_{ik}x_k + C_{ip}y_p + D_{ikp}x_k y_p = Q_i q_0, \\ C_{pi}x_i + E_{pj}y_j + \frac{1}{2}D_{pki}x_k x_i = 0, \\ i,\ k = 1,\ 2,\ \ldots,\ n; \quad p,\ j = 1,\ 2,\ \ldots,\ m. \end{gathered} \tag{3}$$

In the polar co-ordinates with axial symmetry $w = w(r)$, $\varphi = \varphi(r)$, $\alpha = r$, $\beta = \theta$, $ds = 2\pi r dr$, and the operators take the form

$$\Delta = \frac{d^2}{d\bar{r}^2} + \frac{1}{\bar{r}}\frac{d}{dr}, \quad L(\bar{w}, \bar{F}) = \frac{d^2\bar{w}}{d\bar{r}^2} \cdot \frac{1}{\bar{r}}\frac{d\bar{F}}{d\bar{r}} + \frac{1}{r}\frac{d\bar{w}}{d\bar{r}} \cdot \frac{d^2\bar{F}}{d\bar{r}^2}. \tag{4}$$

Substituting $\bar{r}$ by $a\bar{\rho}$ in operators (4), and carrying out the standard transformations (and after division by $2\pi Eh_0^5/a^4$), the system is transformed to its counterpart non-dimensional form. In order to reduce Eq. (3) to the non-dimensional forms the following quantities are introduced:

$$\bar{w} = \frac{w}{h}, \quad \bar{x}_i = \frac{x_i}{h}, \quad \bar{\varphi} = \frac{\varphi}{Eh^3}, \quad y_i = \frac{\bar{y}_i}{Eh^3}, \quad h = \frac{h(\rho)}{\bar{h}}, \quad h = h(0),$$

$$\bar{F} = Eh^2F, \quad \bar{t} = t\tau, \quad \bar{\varepsilon} = \frac{\varepsilon}{\tau}, \quad \tau = \frac{a}{h_0}\sqrt{\frac{a^2\gamma}{Eg}}, \quad \bar{q} = q\frac{Eh_0^4}{a^4},$$

where: t—time, ε—damping coefficient, a—dimension of the square shell, h—thickness of the shell, g—Earth acceleration, γ—material weight density, ν—Poisson's coefficient for the isotropic material ($\nu = 0.3$), E—elasticity modulus, w_0—initial imperfection. Next, bars over the non-dimensional quantities are omitted.

In the case of axially symmetric deformation of the shallow rotational shell of thickness $h = h_0(1 + c\rho)$, the coefficients of system (3) take the following form

$$\begin{aligned} K_{ik} &= \int_0^1 (1+c\rho)\, w_i w_k \rho d\rho, \\ N_{ik} &= \frac{1}{12\left(1-\nu^2\right)} \int_0^1 (1+c\rho)^3 \left[\Delta w_i \Delta w_k - (1-\nu)\, L\left(w_i, w_k\right)\right] \rho d\rho, \\ C_{ip} &= -\int_0^1 \left[\Delta_k \varphi_p + L\left(w_0, \varphi_p\right)\right] w_i \rho d\rho, \\ D_{ikp} &= -\int_0^1 w_i L(w_k, w_p) \rho d\rho, \\ Q_i &= \int_0^1 w_i \rho d\rho, \\ E_{jp} &= -\int_0^1 \frac{1}{1+c\rho} \left[\Delta\varphi_j \Delta\varphi_p - (1+\nu)\, L\left(\varphi_j, \varphi_p\right)\right] \rho d\rho. \end{aligned} \tag{5}$$

Solving the second equation of system (3) regarding y_i, we get

$$y_i = \left[E_{jp}^{-1} C_{ps} + \frac{1}{2}\left(E_{jp}^{-1} D_{pi} x_i\right)_s\right] x_s.$$

Multiplying by K^{-1} the first equation of (3) and using notation $\dot{x}_i = r_i$, the problem is reduced to the first order ODEs of the form

$$\begin{gathered} \dot{r}_i = -\bar{\varepsilon}\, r_i + \left[K_{ik}^{-1} C_{ij} + \left(A_{ik}^{-1} D_{ks} x_s\right)_j\right] \cdot y_j - K_{ik}^{-1} B_{ks} x_s + q_0(\bar{t}) K_{ik}^{-1} Q_k, \\ \dot{x}_i = r_i, \\ i,\ k,\ s = 1, 2, \ldots, n; \quad p,\ j = 1, 2, \ldots, m. \end{gathered} \tag{6}$$

The so far introduced transformation has been possible, since matrices K_{ik}^{-1} and E_{jp}^{-1} exist if the coordinate functions are linearly independent. Equation (6), with

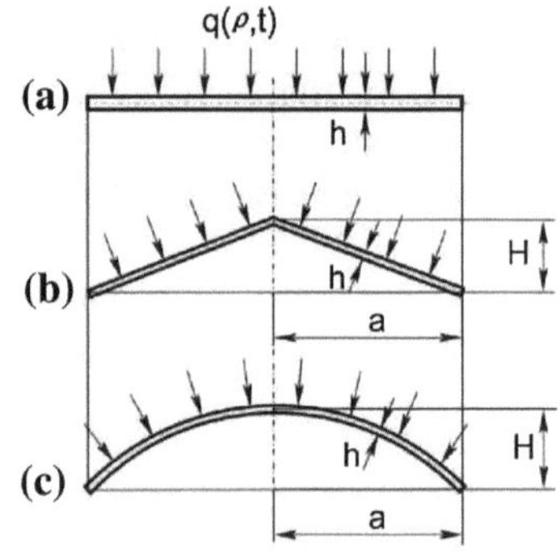

Fig. 2 Structural members surface: **a** plate, **b** cone, **c** sphere

initial conditions $x_i = 0$, $\dot{x}_i = 0$ for $t = 0$, have been solved by the fourth order Runge–Kutta method.

We consider further the axially symmetric deformation of closed shallow rotational shells and circled plates subjected to uniformly distributed load being normal to the mean shell surface. In polar co-ordinates for the axial symmetry we have: $w = w(\rho)$, $F = F(\rho)$, $\alpha = \rho$, $\beta = \theta$, $ds = 2\pi\rho d\rho$, and the thickness is defined by function $h(\rho) = h_0(1 + c\rho)$. The mean shell surface is defined by initial deflection $w_0 = -hR\left(1 - c_1\rho - c_2\rho^2\right)$, $K = H/h_0$, where H is the full shell height over a plane (see Fig. 2). For $c_1 = 1$, $c_2 = 0$ we get a conical shell (Fig. 2b); $c_1 = 0$, $c_2 = 1$ corresponds to a sphere (Fig. 2c); for $k = 0$ we deal with a plate (Fig. 2a). Approximating functions regarding four types of boundary conditions are shown in Table 1.

In order to investigate vibrations of a shallow conical shell, we consider it as a plate ($\Delta_k\varphi \equiv 0$) with initial deflection: $w_0 = -k(1 - \bar{\rho})$, $k = H/h_o$, and we apply the co-ordinate functions given in Table 1. Each of formula (5) can be presented by a sum of integrals of the following form

$$I(x, y) = \int_0^1 \rho^x (1 - \rho^2)^y d\rho = \frac{(2y)!!(x-1)!!}{(x+2y+1)!!},$$

and for four types of the boundary conditions applied, the coefficients of system (6) take the form presented in Table 1.

1. Immovable clamping

$$K_{ik}^{(1)} = \frac{1}{6 + 2i + 2k} + c\frac{(4 + 2i + 2k)!!}{(7 + 2i + 2k)!!},$$

Table 1 System of approximating functions for boundary conditions

$\psi_j(\rho)$	$\varphi_i(\rho)$	
	$(1-\rho^2)^{i+1}$	$(1-\rho^2)^{i}$
ρ^{2j}	movable clamping 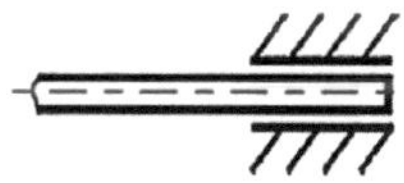	fixed simple support 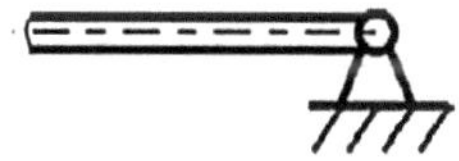
$(1-\rho^2)^{j+1}$	fixed clamping 	movable simple support 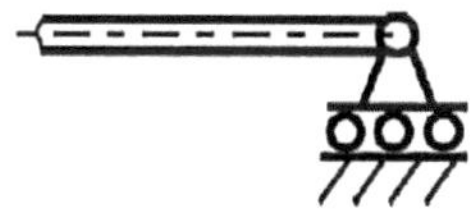

$$\begin{aligned}
N_{ik}^{(1)} &= \frac{4(i+1)(k+1)}{3(1-\nu^2)}\left\{\frac{1}{i+k+1}\left\{\frac{ik}{(i+k)(i+k+1)}\right.\right.\\
&\quad+\left.\frac{3c^2}{2(i+k+2)}\left[\frac{6ik}{(i+k)(i+k-1)}-\frac{1+\nu}{2}\right]\right\}\\
&\quad+\left.\frac{3c(2i+2k-4)!!}{(2i+2k+3)!!}\left[15ik-(1+\nu)(i+k)(i+k-1)\right]\right\},\\
C_{ip}^{(1)} &= -2(p+1)\cdot\frac{H}{h_0}\left(-\int_0^1\left(1-r^2\right)^{i+p+1}dr+2p\int_0^1 r^2\left(1-r^2\right)^{i+p}dr\right),\\
E_{jp}^{(1)} &= 4ip\left[(j+p-1)(1+\nu)-2jp\right]\left(\frac{1}{j+p-1}-\frac{c}{j+p-{}^1/_2}+\frac{c^2}{j+p}\right),\\
D_{ikp}^{(1)} &= 4(i+1)(k+1)p\frac{p!}{(i+k+1)\ldots(i+k+p+1)},\quad Q_i=\frac{1}{2(i+2)}.
\end{aligned}\tag{7}$$

2. Movable clamping

$$K_{ik}^{(2)}=K_{ik}^{(1)},\qquad N_{ik}^{(2)}=N_{ik}^{(1)},$$

$$C_{ip}^{(2)}=2\frac{H}{h_0}\cdot p(2p-1)\cdot\int_0^1 r^{2p-2}\cdot\left(1-r^2\right)^{i+1}dr,\qquad Q_i^{(2)}=Q_i^{(1)},$$

$$\begin{aligned}
E_{jp}^{(2)} &= -16(j+1)(p+1)\left\{\frac{pj}{(j+p+1)(j+p)(j+p-1)}\right.\\
&\quad+\frac{c^2}{2(j+p+1)(j+p+2)}\times\left[\frac{6jp}{(j+p)(j+p-1)}-\frac{1-\nu}{2}\right]
\end{aligned}$$

$$- c\frac{(2j+2p-4)!!}{(2j+2p+3)!!}\left[15jp-(1-\nu)(j+p)(j+p-1)\right]\Bigg\},$$

$$D_{ikp}^{(2)} = -\frac{4(i+1)(k+1)(p+1)}{(i+k+p+1)(i+k+p+2)}. \tag{8}$$

3. Immovable simple support

$$K_{ik}^{(3)} = K_{i-1,k-1}^{(1)},$$

$$N_{ik}^{(3)} = \begin{cases} \frac{2+c(4+3c)}{6(1-\nu)}, & k=i=1, \\ -\frac{c(4+5c)}{15(1-\nu)}, & i=1,k=2,i=2,k=1, \\ N_{i-1,k-1}^{(1)}, & \end{cases} \tag{9}$$

$$C_{ip}^{(3)} = C_{i-1,p}^{(1)}, \quad D_{ikp}^{(3)} = D_{i-1,k-1,p}^{(1)}, \quad Q_{i}^{(3)} = Q_{i-1}^{(1)}, \quad E_{jp}^{(3)} = E_{jp}^{(1)}.$$

4. Movable simple support

$$K_{ik}^{(4)} = K_{i-1,k-1}^{(1)}, \qquad N_{ik}^{(4)} = N_{ik}^{(3)}, \qquad C_{ip}^{(4)} = C_{i-1,p}^{(2)},$$
$$D_{ikp}^{(4)} = D_{i-1,k-1,p}^{(2)}, \qquad Q_{i}^{(4)} = Q_{i-1}^{(1)}, \qquad E_{jp}^{(4)} = E_{jp}^{(2)}. \tag{10}$$

As it has already been mentioned, investigating the spherical shell we treat it as a plate with initial deflection $w_0 = -k(1-r^2)$. For four types of the boundary conditions shown in Table 1, coefficients (6) differ from the case of the conical shell only by C_{ip}:

1. Immovable clamping:

$$C_{ip}^{(1)} = -4\frac{H}{h_0}p\frac{(1+i)!p!}{(i+p+1)!};$$

2. Movable clamping:

$$C_{ip}^{(2)} = 4\frac{H}{h_0}(i+1)(p+1)\frac{(i+p)!}{(i+p+2)!};$$

3. Immovable simple support:

$$C_{ip}^{(3)} = C_{i-1,p}^{(1)};$$

4. Movable simple support:

$$C_{ip}^{(4)} = C_{i-1,p}^{(2)}.$$

The transversal uniformly distributed harmonic load is $q = q_0\sin(\omega_p t)$, where: q_0—amplitude, ω_p—frequency of excitation.

3 Reliability of the Results Obtained

In order to study PDEs governing dynamics of our continuous system we introduce mode functions and we obtain ODEs (3) of infinite dimension. In order to get a correct interpretation of the obtained results, the following remarks should be taken into account. Investigating any continuous system, instead of infinite set of ODEs, we take a truncated system of finite dimension. It is assumed that increasing a number of equations we find a threshold, beginning from each further increase of the equations number does not yield anything new. This approach is also motivated by the occurrence of finite dimension of the system attractor. However, it may happen that with an improper choice of basic functions which serve to reduce PDEs to ODEs, the obtained system of ODEs may have attractors different from that of the original system.

This feature may occur, for instance, in the case of a 2D equation governing the dynamics of heat convection. The Lorenz system Lorenz [7], presenting a three-mode truncation of an approximated PDE, demonstrated complex dynamics including chaos. However, an increase of the mode number yields first an irregular increase of chaos, and hence its decrease. For a sufficiently large number of modes chaos vanishes. The illustrated example shows that in order to get qualitatively true correspondence between the original and truncated dynamics motivated by the use of either Bubnov–Galerkin or Ritz approaches, we need to take into account a sufficient number of modes. Let us investigate the problem of mode number estimation in the Ritz procedure using an example of vibrations of spherical and conical shallow shells being geometrically non-linear and having a constant or non-constant thickness, and bounded along their contour. The applied load uniformly distributed along the shell surface, has the following form

$$q = q_0 \sin(\omega_p t). \tag{11}$$

We consider vibration charts associated with shells of $k = 3$ and $k = 5$ (Figs. 3 and 4, respectively) depending on the magnitude of control parameters $\{q_0, \omega_p\}$ for a different number of partition terms $n = 1, \ldots, 7$.

A further increase of number n in (3) has not changed the charts $\{q_0, \omega_p\}$ qualitatively. For $n = 1$ (Fig. 5) the chart differs from the remaining ones, i.e. it presents only bifurcation zones and harmonic and sub-harmonic vibrations with ω_p and $\omega_p/2$, without any chaotic zones.Increasing n yields new zones of bifurcation and chaos.

Similarly, the chart ($k = 5$) for $n = 1$ strongly differs from the remaining ones, since the increase of n yields different zones becoming similar, i.e. a converging sequence of vibration character is observed. For instance, the sub-harmonic zone is the same for all $n \geq 2$, but for $n = 2$ it is shifted to the right. Chaotic zones become smaller while increasing n, but the separated parts do not change beginning from $n = 4$. Both cases of $k = 5$ and $k = 3$ exhibit better convergence regarding the high frequencies than low frequencies and the frequencies located in the neighborhood of natural frequency.

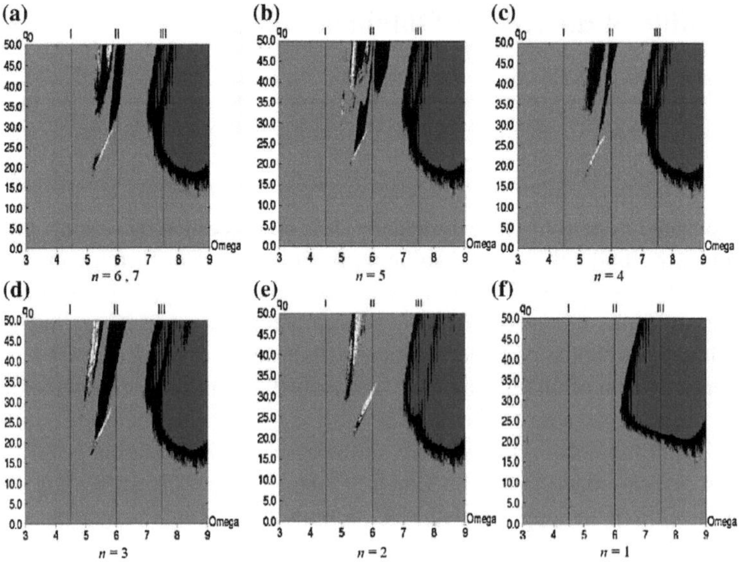

Fig. 3 Charts of control parameters $\{q_0, \omega_p\}$ of the conical shell for $k = 3$. **a** $n = 6.7$. **b** $n = 5$. **c** $n = 4$. **d** $n = 3$. **e** $n = 2$. **f** $n = 1$

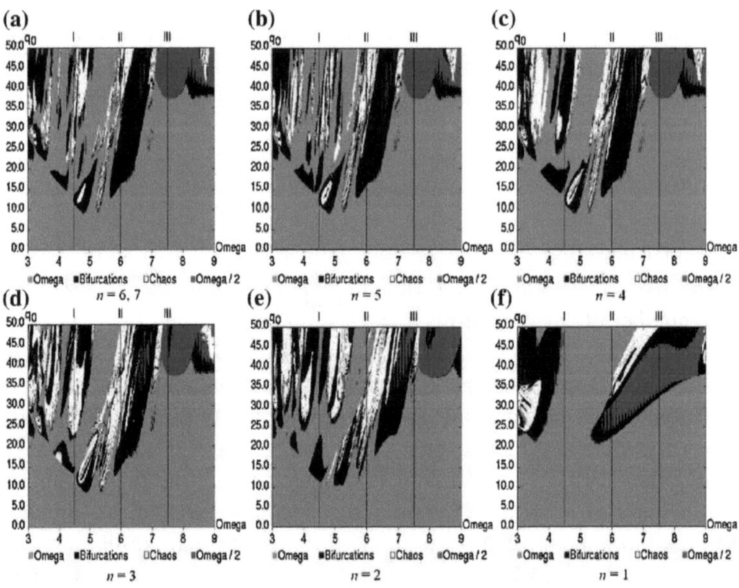

Fig. 4 Charts of control parameters $\{q_0, \omega_p\}$ of the conical shell ($k = 5$). **a** $n = 6.7$. **b** $n = 5$. **c** $n = 4$. **d** $n = 3$. **e** $n = 2$. **f** $n = 1$

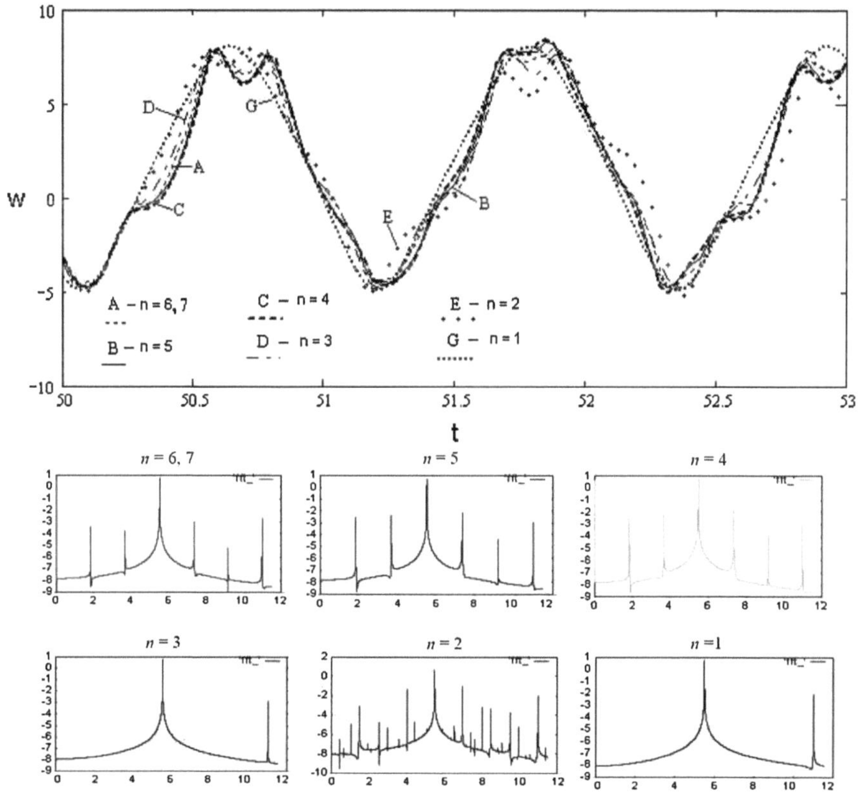

Fig. 5 Relation $w(0, t)$, $t \in [50; 53]$ and $S(\omega)$ versus n for the conical shell ($k = 3$) with constant thickness (moving clamping)

We consider two points: the first one, for $k = 3$, $n = 6, 7$ is located in the bifurcation zone (Fig. 5); the second one, for $k = 5$, $n = 6, 7$ is in the chaotic zone (Fig. 6). In Fig. 5, $w(0, t)$ for $50 \leq t \leq 53$ and power spectra ($S(\omega)$) are reported. Analysis of the results given in Fig. 5 shows that beginning from $n \geq 4$ relations $w(0.5 : t)$ are close to each other, whereas power spectra coincide in full. The results obtained for $n = 1, 2, 3$ differ essentially from those of $n \geq 4$. Hence, we may conclude that beginning from $n \geq 4$, the process of bifurcation is reliable for $k \leq 3$, i.e. there is a convergent sequence, which can be modeled in the following way

$$\left[w^0 - \sum_{i=1}^{n} x_i(t) w_i(\rho) \right] = \min_{t \in [50, 53]}, \quad \left[\varphi^0 - \sum_{i=1}^{n} y_i(t) w_i(\rho) \right] = \min_{t \in [50, 53]},$$

and this is the best approximation to w^0 and φ^0 in the metric H_A.

For $n = 4, 5, 6, 7$ the power spectrum exhibits a period tripling bifurcation, and the attracting orbits have the period three. The occurrence of period 3 orbits yields

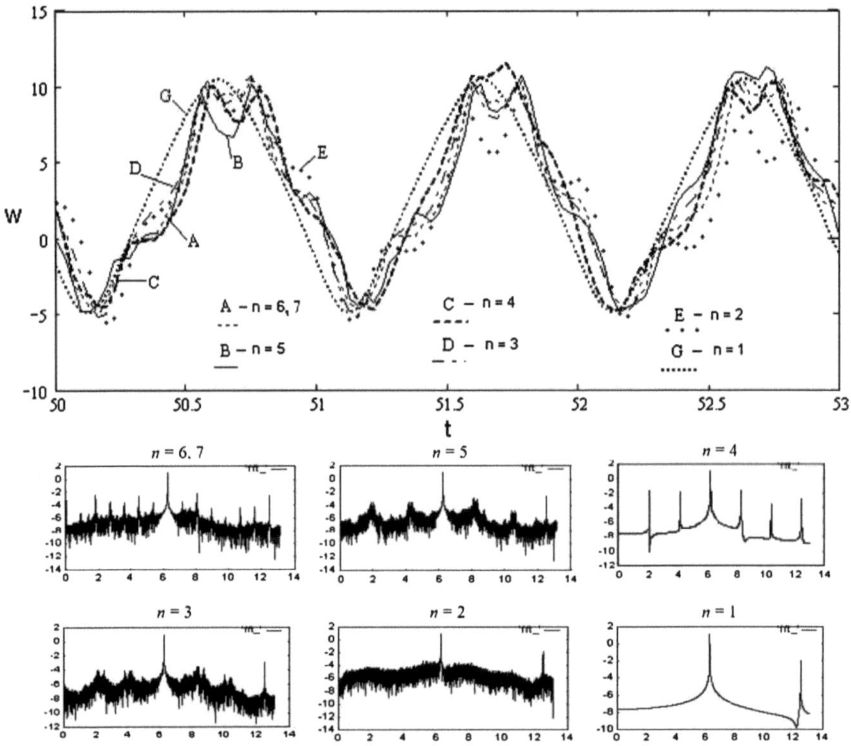

Fig. 6 Relation $w(0, t)$, $t \in [50; 53]$ and $S(\omega)$ versus n for the conical shell ($k = 5$) with constant thickness (moving clamping)

the occurrence of orbits with periods $n = 1, 2, 3, \ldots$. The mentioned remarks are applicable to both real functions and maps of an interval into itself. Here, we aim at analyzing of other orbits which are exhibited by the dynamics of flexible conical shells. In Fig. 5 the same characteristics as in Fig. 6 are shown. However, in this case we do not observe the previously exhibited uniform convergence.

For $n = 1$ we have periodic vibrations with a period $2\pi/\omega$; $n = 4$ corresponds to the period tripling bifurcation (period 3); for $n = 2, 3, 5, 6, 7$ we have chaotic vibrations, but we deal with different types of chaos. Namely, for $n = 2$ we have chaos associated with the excitation frequency. For $n = 3, 5$ we have chaos with frequency 3ω, whereas for $n = 6, 7$ we have chaos associated with frequency 7ω. Owing to this discussion, we may conclude that convergence of the Ritz procedure versus a number of terms of the series (3) essentially depends on the arrow of height k and on the system dynamical regime.

In what follows we investigate convergence of the Ritz procedure versus boundary condition type and a shell geometry using an example of conical shells supported along their edges and having constant (Fig. 7) and non-constant (Fig. 8) thickness ($h = h_0(1 + c\rho)$) for $c = 0.1$, $k = 5$. We consider a point, which for $n = 6, 7$

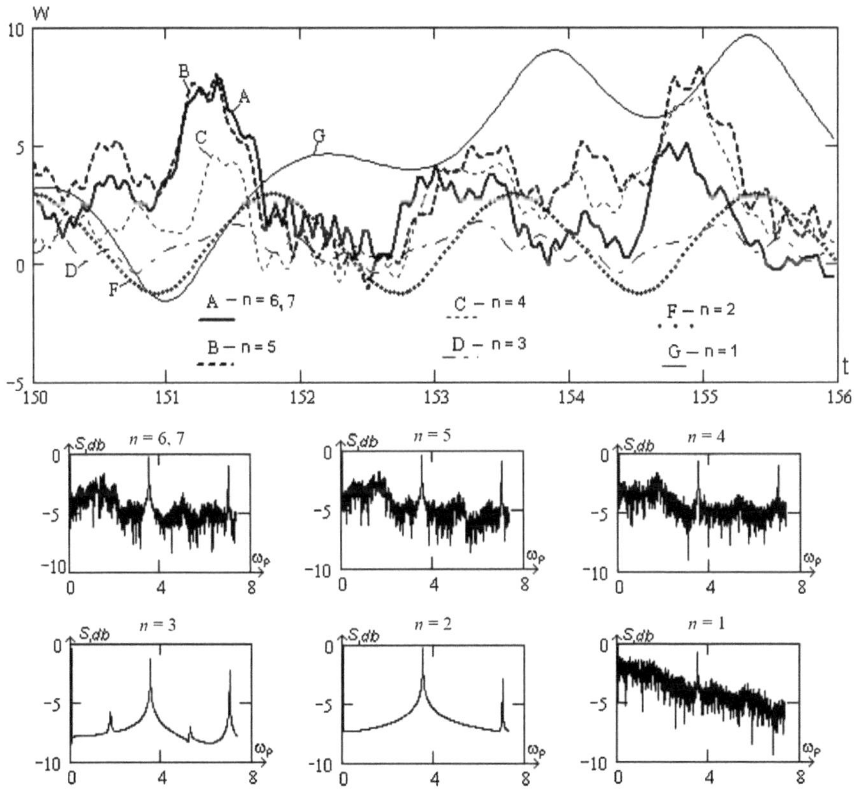

Fig. 7 Relation $w(0, t)$, $t \in [50; 53]$ and $S(\omega)$ versus n for the conical shell ($k = 5$) with constant thickness (movable simple support)

is in a chaotic zone for the following fixed parameters: $q_0 = 2.4$, $\omega_p = 3.5$. In Fig. 7 signals ($w\,(0; t)$, $150 \leq t \leq 156$) and power spectra are reported. For $n = 2$ harmonic excitation with ω_p, $n = 3$ yields the first period doubling bifurcation, whereas for $n = 1, 4, 5, 6, 7$ we have chaos associated with the fundamental shell frequency.

We consider the shell with variable thickness and with the following fixed parameters: $q_0 = 2.4$, $\omega_p = 3.57$. Figure 8 shows signals ($w\,(0; t)$, $150 \leq t \leq 156$) and power spectra. For $n = 2$ and $n = 3$ periodic vibrations with frequency ω_p are shown; $n = 1$ yields sub-harmonic vibrations of $\omega_p/5$, i.e. the first approximation of (3) yields period 5 vibration, whereas for $n = 4, 5, 6$ chaos associated with the natural frequency occurs. In other words, beginning from $n = 4$, a convergent sequence is observed. Further, all results are given for $n = 6$.

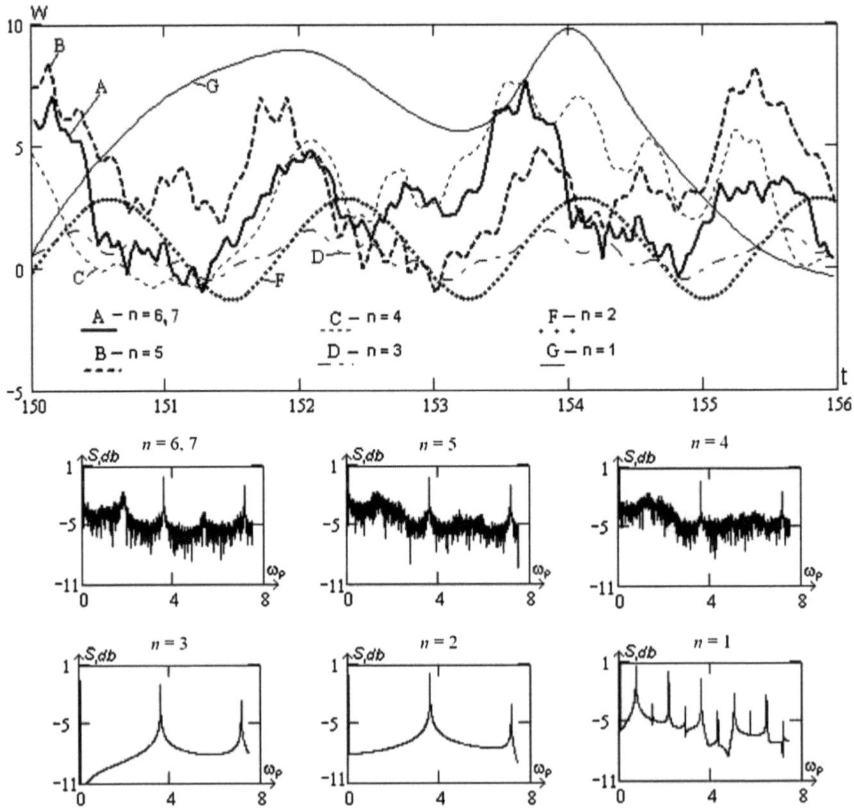

Fig. 8 Relation $w(0, t)$, $t \in [50; 53]$ and $S(\omega)$ versus n for the conical shell ($k = 5$) with non-constant thickness (movable simple support)

4 Spherical Shells with Constant Thickness

We consider vibrations of spherical shallow shells with constant thickness supported on their edges. In Fig. 9 charts of control parameters $\{q_0, \omega_p\}$ for the shell with $k = 1, 1.5, 2, 3, 4, 5$ are reported. The chart showing a transition into chaos exhibits rich dynamics while increasing k coefficient. Analysis of relation $\{q_0, \omega_p\}$ versus k shows that for the plate ($k = 0$), $\{q_0, \omega_p\}$ exhibits only periodic vibrations under constraints $w(0) \leq 5$ and $q_0 \leq 100$. The zones of chaos and bifurcations increase with an increase of k, and for $k = 1$ only two chaotic narrow islands appear within the harmonic zones. Increasing $k \geq 1.5$ yields new zones of bifurcation and chaos.

In order to illustrate the vibrations of spherical shells, Figs. 10 and 11 present characteristics $w_{\max}$ monitored in the shell top versus q_0 for $\omega_p = 5$, $k = 3$ and for $\omega_p = 8$, $k = 5$, respectively.The Lyapunov exponents play an important role in the investigation of dynamical systems. They give a computational qualitative measure

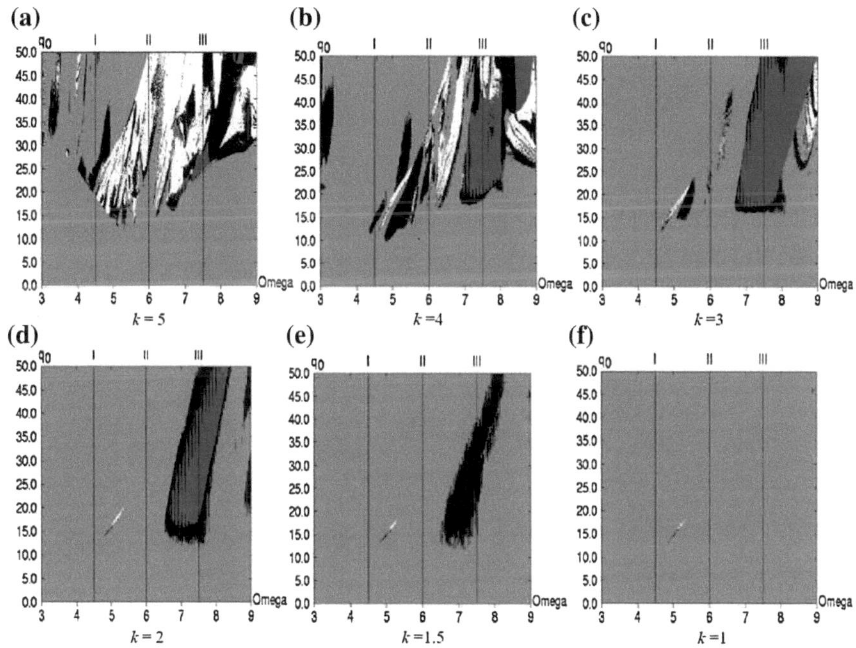

Fig. 9 Charts $\{q_0, \omega_p\}$ of the spherical shell. **a** $n = 5$. **b** $n = 4$. **c** $n = 3$. **d** $n = 2$. **e** $n = 1.5$. **f** $n = 1$

of the stochasticity order. The proposed and developed idea of computation of the series of vibration character of a dynamical system based on the analysis of the power spectrum $S(\omega)$ well coincides with the evolution of the largest characteristic exponent (LE) named here as $\lambda_1(q_0)$. In this work, in order to compute LE the Benettin method is applied. In a chaotic state $\lambda_1 > 0$, and white color represents chaotic zones in the reported scales/charts.

Figure 10 shows in the scale the window L, for which the relation $\lambda_1(q_0)$ $(15 \le q_0 \le 19)$ is given, where four zones of chaos are observed: a $(16.1 \le q_0 \le 16.5)$; $b(16.6 < q_0 < 16.8)$; $c(16.9 \le q_0 \le 17.3)$ and $d(17.9 < q_0 < 18.1)$, with $\lambda_1 > 0$, which well coincide with the scale of vibrations. Observe that in zone d in the corresponding relation $w_{\max}(q_0)$, the local buckling is observed.

The figures include also $w(t)$, $w(\dot{w})$, power spectra, and space forms of vibrations for three points A, B, C for $k = 5$ (point A corresponds to periodic vibrations, B—bifurcations, C—chaos) and two points A, B (point A—periodic vibrations, B—linear combination of two independent frequencies a_1 and b_1 with successive Hopf bifurcations) for $k = 3$, which are shown in relations $w_{\max}(q_0)$.

Analysis of results in either point $w(0, t)$ or in space $w(\rho, t)$ for $50 \le t \le 54$ does not depend on k in the case of periodic vibrations. Bifurcation and chaos essentially depend on k.

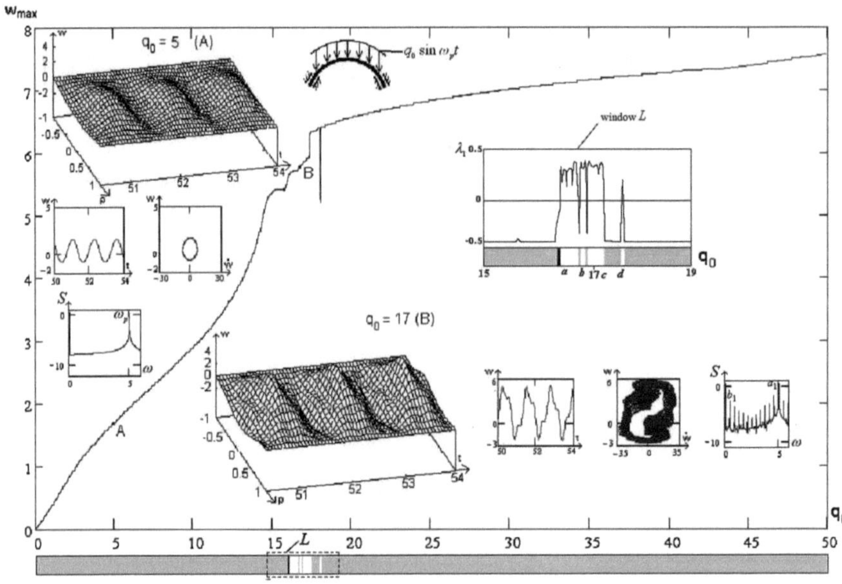

Fig. 10 Relation $w_{max}(q_0)$ for the spherical shell $k = 5$, $\omega_p = 8$ $(q_0 = 5)$

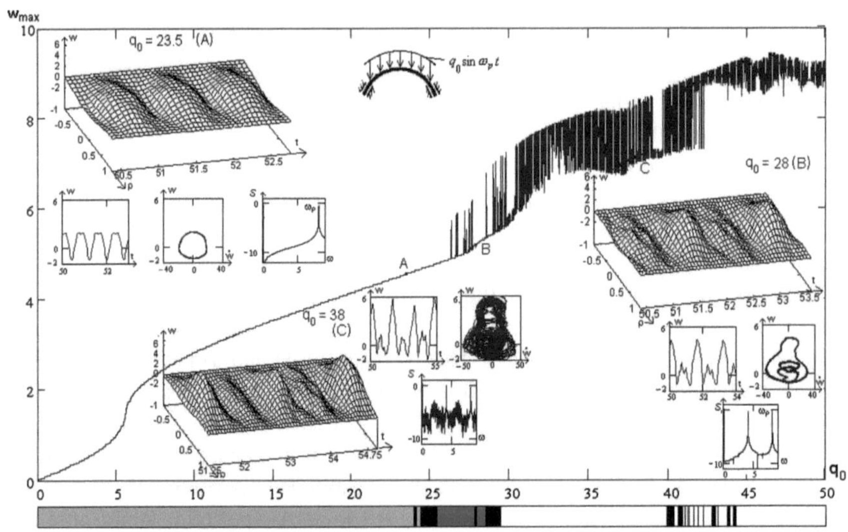

Fig. 11 Relation $w_{max}(q_0)$ for the spherical shell $k = 5$, $\omega_p = 8$ $(q_0 = 23.5)$

5 Conical Shells with Constant Thickness

We consider now vibrations of conical shallow shells of constant thickness and with the movable clamping. Charts $\{q_0, \omega_p\}$ for shells with constant thickness and for $k = 1, 1.5, 2, 3, 4, 5$ are shown in Fig. 12.

The charts of vibration character change qualitatively with an increase of k. Let us compare charts $\{q_0, \omega_p\}$ regarding spherical and conical shells. Analysis of the relation $\{q_0, \omega_p\}$ versus k shows that the influence of shell geometry on the character of vibration increases with an increase of k. For $k = 1$, in both conical and spherical spaces narrow zones of chaos located between $\omega_p = 5$ and $\omega_p = 6$ appear. For $k = 1.5$ both charts exhibit bifurcation zones which are similar. For $k = 2$ zones of rational frequencies are added, but they have a different organization. For $k \geq 3$ the influence of shell geometry implies essential differences in the charts regarding considered spherical and conical shells.

The change of the shell geometry has an impact on the scenarios of transitions into chaos. In the conical shell with $k = 5$, besides zones where a transition from periodic to chaotic vibrations is realized via the so far described scenario for the spherical shells, there are also Feigenbaum scenarios of transitions into chaotic vibrations.

We consider the relation $w_{max}(0)$ versus q_0 for $\omega_p = 5.61$ and scales of bifurcations for conical shells with $k = 3; 5$, as well as vibrations of the shell's surface

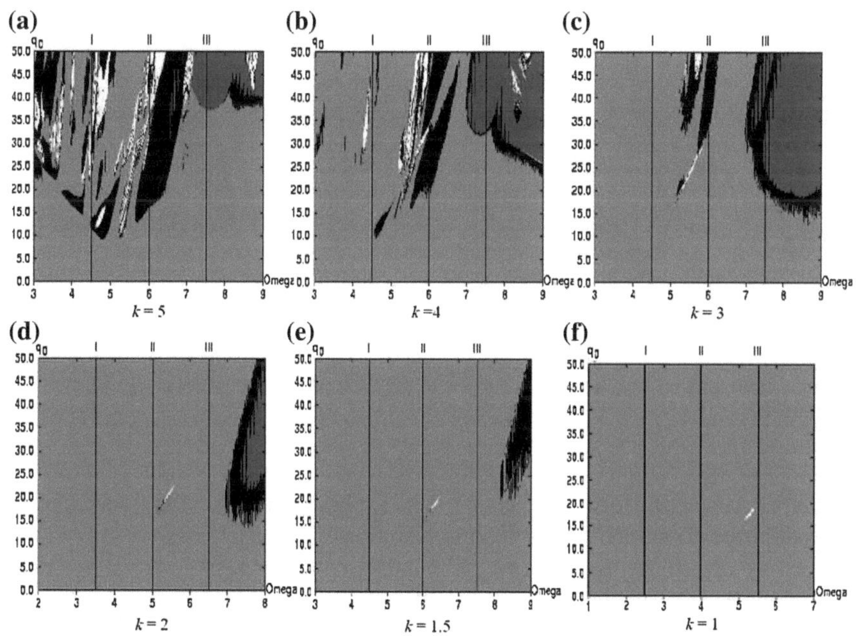

Fig. 12 Charts $\{q_0, \omega_p\}$ of the conical shell (moving clamping) for different k

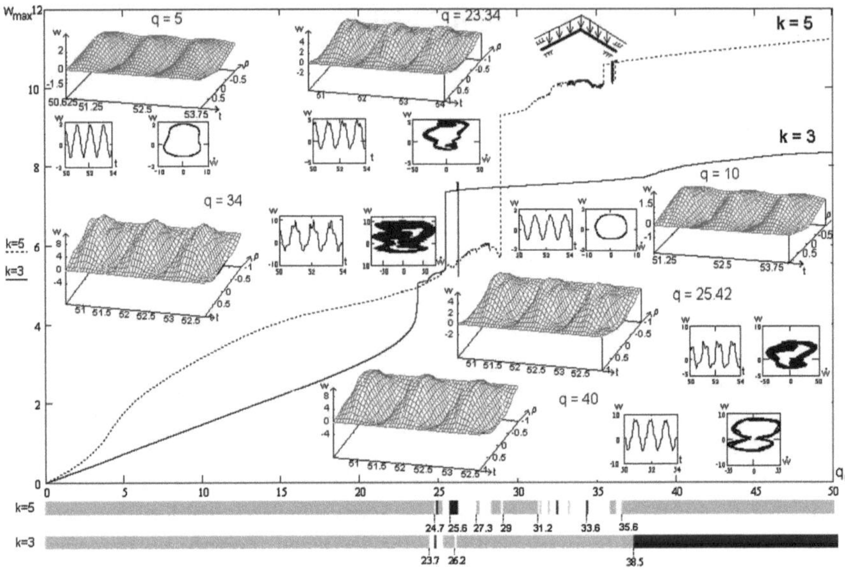

Fig. 13 Relation $w_{max}(q_0)$ for the spherical shell ($\omega_p = 8$)

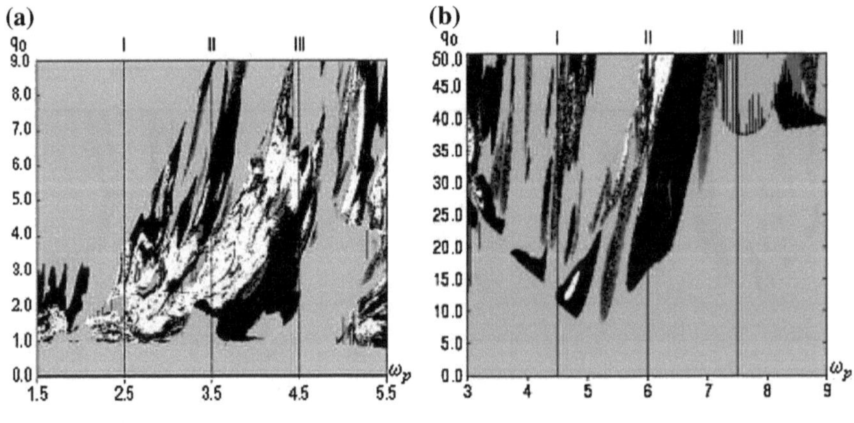

Fig. 14 Charts $\{q_0, \omega_p\}$ of the conical shell with movable simple support (**a**) and with movable clamping (**b**)

in time, signals $w(0, t)$ for $51.25 \leq t \leq 53.75$, phase portraits, vibrations after the series of bifurcations, and finally chaotic vibrations (Fig. 13).

Next, we investigate the influence of boundary conditions on the vibrations character taking as an example the conical shell with $k = 5$ and boundary conditions of movable clamping and movable simple support (Fig. 14). Vibrations corresponding to the movable clamping are more complex than that of the moving simple support.

6 Control of Chaos

We investigate the influence of the shell thickness change on boundary conditions and the shell geometry. We consider shallow conical shells with constant and non-constant thickness $h = (1 + c\rho)$, considering them as plates with initial deflection $w_0 = -k(1 - \rho)$, where $k = H/ho$. For the given type of boundary conditions the approximating functions have the following form

$$w_i(\rho) = \left(1 - \rho^2\right)^i, \qquad \varphi_j(\rho) = \left(1 - \rho^2\right)^{i+1}. \tag{12}$$

We take load $q = q_0 \sin \omega_p t$ and zero value initial conditions.

Figure 15 shows $\{q_0, \omega_p\}$ charts for the constant and non-constant shell thickness for $c = 0.1, -0.1, 0.2$ ($k = 5$). Here the influence of thickness changes on the system state differs essentially from the previously studied case.

For $c = -0.1$ a chaotic zone associated with low frequencies (about 2.5) appears, which does not exist for $c = 0.1$, $c = 0.2$ and $c = 0$, as well as a chaotic zone associated with high frequencies (about 5.5) occurs, which exists for $c = 0$ and does not exist for $c = 0.1, 0.2$. It can be seen that the influence of shell thickness change on its dynamics depends essentially on its geometry and initial conditions.

Figure 16 shows charts of control parameters $\{q_0, \omega_p\}$ for the conical shell ($k = 5$) movably clamped with constant ($c = 0$) and non-constant ($c = 0.1, -0.1$) thickness.The increase of thickness in the shell center ($c = -0.1$) yielded new zones of chaos associated with high frequencies and frequencies close to the natural frequency for $q_0 > 35$. An increase of zones associated with the independent frequencies is also observed. For $c = 0.1$, on the contrary, the zones of chaos and bifurcations essentially decreased.

We analyze spherical shells with $k = 5$. First, we consider spherical shells with the boundary condition of movable clamping and with $k = 5$ of constant and non-constant thickness for $c = 0.1$. In Fig. 17 charts of control parameters $\{q_0, \omega_p\}$ are given.

Here, the influence of thickness on the shell dynamics is more visible than in the previous case. The charts imply that shells with variable thickness have smaller zones of bifurcations and chaotic vibrations.

These results lead to the conclusion that by changing a shape of the transversal shell cross-section, and choosing properly parameters q_0 and ω_p we may control non-linear vibrations of the studied continuous systems.

In order to investigate the vibration character of conical shells versus parameter q_0 we construct $w_{\max}(q_0)$. In Fig. 18 functions $w_{max}(q_0)$ in the shell top are given for $k = 5$ and for $c = 0, 0.1, -0.1$ for $\omega_p = 3.5, 3.57$ and 3.38, respectively. In the relation $w_{max}(q_0)$ for $q_0 = 1$, the first stiff stability loss occurs. In the neighborhood of $q_0 = 2$ there is a zone of the second stiff stability loss, whereas for both critical loads the shell with non-constant stiffness ($c = 0.1$) exhibits higher critical loads in comparison to the case of $c = -0.1$. Critical loads for the constant thickness are located within the mentioned interval. The first critical load occurs for the first Hopf

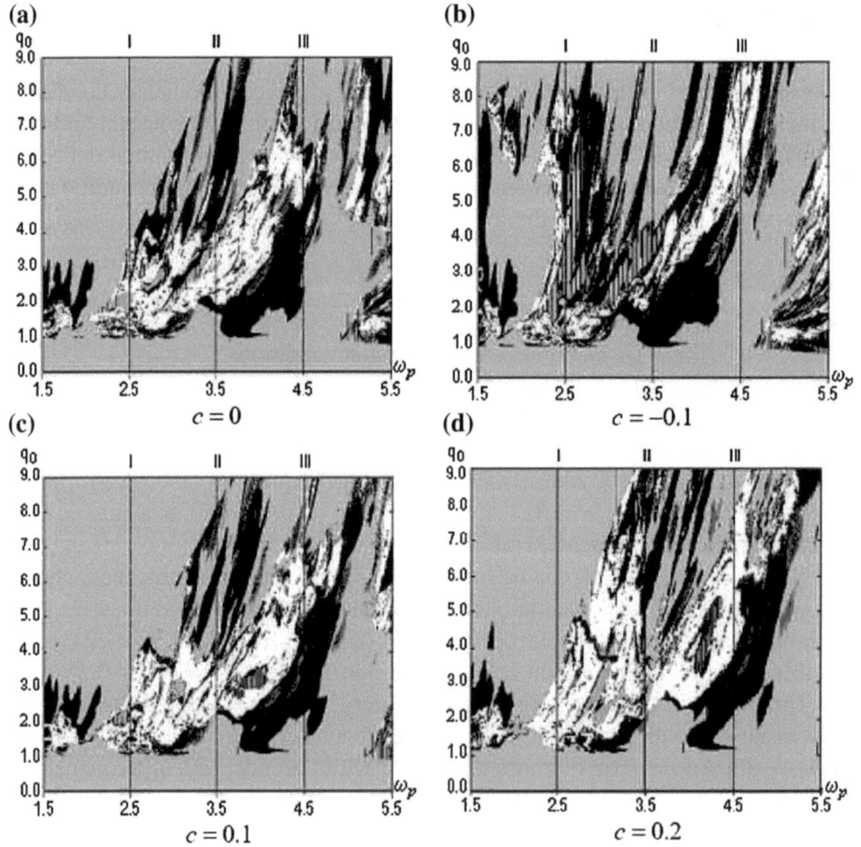

Fig. 15 Charts of control parameters $\{q_0, \omega_p\}$ for conical shells of constant (**a**) and non-constant (**b**, **c**, **d**) thickness ($k = 5$)

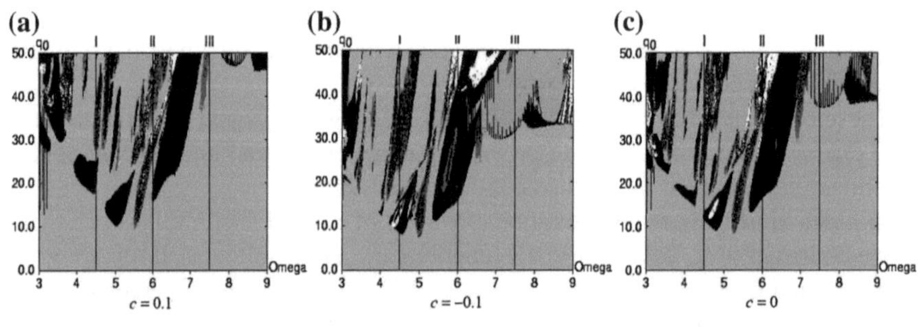

Fig. 16 Charts of control parameters $\{q_0, \omega_p\}$ for spherical shells. **a** $c = 0.1$. **b** $c = -0.1$. **c** $c = 0$

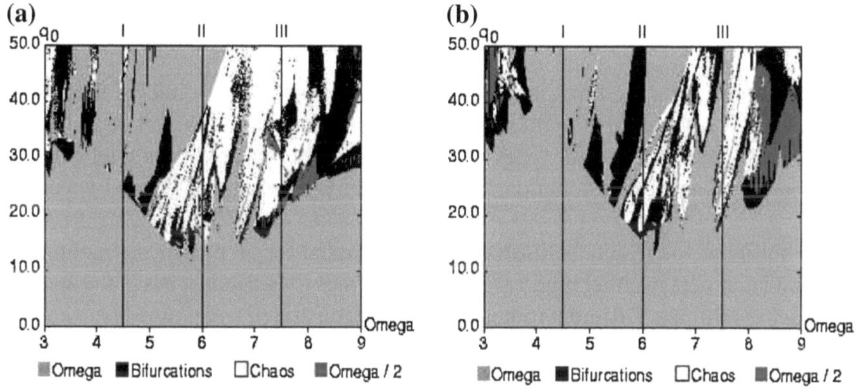

Fig. 17 Charts of control parameters $\{q_0, \omega_p\}$ for spherical shells of constant (**a**) and non-constant (**b**) thickness ($k = 5$)

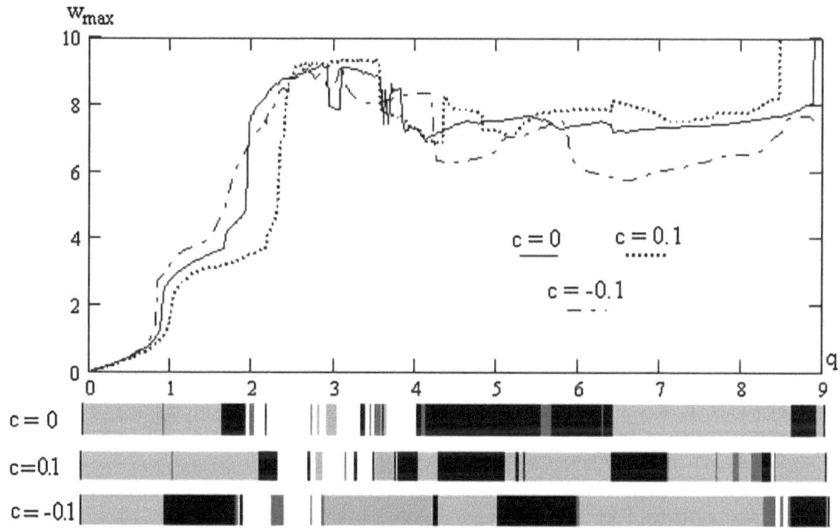

Fig. 18 Relations $w_{max}(q_0)$ and vibration scales for different c of the conical shells

bifurcation ($c = -0.1$) and for the second independent frequency (for $c = 0, 0.1$), which are depicted in the scales by vertical lines in the vicinity of $q_0 = 1$.

7 Concluding Remarks

In this chapter first a shell mathematical model is introduced. The model is reduced to a system of ordinary differential and algebraic equations in both rectangular and polar co-ordinates. Then, four types of boundary conditions are introduced with the associated approximating functions applied to a plate, conical shell and spherical shell.

Reliability of the results is illustrated and discussed for all previously mentioned structural members in either regular or chaotic regimes for various types of the applied boundary conditions. Different routes to chaos are illustrated using time histories and frequency power spectra. Then, non-linear novel dynamical phenomena exhibited by spherical and conical shells are detected, illustrated and discussed versus the k parameter describing a shell slope. Engineering important charts in the $\{q_0, \omega_p\}$ plane are presented for different k, which shows complexity of analyzed shell dynamics, including periodic, quasi-periodic, bifurcational and chaotic dynamics.

The study includes monitoring of the maximum Lyapunov exponent and addresses the problems of shell buckling and stability loss. Then, it is shown how, by using shell thickness c as a control parameter, we may avoid chaotic and bifurcational vibrations of the studied conical and spherical shells which are dangerous from the point of view of engineering applications.

Acknowledgments This paper was financially supported by the National Science Centre of Poland under grant MAESTRO 2, No. 2012/04/A/ST8/00738, for the years 2013-2016.

References

1. Awrejcewicz, J., Krysko, V.A.: Chaos in Structural Mechanics. Springer, Berlin (2008)
2. Awrejcewicz, J., Krysko, V.A., Papkova, I.V., Krysko, A.V.: Routes to chaos in continuous mechanical systems. Part 1: Mathematical models and solution methods. Chaos Solitons and Fractals **45**, 67–708 (2012a)
3. Awrejcewicz, J., Krysko, V.A., Papkova, I.V., Krysko, A.V.: Routes to chaos in continuous mechanical systems. Part 2: Modelling transitions from regular to chaotic dynamics. Chaos Solitons and Fractals **45**, 709–720 (2012b)
4. Awrejcewicz, J., Krysko, V.A., Papkova, I.V., Krysko, A.V.: Routes to chaos in continuous mechanical systems. Part 3: The lyapunov exponents, hyper, hyper-hyper and spatial-temporal chaos. Chaos Solitons and Fractals **45**, 721–736 (2012c)
5. Feng, Z.C., Sethna, P.R.: Global bifurcations in the motion of parametrically excited thin plate. Nonlinear. Dyn. **4**, 389–408 (1993)
6. Han, Q., Hu, H., Yang, G.: A study of chaotic motion in elastic cylindrical shells. Eur. J. Mech. A. Solids **18**, 351–360 (1999)
7. Lorenz, R.N.: Deterministic nonperiodic flow. Atm. Sci. **20**(1), 130–141 (1962)
8. Mikhlin, S.G.: Numerical Realization of Variational Methods (in Russ.). Nauka, Moscow (1966)
9. Shu, X., Han, Q., Yang, G.: The double mode model of the chaotic motion for a large deflection plate. Appl. Math. Mech. **20**, 360–364 (1999)
10. Zhang, W.: Global and chaotic dynamics for a parametrically excited thin plate. J. Sound Vib. **239**, 1013–1036 (2001)

Nonclassical Shell Theories in Ocular Biomechanics

Svetlana M. Bauer and Eva B. Voronkova

Abstract The stress-strain state of a pressurized spherical shell and circular cylinder are studied by means of the exact 3D theory of elasticity and the 2D approximate shell theories of moderate thicknesses. Both the sphere and the cylinder are made of transversally isotropic material. The first problem models the changes in the fluid pressure inside the human eye due to injected additional volume of liquid. The pressurized cylinder is one of simplest model of the human vessels. The algebraic relationships for deflections and stresses are derived. Both described shell theories take into account the variation of the shell thickness, which can be important for soft materials, e.g. the human eye's tissue. The asymptotical analysis of the exact 3D solutions has been performed and the accuracy of the approximate solutions, obtained with the approximate theories is analyzed. The effect of the thickness changes are also discussed.

1 Motivation

Shell-like structures are of very frequent occurrence in biological system, particularly, in a human eye. Shell theory-based mechanical models have been used in recent years as tools to describe, for example, the stress-strain state of the eye shell under encircling band [5, 7]; to build a biomechanical model of the choroidal detachment, to depict the different mechanical aspects of the development of glaucomatous atrophy of the optic nerve fibers and the behavior of Lamina Cribrosa—circular or closed to circular plate, where the optic nerve fibers pass through [5, 6, 10].

S.M. Bauer (✉) · E.B. Voronkova
Department of Theoretical and Applied Mechanics, St. Petersburg State University, Universitetsky Pr. 28, Peterhof, 198504 St. Petersburg, Russia
e-mail: s.bauer@spbu.ru

E.B. Voronkova
e-mail: e.voronkova@spbu.ru

H. Altenbach and G.I. Mikhasev (eds.), *Shell and Membrane Theories in Mechanics and Biology*, Advanced Structured Materials 45, DOI 10.1007/978-3-319-02535-3_4

Estimation of the short time effect of intravitreal injections on the intraocular pressure elevation is one of possible applications of shell-theory based models. Intraocular pressure (IOP), the fluid pressure inside an eye, is a key diagnostic parameter to determine the health of the eye. Intravitreal injections are performed by the amount of fluid (drug) brought into the eye. This procedure are applied widely as a treatment for a variety of eye's diseases. An increase in IOP can be induced by intravitreal injections and leads to undesirable side-effects and complications such as intraocular bleeding and retinal detachment. Relationship between IOP with the intraocular volume (IOV) seems to play an important role in problems concerning intravitreal injections [13]. As it is noted in Lyubimov [15]; Stein [19] the knowledge of the effect of IOP on IOV in a human eye is important to draw a physically correct conclusion from the data of standard measurement procedure used in ophthalmology.

An outer, fibrous, coat of the eye consists of the cornea and sclera. The sclera forms more than 90 % of fibrous eye shell and the sclera is tougher than the cornea [14]. The sclera, as well as cornea, are close to a soft membrane: the moduli of elasticity in the thickness directions are two orders of magnitude smaller than the tangential moduli [12]. For people with normal sight the sclera has a shape close to spherical one. That is why, the simplest model of the eyeball is a spherical shell. However, myopic (shortsighted) and hyperopic (longsighted) eyes have out-of-sphericity shapes.

Following Bauer et al. [8, 9], with the simplest model of the eyeball, we considered ellipsoidal shells of revolution of different shapes (modeling the sclera) as a soft shell (irresistible to flexural deformation), subjected to inner pressure The classical Kirchhoff-Love (KL) shell theory was employed to calculate the change of the inner volume of the shell. The aspherical geometry of the shell can be expressed, for example, by the following equation

$$r^2 + \frac{z^2}{\kappa^2} - \kappa^2 R_a^2 = 0,$$

where the z-axis is the axis of revolution of ellipsoid and also the optical axis of the eye, r is the radial distance from the optical (APA) axis, and R_a is the apical radius of curvature. A value of shape parameter $\kappa < 1$ corresponds to an oblate ellipsoidal shape modelling the hyperopic eye, $\kappa > 1$ to a prolate ellipsoidal shape (the myopic eye), $\kappa = 1$ to a spherical surface (the eye with normal vision).

Figure 1 represents the increasing of shell's inner volume under the pressure 15 mm Hg as a function of the shape parameter κ. We set the initial volume V_0 constant for all values of shape parameter κ. It is important, that the initial geometry (the shape of the shell) influences on the pressure-volume relationship.

However, the mathematical models based on assumptions of irresistibility to flexural deformation cannot explain, for example, the effect of the central cornea thickness on applanation tonometry readings or changes of the readings after refractive surgery [4, 11]. In order to develop the adequate mathematical models one should take into account the strong anisotropy of the eye's tissues, the influence of the bending deformation (while modeling tonometry procedure, for example). One way is

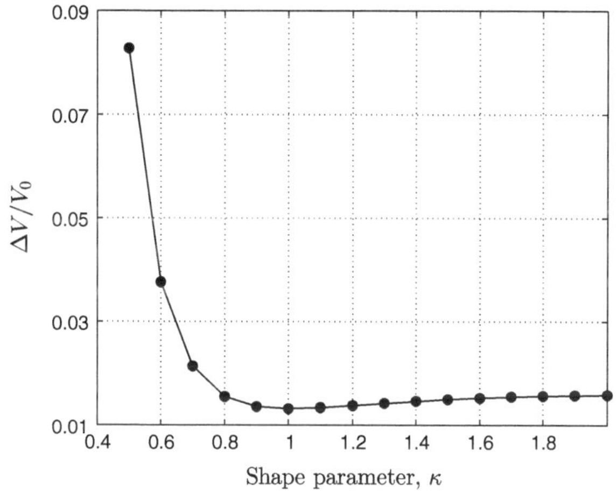

Fig. 1 Relative change of the inner volume ($\Delta V/V_0$). The shape parameter κ varies for constant initial volume V_0. The parameter $\kappa = APA/D$ (the ratio of axial length (APA) to equatorial diameter (D)) increases for a shell modelling a myopic eye, and decreases for a shell modelling a hyperopic eye

to carry out the finite element modeling [4, 11], another possibility is to use the nonclassical shells theories that do not neglect shear deformations and thickness changes [21].

Several non-classical, refined and high-order theories of orthotropic plates and shells have been formulated over the years. Some of these theories are reviewed and discussed in Altenbach [1]; Reddy and Wang [17]; Tovstik and Tovstik [20]. For soft materials, e.g. biological tissues, thickness variation can be important. First order shear deformation theories (for example, the Ambartsumyan theory [2]) still neglect thickness variation.

In the present paper, we treat two simple linear problems for pressurized spherical shells and circular cylinders made of transversally orthotropic material, applying the exact 3D theory and the 2D approximate shell theories of moderate thicknesses described in Palii and Spiro [16]; Rodionova et al. [18]. The theories are based on different principals, but both take into account deformations, rotation and bending of the fibers and, what is more important, their elongations in the direction of the thickness of the shell. The exact 3D solution for isotropic spherical shell can be find, for example, in Atanackovic and Guran [3]. Our concern is an accuracy of the 2D theories for transversally orthotropic shells. Therefore the asymptotical analysis of the exact 3D solutions has been performed and the correctness of the approximate solutions, obtained with refined theories is analyzed.

2 Problem Formulation and General Remarks

We consider a spherical shell and a circular long cylinder both of inner radius R_1 and outer radius $R_2 > R_1$, as shown in Fig. 2. The inner and outer surfaces of the bodies are loaded by the uniform pressures p_1 and p_2, respectively. The sphere, as well as the cylinder, is made of the transversely isotropic around the thickness direction material. A spherical coordinate system $(\rho, \vartheta, \varphi)$ and a cylindrical system of coordinates (ρ, φ, z) are used to define the stress and deformation fields for the sphere and the cylinder, respectively.

The three-dimensional relations between the stresses σ and the strains e of the shells in material principal coordinates (x_1, x_2, x_3), are given by Hooke's law

$$\begin{aligned} e_{11} &= \frac{\sigma_{11}}{E_1} - \frac{\nu_{12}}{E_1}\sigma_{22} - \frac{\nu_{13}}{E_1}\sigma_{33}, \quad e_{12} = \frac{\sigma_{12}}{G_{12}}, \\ e_{22} &= -\frac{\nu_{21}}{E_2}\sigma_{11} + \frac{\sigma_{22}}{E_2} - \frac{\nu_{23}}{E_2}\sigma_{33}, \quad e_{23} = \frac{\sigma_{23}}{G_{23}}, \\ e_{33} &= -\frac{\nu_{31}}{E_3}\sigma_{11} - \frac{\nu_{32}}{E_3}\sigma_{22} + \frac{\sigma_{33}}{E_3}, \quad e_{13} = \frac{\sigma_{13}}{G_{13}} \end{aligned} \tag{1}$$

with E_i being the Young moduli in the ith directions; G_{ij} are the shear moduli in $i - j$ directions; ν_{ij} are Poisson's ratio for the shortening in jth direction under the tension in ith direction and $\nu_{ij}E_j = \nu_{ji}E_i$ $(i, j = 1 \ldots 3)$.

In particular, we denote $x_1 = \vartheta$, $x_2 = \varphi$, and $x_3 = \rho$ for the sphere, and $x_1 = z$, $x_2 = \varphi$, $x_3 = \rho$ for the cylinder.

Taking into account the assumption of transverse isotropy with respect to surfaces orthogonal to the radial $x_3 = \rho$ axis, i.e. assuming the $x_1 - x_2$ surface to be the surface of the isotropy at every point, and inverting (1) we obtain

$$\begin{aligned} &\sigma_{11} = E_{11}e_{11} + E_{12}e_{22} + E_{13}e_{33}, \quad \sigma_{22} = E_{12}e_{11} + E_{11}e_{22} + E_{13}e_{33}, \\ &\sigma_{33} = E_{13}e_{11} + E_{13}e_{22} + E_{33}e_{33}, \quad \sigma_{12} = G_{12}e_{12}, \quad \sigma_{23} = G_{23}e_{23}, \quad \sigma_{13} = G_{13}e_{13}, \end{aligned} \tag{2}$$

where $G_{13} = G_{23}$, and the coefficients E_{ij} and G_{12} are given by

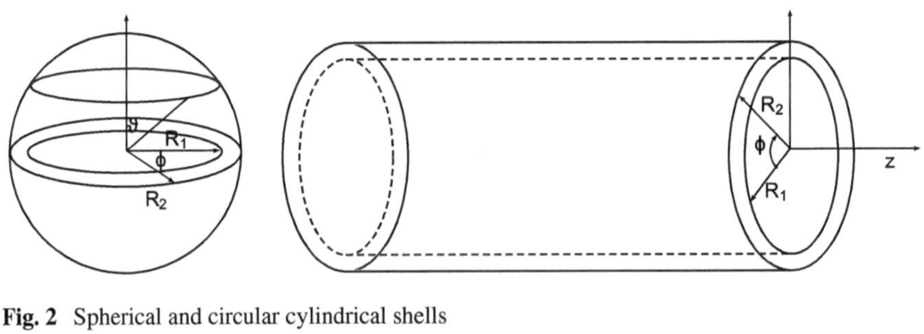

Fig. 2 Spherical and circular cylindrical shells

$$E_{11} = \frac{E_1(E_1 - E_3\nu_{13}^2)}{(1+\nu_{12})(E_1(1-\nu_{12}) - 2E_3\nu_{13}^2)}, \quad E_{12} = \frac{E_1(E_1\nu_{12} + E_3\nu_{13}^2)}{(1+\nu_{12})(E_1(1-\nu_{12}) - 2E_3\nu_{13}^2)},$$
$$E_{13} = \frac{E_1E_3\nu_{13}}{E_1(1-\nu_{12}) - 2E_3\nu_{13}^2}, \quad E_{33} = \frac{E_1E_3(1-\nu_{12})}{E_1(1-\nu_{12}) - 2E_3\nu_{13}^2}, \quad G_{12} = \frac{E_1}{2(1+\nu_{12})}.$$

2.1 3D Equations for Spherical Shell and Circular Cylinder

Due to symmetry the distribution of the stresses depends only on the radial coordinate $x_3 = \rho$ and the equilibrium equations reduce to a single equation for the sphere, as well as, for the cylinder

$$\frac{d\sigma_{33}}{d\rho} + \frac{k}{\rho}(\sigma_{33} - \sigma_{22}) = 0, \tag{3}$$

where $k = 1$ for a cylinder, and $k = 2$ for a sphere.

The inner and the outer surfaces of the shells are subjected to the uniform pressure. Thus, we have

$$\sigma_{33} = -p_i \quad for \quad \rho = R_i, \quad (i = 1, 2). \tag{4}$$

For reasons of symmetry, we suppose for the spherical shell that the displacements in the meridional and the circumferential directions are equal to zero and the radial displacement $u_3(\rho)$ is the only displacement component. For the long cylinder the plain strain conditions are assumed, namely the strain along the longitude axis $z = x_1$ is equal to zero $e_{11} = 0$.

Then the strain-displacements equations for both the sphere and the cylinder can be written as

$$e_{11} = (k-1)\frac{u_3}{\rho}, \quad e_{22} = \frac{u_3}{\rho}, \quad e_{33} = \frac{du_3}{d\rho}. \tag{5}$$

We can rewrite the constitutive Eq. (2) in the form

$$\begin{aligned} \sigma_{11} &= ((k-1)E_{11} + E_{12})\, e_{22} + E_{13}e_{33}, \\ \sigma_{22} &= (E_{11} + (k-1)E_{12})\, e_{22} + E_{13}e_{33}, \\ \sigma_{33} &= kE_{13}e_{22} + E_{33}e_{33}. \end{aligned} \tag{6}$$

By using (5) and (6) in (3) it follows that

$$\frac{d^2u_3}{d\rho^2} + \frac{k}{\rho}\frac{du_3}{d\rho} - M\frac{u_3}{\rho^2} = 0, \tag{7}$$

where parameter $M = k\,(E_{11} + (k-1)(E_{12} - E_{13}))/E_{33}$ depends on the material properties. In particular,

$$M^{\mathrm{s}} = 2\frac{E_{11} + E_{12} - E_{13}}{E_{33}}, \quad M^{\mathrm{c}} = \frac{E_{11}}{E_{33}}.$$

In what follows subscripts 's' and 'c' denote variables associated with the sphere and the cylinder.

Solving (7) together with (4) we obtain the radial displacements u_3 as a function of the coordinate ρ

$$u_3(\rho) = \frac{A_1}{\mu_1}\,\rho^m + \frac{A_2}{\mu_2}\,\rho^{-(m+k-1)}, \tag{8}$$

where

$$m = -\frac{k-1-\sqrt{(k-1)^2+4M}}{2},$$

$$\mu_1 = mE_{33} + kE_{13}, \quad \mu_2 = (m+k-1)E_{33} - kE_{13},$$

$$A_1 = \frac{p_1 R_1^{m+k} - p_2 R_2^{m+k}}{R_2^{2m+k-1} - R_1^{2m+k-1}}, \quad A_2 = (R_1R_2)^{m+k}\,\frac{p_1 R_2^{m-1} - p_2 R_1^{m-1}}{R_2^{2m+k-1} - R_1^{2m+k-1}}.$$

From Hooke's law (6), the strain-displacement Eqs. (5) and (8) one can find the circumferential σ_{22} and the radial σ_{33} stresses for the sphere and the cylinder

$$\sigma_{22} = A_1\frac{m+k-1}{k}\,\rho^{m-1} + A_2\frac{m}{k}\,\rho^{-(m+k)}, \quad \sigma_{33} = A_1\rho^{m-1} - A_2\rho^{-(m+k)} \tag{9}$$

For the spherical shell, the stress σ_{11}^{s} in the meridional direction $x_1 = \vartheta$ is equal to the circumferential stress σ_{22}^{s}.

In the case of the cylindrical shell, the stress in the longitude direction $x_1 = z$ becomes

$$\sigma_{11}^{\mathrm{c}} = A_1\mu_3\,\rho^{m-1} - A_2\mu_4\,\rho^{-m-1}, \tag{10}$$

$$\mu_3 = \frac{mE_{13} + E_{12}}{mE_{33} + E_{13}}, \quad \mu_4 = \frac{mE_{13} - E_{12}}{mE_{33} - E_{13}}.$$

In the case of isotropy,

$$M^s = 2, \;\; \mu_1^s = \frac{E_1}{1-2\nu_{12}} \;\; \mu_2^s = 2\frac{E_1}{1+\nu_{12}}, \;\; m^s = m^c = 1,$$

$$M^c = 1, \;\; \mu_1^c = \frac{E_1}{(1+\nu_{12})(1-2\nu_{12})}, \;\; \mu_2^c = \frac{E_1}{(1+\nu_{12})}, \;\; \mu_3 = \mu_4 = 1.$$

2.2 Asymptotic Solutions

To compare the exact 3D-solutions of the problems with the 2D approximate ones the asymptotic analysis of Eqs. (8–10) was performed. We introduce the dimensionless coordinate in the radial directions ξ

$$\xi = (\rho - R)/h, \quad -1/2 \le \xi \le 1/2,$$

where $h = R_2 - R_1$ and $R = (R_2 + R_1)/2$ are the thickness and the radius of midsurface of the shells.

We use the relative shell thickness $\alpha = h/R$ as the small parameter in the asymptotic expansions. The first terms of the asymptotic expansions of the displacement and the stresses of the shells's middle surface are

$$\sigma_{20} = \frac{p_1 - p_2}{\alpha k}\left[1 - \frac{\alpha k}{2}P^{\pm} + \frac{\alpha^2}{12}\left(k^2 + k - 3 - \frac{M}{2}\right)\right] + O(\alpha^2), \tag{11}$$

$$\sigma_{30} = -\frac{p_1 + p_2}{2}\left[1 - \alpha\frac{k+2}{4}P^{\mp} + \frac{\alpha^2 k}{8}\left(1 - \frac{M}{k}\right)\right] + O(\alpha^3), \tag{12}$$

$$\bar{u}_{30} = \frac{p_1 - p_2}{\alpha\, k E_0}\left[1 - \frac{\alpha k}{2}P^{\pm}\left(1 - \frac{E_{13}}{E_{33}}\right)\right. \tag{13}$$
$$\left. + \frac{\alpha^2}{12}\left(k^2 + k - 3 - \frac{M}{2} - \frac{3k(k+2)E_{13}}{2E_{33}}\right)\right] + O(\alpha^2)$$

with the shorthand notations

$$\sigma_{20} = \sigma_{22}(0), \quad \sigma_{30} = \sigma_{33}(0), \quad \bar{u}_{33} = u_3(0)/R,$$
$$E_0^{s} = \frac{E_1}{1 - \nu_{12}}, \quad E_0^{c} = \frac{E_1}{1 - \nu_{12}^2}, \quad P^{\pm} = \frac{p_1 + p_2}{p_1 - p_2}, \quad P^{\mp} = \frac{p_1 - p_2}{p_1 + p_2}.$$

For the longitudinal stress of the cylinder we have

$$\sigma_{10}^{c} = \sigma_{11}^{c}(0) = \frac{(p_1 - p_2)\nu_{12}}{\alpha}\left[1 - \frac{\alpha}{2}P^{\pm}\left(1 - \frac{\nu_{13}}{\nu_{12}}\right)\right. \tag{14}$$
$$\left. - \frac{\alpha^2}{12}\left(1 - \frac{9\nu_{13}}{2\nu_{12}} + \frac{E_{11}}{2E_{33}}\right)\right] + O(\alpha^2).$$

Note that the classical Kirchhoff-Love (KL) theory gives for the displacements and the tangential stresses the first terms in asymptotic approximation of the 3D solution in (11), (13) and (14). The normal stresses σ_{33} are neglected in the KL theory. For discussed problems the Timoshenko-Reissner (TR) and the Ambartsumyan (Amb) theories gives the same results as the KL-theory in the case of transversely isotropic material.

3 Palii-Spiro Theory

3.1 Basic Assumptions

The refined theory of anisotropic shells of moderate thickness has been introduced in Palii and Spiro [16]. Palii and Spiro have extended the Timoshenko-Reissner (TR) type theory taking into account the effect of transverse normal stress and thickness deformation.

As in the TR theory, the Palii-Spiro (PS) theory supposes, that a rectilinear element normal to the middle surface of a shell remains rectilinear after the deformation. The second assumption of the PS theory gives

$$u_i(\zeta) = u_{i0} + \zeta(\theta_i + \gamma_i), \quad (i = 1, 2), \qquad u_3(\zeta) = u_{30} + F(\zeta), \tag{15}$$

where u_{i0}, u_{30} denote the displacements of a point on the midsurface of the shell in the principal directions β_i and in the ζ direction normal to the midsurface; u_i, u_3 are the displacements of a generic point of the shell; θ_i, γ_i are the rotations of the transverse normals about the β_i-axes and the shear angles in (β_i, ζ) plane, respectively; $F(\zeta)$ is a function to be determine.

According to Palii and Spiro [16], the normal stress distribution along the shell thickness and the thickness stretching $F(\zeta)$ are

$$\sigma_{33} = \frac{1}{(1+\kappa_1\zeta)(1+\kappa_2\zeta)}\left(\frac{m_3}{h} + q_3\frac{\zeta}{h}\right), \tag{16}$$

$$F(\zeta) = \int_0^\zeta \frac{\sigma_{33}}{E_{33}}d\zeta - \zeta\left(\nu_1^*\left(\varepsilon_1 + \frac{\zeta}{2}\tilde{\kappa}_1\right) + \nu_2^*\left(\varepsilon_2 + \frac{\zeta}{2}\tilde{\kappa}_2\right)\right) \tag{17}$$

with κ_1, κ_2 being the principal curvatures in β_1 and β_2 directions, respectively; $\varepsilon_1, \varepsilon_2, \tilde{\kappa}_1\ \tilde{\kappa}_2$ are the strains and the curvature of the middle surface as their defined in the classical shell theory. We employ the notations of Rodionova et al. [18], by defining

$$q_3 = p_1\left(1 - \kappa_1\frac{h}{2}\right)\left(1 - \kappa_2\frac{h}{2}\right) - p_2\left(1 + \kappa_1\frac{h}{2}\right)\left(1 + \kappa_2\frac{h}{2}\right),$$

$$m_3 = -\frac{h}{2}\left[p_1\left(1 - \kappa_1\frac{h}{2}\right)\left(1 - \kappa_2\frac{h}{2}\right) + p_2\left(1 + \kappa_1\frac{h}{2}\right)\left(1 + \kappa_2\frac{h}{2}\right)\right].$$

For a shell made of orthotropic material

$$\nu_1^* = \frac{\nu_{13} + \nu_{12}\nu_{21}}{1 - \nu_{12}\nu_{21}}, \quad \nu_2^* = \frac{\nu_{23} + \nu_{21}\nu_{13}}{1 - \nu_{12}\nu_{21}}.$$

3.2 Formulation for Spherical Shells

Now we derive the governing equations for the transversally isotropic spherical shell of radius R and thickness h, subjected to transversal load (4).

As all variables of interest are the functions of the thickness coordinate ζ, assumptions (15) yield the following strain-displacement relations and the constitutive equations for stress resultants

$$\begin{aligned} u_3^{PS}(\zeta) &= w + F(\zeta), \quad -h/2 \le \zeta \le h/2, \\ e_{11}^{PS} = e_{22}^{PS} &= \varepsilon_1 \left[1 - \frac{\zeta}{R}\left(1 + 2\frac{E_{13}}{E_{33}}\right)\right], \quad \varepsilon_1 = \varepsilon_2 = \frac{w}{R}, \\ T_1^{PS} = T_2^{PS} &= E_0^s h \varepsilon_2 + \frac{E_{13}}{E_{33}} T_0^{PS}, \quad T_0^{PS} = -\frac{h}{2}(p_1 + p_2), \end{aligned} \tag{18}$$

where w is the midsurface displacement, T_1 and T_2 are the meridional and the circumferential stress resultants per unit length.

The tangential normal stresses σ_{11}, σ_{22} can be obtain from Hooke's law (2). For the spherical shell we have

$$\sigma_{11}^{PS} = \sigma_{22}^{PS} = E_0^s e_{22} + \frac{E_{13}}{E_{33}} \sigma_{33}^{PS}(\zeta), \tag{19}$$

where the transverse stress σ_{33}^{PS} is calculated in Eq. (16) at $\kappa_1 = \kappa_2 = R$.

Substituting Eq. (18) into the equilibrium equation of the spherical shell

$$\frac{T_1^{PS}}{R} + \frac{T_2^{PS}}{R} = p_1 - p_2 - \frac{h}{R}(p_1 + p_2), \tag{20}$$

one obtains the displacement w of the middle surface.

Now we can write out the stresses and the displacement of the spherical shell as

$$\begin{aligned} \sigma_{20}^{PS(s)} &= \frac{p_1 - p_2}{2\alpha}\left[1 - \alpha P^{\pm} + \alpha^2 \frac{E_{13}}{E_{33}} + \frac{\alpha^3}{4} P^{\pm} \frac{E_{13}}{E_{33}}\right], \\ \sigma_{30}^{PS(s)} &= -\frac{p_1 + p_2}{2}\left[1 - \alpha P^{\mp} + \frac{\alpha^2}{4}\right], \\ \bar{u}_{30}^{PS(s)} &= w/R = \frac{p_1 - p_2}{\alpha} \frac{1 - \nu_{12}}{2E_1}\left[1 - \alpha P^{\pm}\left(1 - \frac{E_{13}}{E_{33}}\right)\right]. \end{aligned} \tag{21}$$

3.3 Resultant Expressions for Circular Cylindrical Shells

Here we list the resultant expressions for circular cylindrical shell based on the PS theory. As in Sect. (2.1), the plain strain behavior is assumed for transversely isotropic cylindrical shell of radial R and thickness h.

For the longitudinal σ_{11}, the circumferential σ_{22}, the transverse σ_{33} stresses and the displacement of the cylindrical shell we have

$$\begin{aligned}
\sigma_{10}^{PS(c)} &= \frac{(p_1 - p_2)\nu_{12}}{\alpha}\left[1 - \frac{\alpha}{2}p^{\pm}\left(1 - \frac{\nu_{13}}{\nu_{12}}\right) + \frac{\alpha^2}{4}\frac{E_{13}}{E_{33}\nu_{12}}\right],\\
\sigma_{20}^{PS(c)} &= \frac{p_1 - p_2}{\alpha}\left[1 - \frac{\alpha}{2}p^{\pm} + \frac{\alpha^2}{4}\frac{E_{13}}{E_{33}}\right],\\
\sigma_{30}^{PS(c)} &= -\frac{p_1 + p_2}{2}\left(1 - \frac{\alpha}{2}p^{\mp}\right),\\
\bar{u}_{30}^{PS(c)} &= w/R = \frac{p_1 - p_2}{\alpha E_0^c}\left[1 - \frac{\alpha}{2}p^{\pm}\left(1 - \frac{E_{13}}{E_{33}}\right)\right].
\end{aligned} \tag{22}$$

4 Rodionova-Titaev-Chernykh Theory

4.1 Basic Assumptions and Formulation for Spherical Shells

Rodionova, Titaev and Chernykh developed a linear high-order theory of anisotropic shells by expanding all quantities of interest, i.e. strains, stresses and deformations, into series in the thickness directions [18]. The RTCh theory based on following static and kinematic assumptions

- the transverse tangential and normal stresses are distributed along the shell thickness according to the quadratic and cubic laws, respectively;
- along the shell thickness the tangential and normal components of the displacement vector have the polynomial distributions of the third and the second powers, respectively.

For the spherical shell, described in Sect. 3, the displacement u_{33}^{RTC} and the transversal normal stresses σ_{33}^{RTC} are

$$\sigma_{33}^{RTC}(\zeta) = \sigma_{33}^{(0)}P_0(\zeta) + \sigma_{33}^{(1)}P_1(\zeta) + \sigma_{33}^{(2)}P_2(\zeta) + \sigma_{33}^{(2)}P_3(\zeta), \tag{23}$$

$$u_3^{RTC}(\zeta) = w^* P_0(\zeta) + \gamma_3^* P_1(\zeta) + \theta_3^* P_2(\zeta), \tag{24}$$

where P_i, $(i = 0 \ldots 3)$ are the Legendre polynomials

$$P_0(\zeta) = 1, \quad P_1(\zeta) = \frac{2\zeta}{h}, \quad P_2(\zeta) = \frac{6\zeta^2}{h^2} - \frac{1}{2}, \quad P_3(\zeta) = \frac{20\zeta^3}{h^3} - \frac{3\zeta}{h}.$$

Coefficients γ_3^*, θ_3^* in (24) characterize the alteration of the lengths of normals to the middle surface, $(w^* - \theta_3^*/2)$ is a normal displacements of a point on the midsurface.

Representing the strains and stresses as the linear combinations of the Legendre polynomials and using the above hypotheses, one can write the constitutive and the strain-displacement relations in the form

$$\begin{aligned} \varepsilon_1^* &= \varepsilon_2^* = \frac{w^*}{R}, \qquad \gamma_3^* = \frac{T_0^{RTC}}{2E_{33}} - \frac{E_{13}}{E_{33}}\frac{h}{2}\left(\varepsilon_1^* + \varepsilon_2^*\right), \\ \tilde{\kappa}_1^* &= \tilde{\kappa}_2^* = \frac{\gamma_3^*}{R}, \qquad \theta_3^* = \frac{M_0^{RTC}}{hE_{33}} - \frac{E_{13}}{E_{33}}\frac{h}{6}\left(\tilde{\kappa}_1^* + \kappa_2^*\right), \\ T_1^{RTC} &= T_2^{RTC} = E_0^s h \epsilon_2^* + \frac{E_{13}}{E_{33}} T_0^{RTC}, \end{aligned} \tag{25}$$

where

$$M_0^{RTC} = \frac{h^2}{10} q_3, \quad T_0^{RTC} = m_3.$$

By introducing the notations

$$q_0 = p_1 - p_2, \quad m_0 = -\frac{h}{2}(p_1 + p_2),$$

the normal stresses can be written as following [18]

$$\sigma_{33}^{RTC} = \frac{T_0^{RTC}}{h} P_0(\zeta) + \frac{6M_0^{RTC}}{h^2} P_1(\zeta) + \frac{m_0 - T_0^{RTC}}{h} P_2(\zeta) + \left(\frac{q_0}{2} - \frac{6M_0^{RTC}}{h^2}\right) P_3(\zeta) \tag{26}$$

The tangential stresses can be obtain from the Hooke law (2) with expressions for the deformations

$$\begin{aligned} e_{11}^{RTC} &= e_{22}^{RTC} = \left(1 - \frac{\zeta}{R}\right)\left(\frac{w^*}{R} P_0(\zeta) + \frac{\gamma_3^*}{R} P_1(\zeta) + \frac{\theta_3^*}{R} P_2(\zeta)\right), \\ e_{33}^{RTC} &= \frac{2}{h}\gamma_3^* P_0(\zeta) + \frac{6}{h}\theta_3^* P_1(\zeta). \end{aligned} \tag{27}$$

The problem is governed by the equilibrium equation

$$\frac{T_1^{RTC}}{R} + \frac{T_2^{RTC}}{R} = q_3 = p_1\left(1 - \frac{h}{2R}\right)^2 - p_2\left(1 + \frac{h}{2R}\right)^2. \tag{28}$$

Finally, for stresses and the displacements on the midsurface we get

$$\sigma_{10}^{RTC(s)} = \frac{p_1 - p_2}{2\alpha}\left[1 - \alpha P^{\pm} + \frac{\alpha^2}{4}\left(1 - \frac{2}{5}\frac{E_{11} + E_{12}}{E_{33}} + 2\frac{E_{13}}{E_{33}} + \frac{2}{15}\left(\frac{E_{13}}{E_{33}}\right)^2\right) + \alpha^3(\ldots)\right],$$

$$\sigma_{30}^{RTC(s)} = -\frac{p_1 + p_2}{2}\left[1 - \frac{3\alpha}{2}P^{\mp} + \frac{3}{8}\alpha^2\right], \tag{29}$$

$$\bar{u}_{30}^{RTC(s)} = u_3^{RTC(s)}(0)/R = \frac{p_1 - p_2}{\alpha}\frac{1 - \nu_{12}}{2E_1}\left[1 - \alpha P^{\pm}\left(1 - \frac{E_{13}}{E_{33}}\right) + \frac{\alpha^2}{4}\left(1 - \frac{2}{5}\frac{E_{11} + E_{12}}{E_{33}} - 4\frac{E_{13}}{E_{33}} + \frac{2}{15}\left(\frac{E_{13}}{E_{33}}\right)^2\right) + \alpha^3(\ldots)\right].$$

We do not put down the coefficients at α^3 in (29) since they are exceedingly lengthy.

4.2 Stresses and Deformations for Circular Cylindrical Shells

The RTCh theory gives the following expressions for the stresses and the deformations of the cylindrical shell under plain-strain conditions

$$\sigma_{10}^{RTC(c)} = \frac{(p_1 - p_2)\nu_{12}}{2\alpha}\left[1 - \frac{\alpha}{2}P^{\pm}\left(1 - \frac{\nu_{13}}{\nu_{12}}\right) - \frac{\alpha^2}{12}\left(\frac{3}{5}\frac{E_0^c}{E_{33}} - \frac{9\nu_{13}}{2\nu_{12}} - \frac{3}{2}\frac{E_{13}}{E_{33}} + \frac{1}{2}\left(\frac{E_{13}}{E_{33}}\right)^2\right) + \alpha^3(\ldots)\right],$$

$$\sigma_{20}^{RTC(c)} = \frac{(p_1 - p_2)}{\alpha}\left[1 - \frac{\alpha}{2}P^{\pm} - \frac{\alpha^2}{24}\left(\frac{6}{5}\frac{E_0}{E_{33}} - 3\frac{E_{13}}{E_{33}} + \left(\frac{E_{13}}{E_{33}}\right)^2\right) + \alpha^3(\ldots)\right], \tag{30}$$

$$\sigma_{30}^{RTC(c)} = -\frac{p_1 + p_2}{2}\left[1 - \frac{3\alpha}{4}P^{\mp}\right],$$

$$\bar{u}_{30}^{RTC(c)} = \frac{p_1 - p_2}{\alpha E_0^c}\left[1 - \frac{\alpha}{2}P^{\pm}\left(1 - \frac{E_{13}}{E_{33}}\right) - \frac{\alpha^2}{12}\left(\frac{3}{5}\frac{E_0^c}{E_{33}} - 3\frac{E_{13}}{E_{33}} - \frac{1}{2}\left(\frac{E_{13}}{E_{33}}\right)\right) + \alpha^3(\ldots)\right].$$

5 Results and Discussion

For numerical examples the shells properties are taken from literature as material properties of the sclera Iomdina [12]: $E_1 = 3\,\text{MPa}$, $\nu_{12} = 0.45$, $h/R = 1/15$. Calculations have been performed for different $n = E_1/E_3$. The Poisson ration ν_{13}

is selected so that the stain energy of the plate material be positive. We also test the case of thick shells with $h/R = 1/5$.

5.1 Comparison for Normal Stresses

The classical KL theory neglects the thickness changes and the normal stresses σ_{33} in the thickness directions. Both the PS and the RTCh theory proposed a certain law for normal the stress distribution along the shells thickness (16), (26), but it is important to note that no material elastic constants affect the normal stresses in both 2D theories. This is the case for the isotropic spherical and cylindrical shells, but in for a transversely isotropic material the material parameters appears in terms of order α^2 in Eq. (12).

Comparing the expressions for the normal stresses (21) and (22) with the asymptotic terms in (12), we see that on the middle surface the PS theory gives first two correct terms for the normal stress σ_{33} for both test problems. The RTCh theory gives only one correct term for the stress σ_{33}.

Even so, for the case of thin shells ($h/R = 1/15$), both refined theories give a good agreement with the exact 3D solution of the normal stress distribution for Young's moduli ratio up to $E_1/E_3 = 100$ (see Fig. 3). For a thick ($h/R = 1/5$) shell the PS theory gives excellent accuracy for a case of isotropy, and both theories are inaccurate for transverse isotropy with $E_1/E_3 = 100$ (Fig. 3).

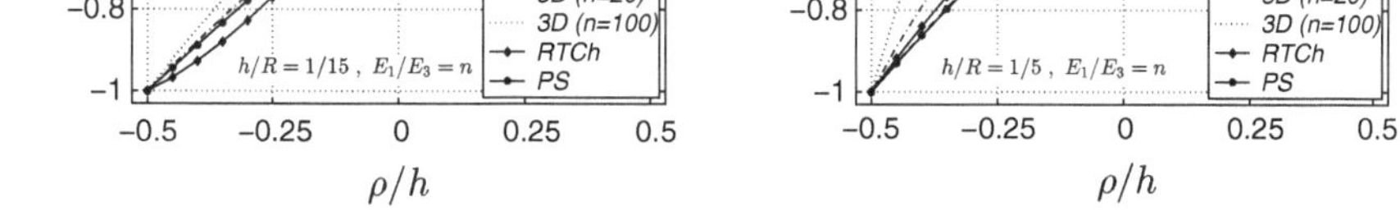

Fig. 3 Distribution of normal stresses in the spherical shell with $h/R = 1/15$ (*left*) and $h/R = 1/5$ (*right*). *Solid line* with *disk markers* describes the PS-theory; with *diamond markers* describes the RTCh theory; *dashed*, *dot-dashed* and *dotted lines* describes the 3D solution at different values of E_1/E_3

5.2 Comparison for Displacements

Both 2D theories give two first correct terms comparing with the asymptotic terms the deflection of the middle surface for the spherical shell along with the circular cylinder (Eqs. 21–22 and 29–30). The RTC theory gives one of three terms of the third order in Eq. (13).

For the thick spherical shell ($h/R = 1/5$) the normalized midplane deflection according to the presented theories and those obtained by the exact 3D theory are plotted in Fig. 4. We denote the KL approximation of the mid-surface deflection as $u_0 = p(1 - \nu_{12})R^2/2E_1h$. In the case of isotropy (Fig. 4, left), the RTCh theory almost exact coincide with the 3D solution on the middle surface as well as along the thickness of the shell. In the case of transversal isotropy with $E_1/E_3 = 20$ the RTCh theory gives excellent accuracy on the middle surface but diverges with the exact solution while moving to the shell's inner and outer surfaces (Fig. 4).

The large difference between the PS and 3D solutions for the displacements is 3 and 19 % for $E_1/E_3 = 1$ and $E_1/E_3 = 20$, respectively. Figure 5 shows the normalized deflection of the thin cylindrical shell at $E_1/E_3 = 100$.

5.3 Thickness Variation and Change of the Inner Volume

For soft materials, e.g. biological tissues, thickness variation can be important. Results, obtained with the help of shell theories, which neglect thickness variation, can be inaccurate or inadequate compered with experimental data.

We calculate the thickness variation both for thin and thick shells. Figure 6 presents the effect of Young's moduli ratio on the thickness variation. For a thin shell the numerical results obtained with the 2D refined theories are almost perfectly superimposed to the exact 3D results up to $E_1/E_3 = 50$. For a thick shell the refined

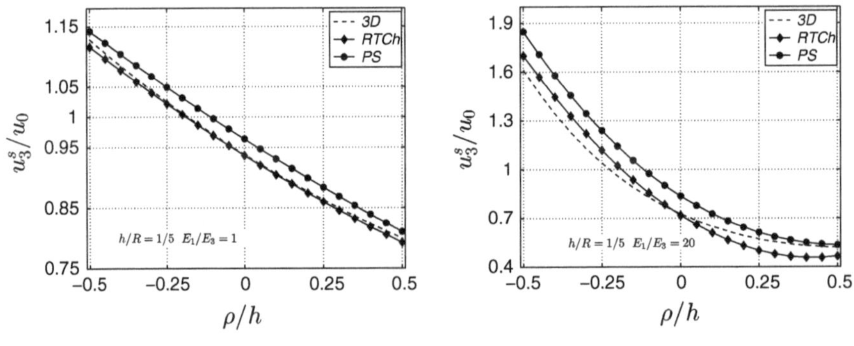

Fig. 4 Normalized deflection of the thick isotropic (*left*) and transversely isotropic (*right*) with $E_1/E_3 = 20$ spherical shell. *Dashed line* corresponds to the 3D exact results, *solid line* with *disk markers* refers to the PS-theory, *solid line* with *diamond markers*—to the RTCh theory

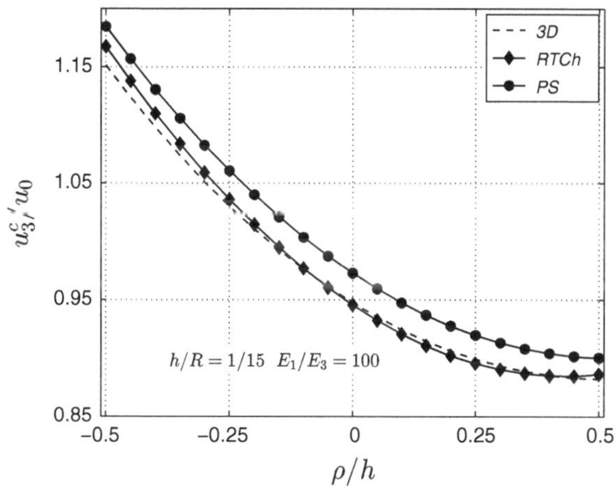

Fig. 5 Normalized deflection of the thin cylindrical shell at $E_1/E_3 = 100$, $\nu_{12} = 0.45$, $\nu_{13} = 0.1$. *Dashed line* corresponds to the 3D exact results, *solid line* with *disk markers* refers to the PS-theory, *solid line* with *diamond markers*—to the RTCh theory. u_0 is the KL approximation of the mid-surface deflection

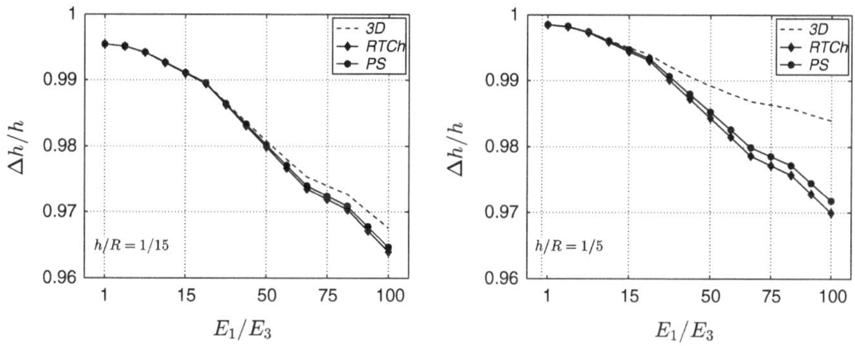

Fig. 6 Relative thickness variation for the pressurized spherical shell at $h/R = 1/15$ (*left*) and $h/R = 1/5$ (*right*). *Dashed line* corresponds to the 3D exact results, *solid line* with *disk markers* refers to the PS-theory, *solid line* with *diamond markers*—to the RTCh theory

theories are inaccurate in prediction of a thickness reduction starting from value of E_1/E_3 equal to 20.

Figure 7 shows the effect of injected volumes on IOP variation. We take the change in intraocular volume equal to 0.1 ml and calculate the corresponding changes in inner pressure for different mechanical properties of the eyeball tissue. The KL-theory, as well as the TR and Amb theories, gives the changes in inner pressure value of 26.7 mm Hg and overestimates the exact results up to 10 mm Hg.

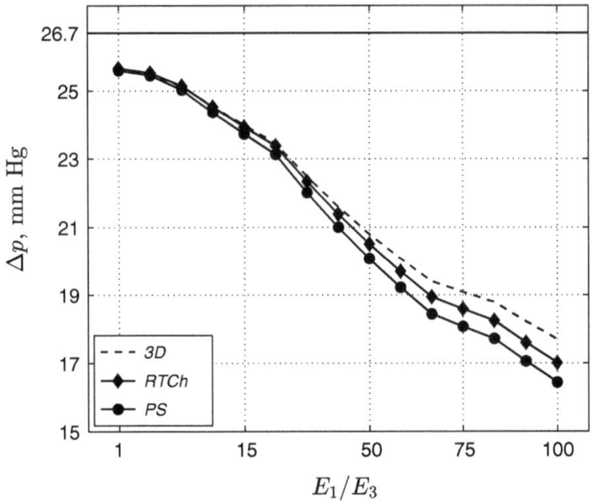

Fig. 7 Effect of absolute changes of IOV on IOP variation. Absolute change of IOV $\varDelta V = 0.1$ ml. relative thickness of the shell is $h/R = 1/15$. Value of 26.7 mm Hg corresponds to the results obtained by the KL-theory (with no thickness variation). *Dashed line* corresponds to the 3D exact results, *solid line* with *disk markers* refers to the PS-theory, *solid line* with *diamond markers*—to the RTCh theory

6 Conclusions

The stress-strain states of a pressurized transversally isotropic spherical shell and a circular cylinder are studied by means of two refined theories for transversally isotropic shells of moderate thickness. The algebraic relationships for deflections and stresses obtained by the approximate theories are compared with the 3D exact solutions.

Both theories give acceptable fit to the 3D solutions for the displacement, the stresses and thickness variations for shells with $h/R = 1/15$. For a thick shell ($h/R = 1/5$), results obtained by the PS and RTCh theories are accurate but only for shells made of isotopic material. Note that in general case, the RTC theory has the 14th order while the PS theory is of the 10th order, which makes a calculation procedure simpler.

In the first approximation, the human eye can be modeled as a spherical shells. However, the myopic and hyperopic eyes have out-of-spherical (ellipsoidal) form. Refined shell theories, discussed here, are well agreed with the exact 3D solutions. Therefore it is reasonable to use these theories for models that describes the human eye more precisely, for example, as the multilayered transversally isotropic spherical shells or/and ellipsoidal shells.

References

1. Altenbach, H.: Theories for laminated and sandwich plates. An overview. Mech. Compos. Mater. **34**, 333–349 (1998)
2. Ambartsumyan, S.: Theory of Anisotropic Plates. Technomic Publishing, Stamford (1970)
3. Atanackovic, T.M., Guran, A.: Theory of Elasticity for Scientist and Engineers. Birkhäuser, Berlin (2000)
4. Bauer, S.M., Kachanov, A.B., Semenov, B.N., Sliesoraityte, I.: The effect of cornea thickness on measurements of intraocular pressure by applanation methods. In: Proceeding of the Ocular Biomechanics Conference, pp. 69–73 (2007)
5. Bauer, S.M., Zimin, B.A., Tovstik, P.E.: Simplest Models of Plates and Shells in Ophthalmology (in Russ.). St. Petersburg University Press, St. Petersburg (2000)
6. Bauer, S.M., Voronkova, E.B.: On the deformation of the lamina cribrosa under intraocular pressure. Russ. J. Biomech. **5**(1), 73–82 (2001)
7. Bauer, S.M., Mironov, A.N.: On the mathematical simulation of the stress-strain state of the eye shell undergoing the scleral buckling procedure. Acta. Bioeng. Biomech. **4**, 726–727 (2002)
8. Bauer, S.M., Lyubimov, G.A., Tovstik, P.E.: On the mathematical simulation of the measuring of the intraocular pressure by Maklakov method. Technische Mechanik **24**, 231–235 (2004)
9. Bauer, S.M., Lyubimov, G.A., Tovstik, P.E.: Mathematical modelling of Maklakoffs method for mesuring the intraocular pressure. Fluid Dyn. **40**, 20–33 (2005)
10. Bauer, S.M.: Mechanical models of the development of glaucoma. In: Guran, A., Smirnov, A.L., Steigmann, D.J., Vaillancourt, R. (eds.) Advances in Mechanics of Solids in Memory of Prof. E.M. Haseganu, pp. 153–178. World Scientific Publishing, Singapore (2006)
11. Bauer, S.M., Karamshina, L.A., Kachanov, A.B.: Mechanical models of the measurements of intraocular pressure by Goldmann and Maklakov applanation tonometers after refractive surgery (in Russ.). Russ. J. Biomech. **16**(3(57)), 25–32 (2012)
12. Iomdina, E.N.: Mechanical properties of the human eye tissues. Modern Problems of Biomechanics, pp. 183–200. Moscow University Press, Moscow (2006)
13. Kotliar, K., Maier, M., Bauer, S., Feucht, N., Lohmann, C., Lanzl, I.: Effect of intravitreal injections and volume changes on intraocular pressure: clinical results and biomechanical model. Acta. Ophthamol Scand. **85**, 777–781 (2007)
14. Lang, G.K.: Ophthalmology. Thieme, Stuttgart (2000)
15. Lyubimov, G.A.: Opportunities of the elastometry method for investigating for elastic properties of the eyeball shell. Fluid Dyn. **45**, 169–176 (2010)
16. Palii, O.M., Spiro, V.E.: Anisotropic Shells in Shipbuildings. Theory and Analysis (in Russ.). Sudostroenie, Leningrad (1977)
17. Reddy, J.N., Wang, C.M.: An overview of the relationships between of the classical and shear deformation plate theories. Compos. Sci. Technol. **60**, 2327–2335 (2000)
18. Rodionova, V.A., Titaev, V.F., Chernykh, K.F.: Applied Theory of Anisotropic Plares and Shells (in Russ.). St. Petersburg University Press, St. Petersburg (1996)
19. Stein, A.A.: Pressure-volume dependence for eyeball under external load. Fluid Dyn. **45**, 177–186 (2010)
20. Tovstik, P.E., Tovstik, T.P.: On the 2d models of plates and shells including the transversal shear. Z. Angew. Math. Mech. **87**, 160–171 (2007)
21. Voronkova, E.B., Bauer, S.M., Eriksson, A.: Nonclassical theories of shells in application to soft biological tissue. In: Altenbach, H., Eremeyev, V.A. (eds.) Shell-like Structures, Advanced Structured Materials, vol. 15, pp. 647–654. Springer, Berlin (2011)

Linear Oscillations of Suspended Graphene

Igor Berinskii and Anton Krivtsov

Abstract Due to their excellent mechanical properties and extra high electroconductivity, suspended graphene sheets recently were proposed as perspective working elements of nanosystems. This work is devoted to derivation of natural frequencies of such sheets. Two different approaches are proposed. The first one is based on representation of the graphene sheet as a thin rectangular membrane. In this case the transversal oscillations are described with the classical one-dimensional wave equation. Evaluation of the tension force in the membrane is performed basing on the misfit between the graphene and silicon substrate crystal lattices. As a result, the natural frequencies are found as the functions of the membrane length. Another approach is to represent a graphene sheet as a thin plate. In this case a bending rigidity of graphene has to be taken into account. As a result, it is shown that the bending rigidity is more significant for the short resonators and leads to the higher frequencies in comparison the long resonators.

1 Introduction

This work is devoted to the suspended graphene oscillations. Graphen is a monoatomic thin film with unique elastic, electrical, optical and thermal properties [7]. Thin films are used in novel nanoelectromechanical systems (NEMS), e.g. nanoresonators [6, 8]. Recently, graphene sheet were proposed for using in such systems. Up to now, the existing graphene-based experimental set-ups [4, 5, 13] have not achieved as high oscillation properties as silicon-based resonators. However, the technical problems

I. Berinskii (✉) · A. Krivtsov
Department of Theoretical and Applied Mechanics,
St. Petersburg State Polytechnical University, Polytechnicheskaya 29,
195251 St. Petersburg, Russia
e-mail: iberinsk@gmail.com

I. Berinskii · A. Krivtsov
Laboratory for Discrete Models in Mechanics,
Institute for Problems in Mechanical Engineering, Bol'shoy Pr. V.O. 61,
199178 St. Petersburg, Russia
e-mail: akrivtsov@bk.ru

H. Altenbach and G.I. Mikhasev (eds.), *Shell and Membrane Theories in Mechanics and Biology*, Advanced Structured Materials 45, DOI 10.1007/978-3-319-02535-3_5

will be hopefully solved soon, and the advantages of graphene systems will be used for practical applications. First of all, they can be much easier miniaturized to the nanoscale. As single-atomic graphene layer is a thinnest material at all, it is much more sensitive to external effects than other films. It allows to use them as mass sensors for nanoparticles.

This paper shows that graphene-based resonators can be used as the sources of the high-frequency (up to several THz) oscillations. The transversal linear oscillations of graphene sheet stretched over the trench in silicon oxide substrate are considered. Two models are used to calculate the natural frequencies. First model represents graphene sheet as a membrane with a constant tension. The second one represents graphene as a thin plate. This model takes a bending rigidity of graphene into account in addition to the tensile stiffness.

There are several works devoted to calculation of the natural frequencies of graphene (see e.g. [1, 9, 10]). Most of them do not take the bending rigidity of graphene into account. Recently, some authors paid their attention to this property of monoatomic graphene sheets. As it will be shown later in this work, the bending rigidity plays a significant role in case of the short resonators.

2 Membrane Model

2.1 Basic Equations

Let us consider the graphene sheet suspending over the trench in SiO_2 substrate. Suppose that it can be approximately represented as a flexible membrane fixed at two supports. Let us consider the uniform deformations along the preferential direction in a plane orthogonal to the direction of oscillations. In this case we can write the equations of small oscillation of the membrane as

$$\frac{Hm}{S_0}\ddot{w} = T\varphi', \quad \varphi = w', \tag{1}$$

here T is a tension force, φ is an angle of membrane element rotation, w is a membrane bending deflection, H is a sheet width, ρ_0 is a linear density of graphene sheet (mass of a unit of area). A point and an accent correspond to the time (t) and longitudinal (x) derivatives respectively. For Eq. (1) is was used that $\rho_0 H \mathrm{d}x$ $\mathrm{d}x$. Relations (1) give the equations of transversal oscillations of the membrane

$$\ddot{w} - c^2 w'' = 0, \quad c^2 = \sigma/\rho_0, \tag{2}$$

where $\sigma = T/H$ is a longitudinal distributed force in graphene sheet.

Let us note that the thickness of the membrane was not included into Eqs. (1) and (2). This fact has a simple explanation. The thickness of an single-atomic layer can not be determined uniquely hence it is non-objective and can not be included into the equations of dynamics [12]. Moreover, the width of the graphene sheet H also is not used in Eq. (2).

The boundary equations are following: $w = 0$ at $x = 0$ and $x = l$. Here l is a length of the membrane. Let us suppose that

$$w(x, t) = W(x)e^{i\omega t},$$

then

$$W'' + \kappa^2 W = 0, \quad \kappa = \omega/c. \tag{3}$$

Its common solution may be written as

$$W = A_1 \cos \kappa x + A_2 \sin \kappa x. \tag{4}$$

A satisfaction to the boundary conditions let us find $\kappa = \pi n/l$, where $n = 1, 2, 3, \ldots$ After κ is found, the natural frequencies can be determined by formulae

$$\omega = \kappa c = \frac{\pi n}{l}\sqrt{\frac{\sigma}{\rho_0}}. \tag{5}$$

2.2 Tension of the Sheet

Let us evaluate a tension in the graphene sheet. Let us believe that a tension is connected with a misfit of graphene and substrate (SiO_2) crystal lattices. It may give a maximum elongation of the graphene sheet on the order of $a/2$. Then a corresponding tension stress may be approximately found as

$$\sigma = \frac{a}{2l}E, \tag{6}$$

where E is a two-dimensional Young modulus of graphene measured in N/m.

A surface density of graphene sheet can be found as

$$\rho_0 = m/S_0, \quad S_0 = \frac{3\sqrt{3}}{4}a^2, \tag{7}$$

where S_0 is a area pear atom in graphene sheet (a half of the elementary cell); a is a distance between the nearest atoms. The substitution of Eqs. (6) and (7) to Eq. (5) gives

$$\omega = k_E \left(\frac{a}{l}\right)^{3/2} \omega_E, \quad k_E = \frac{\pi n}{2}\sqrt{\frac{3}{2}\sqrt{3}}, \quad \omega_E = \sqrt{\frac{E}{m}}. \tag{8}$$

Here k_E is a dimensionless ratio, ω_E has the same unit of measure as a frequency. For the first natural frequency $k_E \approx 2.53$, (8) gives a following evaluation for the frequency of oscillations

$$\omega = 2.53 \left(\frac{a}{l}\right)^{3/2} \omega_E. \tag{9}$$

2.3 *Frequencies Calculation*

Let us evaluate a first frequency of the transversal graphene sheet oscillations on a base of previous results. According to [11], a graphene 2D Young modulus is $E = 350\,\mathrm{N/m}$. A carbon atom mass $m = 1.99 \cdot 10^{-26}\,\mathrm{kg}$. Then, using calculations with formula (8) one can obtain for the frequency ω_E

$$\omega_E = 1.33 \cdot 10^{14}, \quad \nu_E = \frac{1}{2\pi}\omega_E = 21.1\ \mathrm{THz}. \tag{10}$$

Then let us find the frequencies of the graphene sheets with different lengths taking into account that a nearest distance between atoms in graphene is $a = 0.142\,\mathrm{nm}$ [11]. The results if the calculations are given in Table 1. An average distance between the rows of atoms in graphene differs dependently on the direction. It changes from $0.75\,a$ to $\frac{\sqrt{3}}{2}\,a \approx 0.87\,a$. Due to this, an average distance $0.8\,a = 0.114\,\mathrm{nm}$ was used to calculate an approximate number of the atomic rows along the graphene layer as $N = l/(0.8a)$. It is shown in a second column of the table.

3 Plate Model

3.1 *Basic Equations*

Let us consider a graphene sheet laying over a trench in a silicon oxide substrate as a plate on the two supports. Similar to the previous part, let us consider the uniform deformations along the preferential direction in a plane orthogonal to the direction of oscillations. In this case a problem of the plate small oscillations can be reduced to the problem of Bernoulli-Euler beam oscillations:

Table 1 Natural frequencies of the graphene sheets (membrane model)

l (nm)	N	ν
1	8,803	90 MHz
100	880	2.9 GHz
10	88	90 GHz
1	9	2.9 THz

$$H\rho_0\ddot{w} = N', \qquad N = -M', \quad M = DH\varphi', \quad \varphi = w'. \tag{11}$$

Here N is a transversal force, M is a bending moment, φ is an angle of the plate element rotation, w is a plate deflection, H is a graphene sheet width, ρ_0 is a linear density of graphene sheet (mass of the unit of the square), D is a bending rigidity; a dot and a stroke correspond to the time and x coordinate respectively. When obtaining the Eq. (11) it was used that $\rho_0 H dx$ is a mass of the element with a width H and a length dx, and a bending stiffness D is a ratio between a distributed bending moment M/H and the angular deformation φ'. Equation (11) give the equation of the bending oscillations of the plate

$$\ddot{w} + b^4 w'''' = 0, \quad b^4 = D/\rho_0. \tag{12}$$

A thickness of the plate was not included into the Eqs. (11) and (12) as well as the thickness of the membrane was not included in the oscillations equation in the previous part. The explanation is the same: atomic sheet thickness can not be determined uniquely hence it is non–objective and can not be included into the equations of dynamics (11) and (12). The width of the layer H also was not included in Eq. (12).

The boundary conditions at $x = 0$ and $x = l$ where l is a plate length are the following:

$$w = 0, \qquad w' = 0 \text{ (solid support)}, \quad w'' = 0 \text{ (joint)}, \tag{13}$$

relatively for the solid support or for the joint.

Let us allow $w(x,t) = W(x)\,e^{i\omega t}$. Then one can obtain

$$W'''' - \kappa^4 W = 0, \quad \kappa^2 = \omega/b^2. \tag{14}$$

Its general solution may be found as

$$W = A_1 \cos\kappa x + A_2 \sin\kappa x + A_3 \cosh\kappa x + A_4 \sinh\kappa x. \tag{15}$$

After the satisfaction to the boundary conditions one can findκ as $\kappa = \gamma/l$. Here a dimensionless ratio γ is a solution of the following equations

$$\sin\gamma \sinh\gamma = 0 \text{ (joint)}, \qquad \cos\gamma \cosh\gamma = 1 \text{ (solid support)}. \tag{16}$$

From the solutions of these equations (exact solution for the joint bearing and an approximate one for the solid support) it follows

$$\sin\gamma = 0 \quad \Rightarrow \quad \gamma = \pi n \text{ (joint)} \tag{17}$$

or

$$\cos\gamma = 0 \quad \Rightarrow \quad \gamma = \frac{\pi}{2} + \pi n \text{ (solid support)}, \tag{18}$$

where $n = 1, 2, 3\ldots$ The maximum error 1.8 % follows from the approximate solution for the first natural frequency. For the other frequencies the approximate solution is very close to the exact one. The exact solution of equations (16) give the following values for the first natural frequency

$$\gamma = 3.145193\ \text{(joint)}, \quad \gamma = 4.730041\ \text{(solid support)}. \tag{19}$$

After γ is found the natural frequencies are determined from (12) to (14)

$$\omega = \kappa^2 b^2 = \frac{\gamma^2}{l^2}\sqrt{\frac{D}{\rho_0}}. \tag{20}$$

There is an important conclusion following from the last formula. It can be noticed that the natural frequencies are inversely as the square of the plate length. It gives an opportunity to change a frequency of the graphene oscillator in a very wide range. For instance, if the length of the plate increases 10 times then the frequency increases 100 times.

3.2 Evaluation of the Bending Oscillations Frequency

The bending rigidity of the graphene layer is given by formula [3]

$$D = \frac{\sqrt{3}}{6}\,\frac{3c_T + c_B}{c_T + c_B}\,c_B, \tag{21}$$

where c_T and c_B are the torsional and binding stiffnesses respectively. We use this parameters basing on the concept of mechanical representation of the carbonic bond [2]. Unfortunately, so far there are no experiments to determine c_T and c_B. Hence, we will use the relations between these parameters and c_D, where c_D is a transversal stiffness of carbonic bond in graphene.

$$c_T = \frac{1}{12}\,\frac{c_D a^2}{1+\nu}, \quad c_B = \frac{1}{12}\,c_D a^2, \tag{22}$$

where a is a bond length, ν is a dimensionless parameter of interaction. Let us substitute the (22) to formula (21). Then

$$D = \frac{\sqrt{3}}{36}\,k(\nu) c_D a^2, \quad k(\nu) = \frac{1}{2}\,\frac{4+\nu}{2+\nu}. \tag{23}$$

There is a limit $-1 \le \nu \le \frac{1}{2}$ for parameter ν which gives for $k(\nu)$

$$\frac{9}{10} \le k(\nu) \le \frac{3}{2}. \tag{24}$$

Hence, this parameter does not changes significantly and can be taken approximately equal to 1.

A linear density of graphene can be found as

$$\rho_0 = m/S_0, \quad S_0 = \frac{3\sqrt{3}}{4}\, a^2; \tag{25}$$

where S_0 is a square per one atom in graphene list (half of elementary cell square); a is an interatomic distance.

Substitution of the formulae (23) and (25) into (6.22) allows to find the following expression for the bending oscillations frequencies of graphene list

$$\omega = k_D \frac{a^2}{l^2}\, \omega_D, \qquad k_D = \frac{\gamma^2}{4}\sqrt{k(\nu)}, \quad \omega_D = \sqrt{\frac{c_D}{m}}. \tag{26}$$

Here k_D is a dimensionless parameter, ω_D is a frequency of the transversal oscillations of the atom held by the bond with a stiffness c_D. Taking the Eqs. (19) and (24) into account for the solid support boundary conditions one will obtain

$$4.66 \le k_D \le 8.39. \tag{27}$$

Let us take for simplicity $k_D = 6$. Then the following approximate estimation for the graphene layer

$$\omega = 6\frac{a^2}{l^2}\, \omega_D. \tag{28}$$

3.3 Frequencies Calculation

Let us evaluate the first natural frequency of the bending oscillations of graphene layer using the results obtained above. Following [11], the transversal stiffness of the bond between the carbon atoms in graphene is $c_D = 402$ N/m. Mass of the carbon atom is $m = 1.99 \cdot 10^{-26}$ kg. Then using (26) one could calculate a frequency of the transversal oscillations of the carbon atom

$$\omega_D = 1.42 \cdot 10^{14}\ \text{s}^{-1}, \quad \nu_D = \frac{1}{2\pi}\, \omega_D = 22.6\ \text{THz}. \tag{29}$$

Here and after symbol ω denotes the cyclic frequency, ν is an ordinary frequency (inversely to the oscillations period) which is 2π times lower than the cyclic one.

Table 2 Natural frequencies of the graphene sheets (plate model)

l (nm)	N	ν (MHz)
1	8,803	2.7
100	880	273
10	88	27
1	9	2.7

Now let us use (28) to calculate a frequency of the graphene layer oscillations $\nu = \omega/(2\pi)$. Let us take a distance between the nearest carbon atoms in graphene as $a = 0.142$ nm [11]. The results of the oscillations are given in Table 2. A second column shows an approximate number of the rows of atoms along the sheet length. It can be found as $N = l/(0.8a)$ as it was done for Table 1.

Following (28), the frequency is inversely as a square of the graphene layer length. Hence, change of the length lead to the much more significant change of the frequency. According to this fact, one can variate the graphene frequency in a very large limit. For instance, as it follows from Table 2 decrease of the resonator length from 1 mkm to 1 nm lead to the increase of the frequency from 2.7 MHz to 2.7 THz. Thus, the short resonators (1 nm or less) can give the frequencies of terahertz range.

4 Conclusions

Figure 1 shows a comparison of the calculation results obtained with the plate and membrane models. The graphs show that for the lengths more than 1 nm the membrane model gives higher frequencies. If the lengths are less than 1 nm then a plate

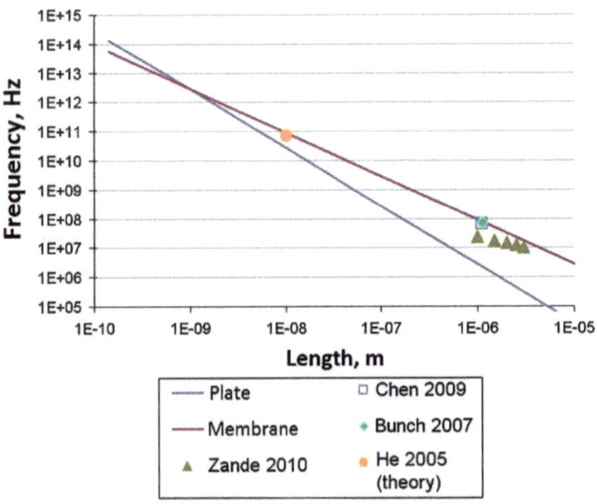

Fig. 1 Comparison of the frequencies of graphene oscillations obtained by different methods

model gives higher frequencies. This means that the plate type of the oscillations dominates for the shortest graphene sheets having up to 10 rows of atoms. Let us note that this result was obtained for the sufficiently high tension of graphene. If the tension is low, then plate type of the oscillations can be realized even for the graphene sheets longer that several nanometers.

A difference between results obtained with membrane and plate models is connected with a bending rigidity influence. The shorter is graphene layer, the more significant bending rigidity is. Thus, a plate model should work better for the shorter resonators but in the case of long resonators the bending rigidity of graphene can be neglected.

Acknowledgments The work is supported by RFBR grant 14-01-802 and the President's of Russian Federation grant MK–4873.2014.1.

References

1. Alyokhin, V.V., Annin, B.D., Babichev, A.V., Korobeynikov, S.N.: Free vibrations and buckling of graphene sheets. Dokl. Phys. **58**(11), 487–490 (2013)
2. Berinskii, I.E.: Rod model of graphene crystal lattice (in Russ.). Nauchno-Technicheskie vedomosti SPBSPU, vol. 3, p. 1320 (2010)
3. Berinskii, I.E., Krivtsov, A.M., Kudarova, A.M.: Determination of bending rigidity of graphene sheet (in Russ.). Fizicheskaya Mezomekhanika **17**(1), 57–65 (2014)
4. Bunch, J.S., van der Zande, A.M., Verbridge, S.S., Frank, I.W., Tanenbaum, D.M., Parpia, J.M., Craighead, H.G., McEuen, P.L.: Electromechanical resonators from graphene sheets. Science **315**(5811), 490–493 (2007)
5. Chen, C., Rosenblatt, S., Bolotin, K.I., Kalb, W., Kim, P., Kymissis, I., Stormer, H.L., Heinz, T.F., Hone, J.: Performance of monolayer graphene nanomechanical resonators with electrical readout. Nat. Nanotechnol. **4**, 861–867 (2009)
6. Eom, K., Park, H.S., Yoon, D.S., Kwon, T.: Nanomechanical resonators and their applications in biological/chemical detection: nanomechanics principles. Phys. Rep. **503**, 115–163 (2011)
7. Geim, A.K., Novoselov, K.S.: The rise of graphene. Nat. Mater. **6**(3), 183–191 (2007)
8. Greenberg, Y., Pashkin, Y.A., Il'ichev, E.: Nanomechanical resonators. Phys. Usp. **55**(4), 382–407 (2012)
9. Gupta, S.S., Batra, R.C.: Elastic properties and frequencies of free vibrations of single-layer graphene sheets. J. Comput. Theor. Nanosci. **7**, 2151–2164 (2010)
10. He, X.Q., Kitipornchai, S., Liew, K.M.: Resonance analysis of multi-layered graphene sheets used as nanoscale resonators. Nanotechnol. **16**, 2086–2091 (2005)
11. Ivanova, E.A., Krivtsov, A.M., Morozov, N.F.: Derivation of macroscopic relations of the elasticity of complex crystal lattices taking into account the moment interactions at the microlevel. J. Appl. Math. Mech. **71**(4), 543–561 (2007)
12. Krivtsov, A.M., Morozov, N.F.: On mechanical characteristics of nanocrystals. Phys. Solid State **44**(12), 2260–2265 (2002)
13. van der Zande, A.M., Barton, R.A., Alden, J.S., Ruiz-Vargas, C.S., Whitney, W.S., Pham, P.H.Q., Park, J., Parpia, J.M., Craighead, H.G., McEuen, P.L.: Large-scale arrays of single-layer graphene resonators. Nano Lett. **10**, 4869–4873 (2010)

On Discrete-Kirchhoff Plate Finite Elements: Implementation and Discretization Error

Boštjan Brank, Adnan Ibrahimbegović and Uroš Bohinc

Abstract We present neat and efficient computer code implementation of two types of Discrete-Kirchhoff plate finite elements—the Discrete-Kirchhoff triangle and the Discrete-Kirchhoff quadrilateral—which can be used to model numerous thin plate problems in mechanics and biology. We also present an implicit a posteriori discretization error indicator computation, based on the superconvergent patch recovery technique, for the Discrete-Kirchhoff plate finite elements. This error indicator can drive an adaptive meshing algorithm providing the most suitable finite element mesh. For an illustration, some numerical results are given.

1 Introduction

There exist numerous applications in engineering, technology and biology, which call for the thin plate finite element (FE) modeling. Frequently used thin plate finite elements are the Discrete-Kirchhoff (DK) elements (e.g. [6]), which deliver higher-order approximations of curvatures, even for FEs with small number of nodes. The plate elements of this kind are part of many commercial FE codes, either as 3-node triangle (DKT) or as 4-node quadrilateral (DKQ).

B. Brank (✉)
Faculty of Civil and Geodetic Engineering, University of Ljubljana, Jamova Cesta 2, 1000 Ljubljana, Slovenia
e-mail: bostjan.brank@fgg.uni-lj.si

A. Ibrahimbegović
Ecole Normale Supérieure de Cachan, LMT, Cachan, France
e-mail: adnan.ibrahimbegovic@ens-cachan.fr

U. Bohinc
Slovenian National Building and Civil Engineering Institute, Dimičeva 12, 1000 Ljubljana, Slovenia
e-mail: uros.bohinc@zag.si

H. Altenbach and G.I. Mikhasev (eds.), *Shell and Membrane Theories in Mechanics and Biology*, Advanced Structured Materials 45, DOI 10.1007/978-3-319-02535-3_6

The DK elements are the result of a long line of developments on the Kirchhoff plate elements, e.g. [10, 11], requiring C^1 continuity of the transverse displacement interpolation. For many reasons, the Kirchhoff plate elements were abandoned and preference was given to the Reissner-Mindlin (RM) thick plate elements, see e.g. [13, 29] and references therein, using only C^0 interpolations for the transverse displacement and the independent rotations. The RM plate elements turned to be very sensitive to the shear locking (e.g. [29]). The DK element was introduced within the RM plate theory framework, trying to deal with the shear locking by completely removing the shear deformation. The consistent mixed formulation for the DKQ was proposed in [14] and that approach was extended to the DKT in [8]. In this work we show that both the DKT and the DKQ can be implemented neatly into the FE computer code by the same algorithm.

In constructing the FE discretization error estimates for the DK elements, the preference is given to an implicit a posteriori version, see e.g. [1, 2, 17], that starts from the computed solution of the FE produced set of equilibrium equations. The bending moments of an increased accuracy provide the corresponding error estimates by comparison against the FE solution. The enhanced solution might be obtained by the superconvergent patch recovery (SPR) (e.g. see [27–29]), which is the least square fit of the FE results computed in the superconvergence points (roughly, the reduced numerical integration points of a particular element). In this work we present an application of the SPR for the DK plate elements.

2 Kirchhoff Plate Bending Model

Consider an elastic plate (Fig. 1), with the middle-plane Ω that is placed in the xy plane, the middle-plane boundary Γ_p, and the thickness t, $z \in [-t/2, t/2]$. At the middle-plate boundary point, two vectors are defined: $\mathbf{s} = [-n_y, n_x]^T$, and $\mathbf{n} = [n_x, n_y]^T$, where $\mathbf{n}$ is the outward unit exterior boundary normal. Coordinates s and n change, respectively, along the boundary and in the direction of the normal. The plate is loaded by a distributed area loading f, a line boundary moment in the s direction $\overline{m}_s$ on $\Gamma_N \subset \Gamma_p$ and a line boundary effective force in the z direction $\overline{q}_{\text{ef}}$ on $\Gamma_N \subset \Gamma_p$. The Kirchhoff plate model is built upon the kinematic constraint that fixes through-the-thickness fibre as perpendicular to the plate middle surface. This allows to define the fibre rotations $\boldsymbol{\theta} = [\theta_x, \theta_y]^T$ as the derivatives of the transverse displacement w:

$$\nabla^K w - \boldsymbol{\theta} = \mathbf{0}, \quad \nabla^K = \left[\frac{\partial}{\partial y}, -\frac{\partial}{\partial x}\right]^T \tag{1}$$

The above constraint sets the shear strains to zero. Thus, the only generalized deformations are the middle-plane curvatures $\boldsymbol{\kappa} = [\kappa_{xx}, \kappa_{yy}, 2\kappa_{xy}]^T$:

$$\kappa_{xx} = \frac{\partial^2 w}{\partial x^2} = -\frac{\partial \theta_y}{\partial x}, \ \kappa_{yy} = \frac{\partial^2 w}{\partial y^2} = \frac{\partial \theta_x}{\partial y}, \ \kappa_{xy} = \frac{\partial^2 w}{\partial x \partial y} = \frac{1}{2}\left(\frac{\partial \theta_x}{\partial x} - \frac{\partial \theta_y}{\partial y}\right) \tag{2}$$

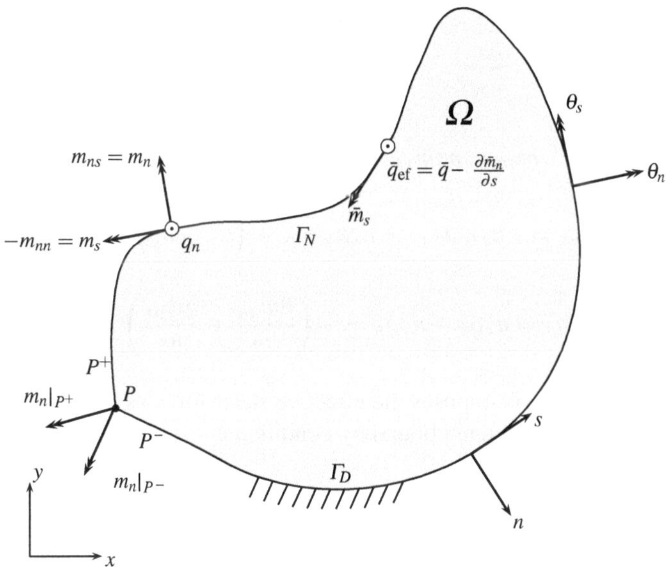

Fig. 1 Notation for plate problem

The curvatures relate to the bending moments $\mathbf{m} = [m_{xx}, m_{yy}, m_{xy}]^T$ as:

$$\mathbf{m} = \underbrace{D\left[\left[1\ \nu\ 0\right]^T, \left[\nu\ 1\ 0\right]^T, \left[0\ 0\ \frac{1-\nu}{2}\right]^T\right]}_{\mathbf{C}_B}\kappa, \quad D = \frac{Et^3}{12(1-\nu^2)}, \tag{3}$$

where E is Young's modulus and ν is Poisson's ratio. According to the kinematic constraint, there cannot be a similar role for the shear constitutive equations. Rather, the shear forces have to be computed from the corresponding moment equilibrium equations:

$$q_x = -\left(\frac{\partial m_{xx}}{\partial x} + \frac{\partial m_{xy}}{\partial y}\right), \quad q_y = -\left(\frac{\partial m_{yy}}{\partial y} + \frac{\partial m_{xy}}{\partial x}\right) \tag{4}$$

The last equations can be inserted into the force equilibrium equation, in order to provide a single equilibrium equation in terms of moments:

$$\frac{\partial q_x}{\partial x} + \frac{\partial q_y}{\partial y} + f = 0 \Longrightarrow \frac{\partial m_{xx}^2}{\partial x^2} + 2\frac{\partial^2 m_{xy}}{\partial x \partial y} + \frac{\partial m_{yy}^2}{\partial y^2} = f \tag{5}$$

The following transformations, applied at the plate boundary, define the boundary rotations, the boundary moments and the boundary shear force:

$$\theta_n = \frac{\partial w}{\partial s} = -n_y \underbrace{\frac{\partial w}{\partial x}}_{-\theta_y} + n_x \underbrace{\frac{\partial w}{\partial y}}_{\theta_x}, \quad -\theta_s = \frac{\partial w}{\partial n} = n_x \underbrace{\frac{\partial w}{\partial x}}_{-\theta_y} + n_y \underbrace{\frac{\partial w}{\partial y}}_{\theta_x}, \tag{6}$$

$$m_{nn} = n_x^2 m_{xx} + n_y^2 m_{yy} + 2 n_x n_y m_{xy}, \tag{7}$$

$$m_{ns} = -n_x n_y m_{xx} + n_x n_y m_{yy} + \left(n_x^2 - n_y^2\right) m_{xy}, \tag{8}$$

$$q_n = n_x q_x + n_y q_y = -\left(\frac{\partial m_{nn}}{\partial n} + \frac{\partial m_{ns}}{\partial s}\right), \tag{9}$$

The kinematic hypothesis imposes the effective shear force as the only appropriate way for imposing the Neumann boundary conditions:

$$q_{\text{ef}} = q_n - \frac{\partial m_{ns}}{\partial s} \quad \text{on} \quad \Gamma_p, \qquad \overline{q}_{\text{ef}} = \overline{q} - \frac{\partial \overline{m}_n}{\partial s} \quad \text{on} \quad \Gamma_N, \tag{10}$$

In general, the middle-plane boundary can be split into the Neumann Γ_N, where we impose the normal moment and the effective shear force (m_{nn}, q_{ef}), and the Dirichlet Γ_D, where we impose the displacement and its normal derivative ($w, \frac{\partial w}{\partial n}$). We can also have a mixed case or Navier boundary Γ_M, where we impose the displacement and the normal moment (w, m_{nn}).

The strong form of the boundary value problem can be stated as: given distributed load f, imposed boundary shear and moment $\overline{q}_{\text{ef}}, \overline{m}_s$, and imposed displacement and rotation $\overline{w}, \overline{\theta}_s$, find w, such that

$$\begin{aligned}
&\Delta\Delta w = \frac{f}{D}\,;\ \Delta = \frac{\partial^2}{\partial x^2} + \frac{\partial^2}{\partial y^2} \quad \text{in} \quad \Omega, \\
&w = \overline{w}\,;\ -\frac{\partial w}{\partial n} = \overline{\theta}_s \quad \text{on} \quad \Gamma_D, \\
&w = \overline{w}\,;\ m_{nn}(w) = -\overline{m}_s \quad \text{on} \quad \Gamma_M, \\
&q_{\text{ef}}(w) = \overline{q}_{\text{ef}}\,;\ m_{nn}(w) = -\overline{m}_s \quad \text{on} \quad \Gamma_N
\end{aligned} \tag{11}$$

and for any corner point P on Γ_N: $-m_{ns}|_{P-} + m_{ns}|_{P+} = -\overline{m}_n|_{P-} + \overline{m}_n|_{P+}$.

The corresponding variational formulation, considering the weak form of plate equilibrium equation along with strong form of kinematics and constitutive equations, can be written as

$$a(w, v) = l(v), \qquad w \in \mathcal{V}, \qquad \forall v \in \mathcal{V}_0, a(w, v) = l(v), \qquad w \in \mathcal{V}, \qquad \forall v \in \mathcal{V}_0, \tag{12}$$

$$
\begin{aligned}
a(w, v) &= \int_{\Omega} \kappa^T(v)\, \mathbf{C}_B \kappa(w)\, \mathrm{d},\Omega, \\
l(v) &= \int_{\Omega} v f\, \mathrm{d}\Omega + \sum_{\Gamma_{IJ}} \left(\int_{\Gamma_{IJ}} (v\overline{q}_{ef} + \theta_s \overline{m}_s)\, \mathrm{d}s + [v\overline{m}_n]_I^J \right)
\end{aligned}
\tag{13}
$$

Here, I, J are the corner points, $\Gamma_{IJ} \subset \Gamma_N$, $\sum_{\Gamma_{IJ}} = \Gamma_N$, $\mathscr{V}$ is the trial space, and $\mathscr{V}_0$ is the test space. For the future use, let us rewrite the weak form (12) as

$$
a(\mathbf{u}, \mathbf{v}) = l(\mathbf{v}),\ \mathbf{u} \in \boldsymbol{V},\ \forall \mathbf{v} \in \boldsymbol{V}_0, \tag{14}
$$

where

$$
\mathbf{u} = \left[w, \theta_x = \frac{\partial w}{\partial y}, \theta_y = -\frac{\partial w}{\partial x} \right]^T \quad \text{and} \quad \mathbf{v} = \left[v, \phi_x = \frac{\partial v}{\partial y}, \phi_y = -\frac{\partial v}{\partial x} \right]^T
$$

The trial and the test spaces are given as ($\mathscr{H}^i$ is the ith order Sobolev space):

$$
\boldsymbol{V} = \left\{ \begin{matrix} w \in \mathscr{H}^2(\Omega) \\ \theta \in \left[\mathscr{H}^1(\Omega)\right]^2 \end{matrix} \;\middle|\; \begin{matrix} \theta_x = \dfrac{\partial w}{\partial y}, \theta_y = -\dfrac{\partial w}{\partial x}, \\ w = \overline{w}, -\dfrac{\partial w}{\partial n} = \theta_s = \overline{\theta}_s \text{ on } \Gamma_D;\ w = \overline{w} \text{ on } \Gamma_M \end{matrix} \right\} \tag{15}
$$

$$
\boldsymbol{V}_0 = \left\{ \begin{matrix} v \in \mathscr{H}^2(\Omega) \\ \phi \in \left[\mathscr{H}^1(\Omega)\right]^2 \end{matrix} \;\middle|\; \begin{matrix} \phi_x = \dfrac{\partial v}{\partial y}, \phi_y = -\dfrac{\partial v}{\partial x}, \\ v = 0, -\dfrac{\partial v}{\partial n} = \phi_s = 0 \text{ on } \Gamma_D;\ v = 0 \text{ on } \Gamma_M \end{matrix} \right\} \tag{16}
$$

3 Implementation of the Discrete-Kirchhoff Finite Elements

Prior designing the DK interpolations, we will replace (14) with a mixed variational formulation in order to reduce the displacement continuity requirements:

$$
\begin{aligned}
&a_\gamma(\mathbf{u}, \gamma, \mathbf{q}; \mathbf{v}) = l(\mathbf{v}),\ a_\gamma(\mathbf{u}, \gamma, \mathbf{q}; \mathbf{v}) = \int_{\Omega} \left\{ \kappa^T(\mathbf{v})\, \mathbf{m}(\mathbf{u}) + [\gamma(\mathbf{v}) - \gamma^*]^T \mathbf{q} \right\} \mathrm{d}\Omega, \\
&\int_{\Omega} \left\{ \mathbf{q}^{*,T} [\gamma(\mathbf{u}) - \gamma]^T \right\} \mathrm{d}\Omega = 0
\end{aligned}
\tag{17}
$$

Here $\mathbf{u}$ and $\mathbf{v}$ are kinematic variables and their variations,

$$
\gamma(\mathbf{u}) = \left(\frac{\partial w}{\partial x} + \theta_y, \frac{\partial w}{\partial y} - \theta_x \right)^T \quad \text{and} \quad \gamma(\mathbf{v}) = \left(\frac{\partial v}{\partial x} + \phi_y, \frac{\partial v}{\partial y} - \phi_x \right)^T
$$

are displacement-based shear strains and their variations, $\mathbf{q}$ and $\mathbf{q}^*$ are shear forces and their variations, γ and γ^* are independent shear strains and their variations. By assuming the shear force is defined independently in each element, we can recover the variational consistency of the assumed shear strain formulation by enforcing the validity of the variational equation (17) in each element.

To show how to construct the DK interpolations, let us first consider each edge of the DK element as the 2-node planar beam FE of length L. We define the coordinates $s^B = \xi^B \frac{L}{2} + \frac{L}{2} \in [0, L], \xi^B \in [-1, +1]$. Along each edge, we choose the following interpolations for displacement and rotation:

$$w_h^B = \sum_{I=1}^{2} w_I N_I^B + \sum_{I=3}^{4} w_I N_I^B, \quad \theta_h^B = \sum_{I=1}^{2} \theta_I N_I^B + \theta_3 N_3^B \tag{18}$$

where N_I^B are hierarchical Lagrangian functions

$$N_1^B = \frac{1}{2}(1 - \xi^B), \; N_2^B = \frac{1}{2}(1 + \xi^B), \; N_3^B = \left(1 - \left(\xi^B\right)^2\right), \; N_4^B = \xi^B \left(1 - \left(\xi^B\right)^2\right)$$

Only two nodal displacements, w_1, w_2, and two nodal rotations, θ_1, θ_2, will remain as acceptable parameters. The other parameters are condensed out by enforcing the kinematic hypothesis

$$\frac{\partial w_h^B}{\partial s^B} - \theta_h^B = 0,$$

resulting with:

$$w_3 = \frac{L}{8}(\theta_1 - \theta_2), \quad w_4 = \frac{L}{4}\left(\frac{w_2 - w_1}{L} - \frac{1}{2}(\theta_1 + \theta_2)\right), \quad \theta_3 = \frac{6}{L} w_4 \tag{19}$$

By placing (19) into (18), we finally recover a four-parameter interpolation along ξ^B. With this interpolation the DK element will have the zero transverse shear imposed along each edge, which for assumed strain interpolation further implies zero shear strains everywhere within the element domain. This result is directly applicable to the DK element by recognizing that $\theta_h^B = \mathbf{n}^T \theta_{h,e} = \theta_{n,h,e}$ and that the kinematic constraint

$$\frac{\partial w_{h,e}}{\partial s} - \theta_{n,h,e} = 0$$

implies for each edge of element e. We obtain the following DK approximations of displacement and rotations represented by the sum of nodal, I, and edge, IJ, contributions

$$\begin{aligned} w_{h|e} = w_{h,e} &= \sum_{I} w_I N_I + \sum_{IJ} \left(w_{3,IJ} N_{IJ} + w_{4,IJ} M_{IJ}\right), \\ \theta_{h|e} = \theta_{h,e} &= \sum_{I} \theta_I N_I + \sum_{IJ} \theta_{3,IJ} N_{IJ} \end{aligned} \tag{20}$$

Here, $w_{3,IJ}$, $w_{4,IJ}$ and $\theta_{3,IJ}$ are in analogy with the parameters in (19)

$$\begin{aligned} w_{3,IJ} &= \frac{L_{IJ}}{8}\mathbf{n}_{IJ}^T(\theta_I - \theta_J), \quad w_{4,IJ} = \frac{L_{IJ}}{4}\left(\frac{w_J - w_I}{L_{IJ}} - \frac{1}{2}\mathbf{n}_{IJ}^T(\theta_I + \theta_J)\right), \\ \theta_{3,IJ} &= \frac{6}{L_{IJ}} w_{4,IJ}\mathbf{n}_{IJ}, \end{aligned} \tag{21}$$

where $\mathbf{n}_{IJ} = [n_{x,IJ}, n_{y,IJ}]^T$ is the outward unit normal of an edge between nodes I and J, and L_{IJ} is the length of that edge. The interpolations N_{IJ}, M_{IJ} will be presented below. The above DK interpolations can also be given in the matrix notation

$$w_{h,e} = \sum_I \mathbf{N}_{w,I}\mathbf{u}_I, \quad \theta_{h,e} = \sum_I \mathbf{N}_{\theta,I}\mathbf{u}_I, \quad \kappa_h = \sum_I \mathbf{B}_{\kappa,I}\mathbf{u}_I, \tag{22}$$

where vector $\mathbf{u}_I = [w_I, \theta_{xI}, \theta_{yI}]^T$ collects the degrees of freedom of node I. Explicit forms of $\mathbf{N}_{w,I}$, $\mathbf{N}_{\theta,I}$ and $\mathbf{B}_{\kappa,I}$ can be obtained from (20), (21) and (2).

By using the so defined DK interpolations for both trial, $\mathbf{u}_{h,e} = [w_{h,e}, \theta_{h,e}^T]^T$, and test functions, $\mathbf{v}_{h,e} = [v_{h,e}, \phi_{h,e}^T]^T$, in the weak form (12), we obtain

$$\begin{aligned} a_e(\mathbf{u}_{h,e}, \mathbf{v}_{h,e}) &= \int_{\Omega_{h,e}} \kappa_h^T(\mathbf{v}_{h,e})\, \mathbf{m}_h(\mathbf{u}_{h,e})\, \mathrm{d}\Omega = \sum_{I,J} \mathbf{v}_I^T\, \mathbf{K}_{IJ}^e\, \mathbf{u}_J, \\ \mathbf{K}_{IJ}^e &= \int_{\Omega_{h,e}} \mathbf{B}_{\kappa,I}^T \mathbf{C}_B \mathbf{B}_{\kappa,J}\, \mathrm{d}\Omega \end{aligned} \tag{23}$$

where $\mathbf{v}_I = [v_I, \phi_{xI}, \phi_{yI}]^T$, and $[\mathbf{K}_{IJ}^e]$ is the element stiffness matrix. The element consistent load vector $\mathbf{f}^e = \left[\mathbf{f}_I^{e,T}\right]^T$ follows from:

$$\begin{aligned} l_e(\mathbf{v}_{h,e}) &= \int_{\Omega_{h,e}} v_{h,e}\, f\, \mathrm{d}\Omega + \sum_{\Gamma_{IJ}}\left[\int_{\Gamma_{IJ}} (v_{h,e}\, \overline{q}_{\mathrm{ef}} + \phi_{s,h,e}\, \overline{m}_s)\, \mathrm{d}s + [v_{h,e}\, \overline{m}_n]_I^J\right] \\ &= \sum_I \mathbf{v}_I^T \mathbf{f}_I^e = \sum_I \mathbf{v}_I^T (\mathbf{f}_{f,I}^e + \mathbf{f}_{t,I}^e + \mathbf{f}_{c,I}^e) \end{aligned} \tag{24}$$

where $\phi_{s,h,e} = \mathbf{s}_{IJ}^T \phi_{h,e}$, $\mathbf{s}_{IJ} = \left[-n_{y,IJ}, n_{x,IJ}\right]^T$, $\Gamma_{IJ} \subset \Gamma_{N,h}$, and

$$\begin{aligned} \mathbf{f}_{f,I}^e &= \int_{\Omega_{h,e}} f\, \mathbf{N}_{w,I}\, \mathrm{d}\Omega, \quad \mathbf{f}_{t,I}^e = \sum_{\Gamma_{IJ}} \int_{\Gamma_{IJ}} (\overline{q}_{\mathrm{ef}} \mathbf{N}_{w,I} + \overline{m}_s\, \mathbf{s}_{IJ}^T \mathbf{N}_{\theta,I}) \mathrm{d}s, \\ \mathbf{f}_{c,I}^e &= [\overline{m}_n|_{I^+} - \overline{m}_n|_{I^-}, 0, 0]^T, \quad \forall\, I \in \Gamma_{N,h} \end{aligned} \tag{25}$$

The trial and the test spaces for a plate problem, discretized by the DK finite element, can be written as:

$$\mathbf{V}_h^{DK} = \left\{ \begin{array}{l} w_h \in \mathscr{U} \mid w_{h|e} \in p\,(e)\ \forall e \in \mathscr{C}_h, \\ \theta_h \in \mathbf{U} \mid \theta_{h|e} \in [p_1\,(e)]^2\ \forall e \in \mathscr{C}_h, \\ \mathbf{n}_{IJ}^T \left(\nabla^K w_{h|e} - \theta_{h|e}\right) |_{\Gamma_{IJ}} = 0\ \forall \Gamma_{IJ} \in e \wedge \forall e \in \mathscr{C}_h \end{array} \right\} \tag{26}$$

$$\mathbf{V}_{0,h}^{DK} = \left\{ \begin{array}{l} v_h \in \mathscr{U}_0 \mid v_{h|e} \in p\,(e)\ \forall e \in \mathscr{C}_h, \\ \phi_h \in \mathbf{U}_0 \mid \phi_{h|e} \in [p_1\,(e)]^2\ \forall e \in \mathscr{C}_h, \\ \mathbf{n}_{IJ}^T \left(\nabla^K v_{h|e} - \phi_{h|e}\right) |_{\Gamma_{IJ}} = 0\ \forall \Gamma_{IJ} \in e \wedge \forall e \in \mathscr{C}_h \end{array} \right\} \tag{27}$$

where $p\,(e)$ and $p_1\,(e)$ are the spaces of polynomials of degrees at most p and p_1, respectively, on element e with edges Γ_{IJ} and nodes I. For the DKQ $p = 4$ and for the DKT $p = 3$, $p_1 = p - 1$. The $\mathscr{C}_h$ represents the collection of all the elements of the mesh, and

$$\begin{aligned} \mathscr{U} &= \left\{ w \in \mathscr{H}^1\,(\Omega) \mid w = \overline{w} \text{ on } \Gamma_D \wedge \Gamma_M \right\}, \mathbf{U} = \left\{ \theta \in \left[\mathscr{H}^1\,(\Omega) \right]^2 \mid \theta_s = \overline{\theta}_s \text{ on } \Gamma_D \right\}, \\ \mathscr{U}_0 &= \left\{ v \in \mathscr{H}^1\,(\Omega) \mid v = 0 \text{ on } \Gamma_D \wedge \Gamma_M \right\}, \mathbf{U}_0 = \left\{ \phi \in \left[\mathscr{H}^1\,(\Omega) \right]^2 \mid \phi_s = 0 \text{ on } \Gamma_D \right\} \end{aligned} \tag{28}$$

The DKT has 3 vertex nodes; the shape functions in the area coordinates are:

$$N_I = \zeta_I; \quad N_{IJ} = 4\zeta_I\zeta_J; \quad M_{IJ} = 4\zeta_I\zeta_J(\zeta_J - \zeta_I) \tag{29}$$

The 4-point triangular scheme for the numerical integration is used to compute the stiffness matrix and the load vector for the DKT. The DKQ has 4 vertex nodes; the shape functions are given in the coordinates $(\xi, \eta) \in [-1, +1] \times [-1, +1]$:

I	1	2	3	4
IJ	12	23	34	41
r	ξ	η	ξ	η
s	η	ξ	η	ξ
p_r	+1	+1	−1	−1
p_s	+1	−1	−1	+1

$$\begin{aligned} N_I &= (1 - p_r r)(1 - p_s s)/4; \\ N_{IJ} &= (1 - r^2)(1 - p_s s)/2; \\ M_{IJ} &= p_r r(1 - r^2)(1 - p_s s)/2 \end{aligned} \tag{30}$$

The 2×2 Gauss scheme is used for the numerical integration of the stiffness matrix and the load vector for the DKQ.

4 Discretization Error Estimation

In the Bubnov-Galerkin FE method, the optimal approximation $\mathbf{u}_h \in \boldsymbol{V}_h \subset \boldsymbol{V}$ is the one for which

$$a(\mathbf{u}_h, \mathbf{v}_h) = l(\mathbf{v}_h) \qquad \forall \mathbf{v}_h \in \boldsymbol{V}_{0,h} \subset \boldsymbol{V}_0 \tag{31}$$

The exact solution $\mathbf{u} \in \boldsymbol{V}$ also satisfies (31)

$$a(\mathbf{u}, \mathbf{v}_h) = l(\mathbf{v}_h) \qquad \forall \mathbf{v}_h \in \boldsymbol{V}_{0,h} \subset \boldsymbol{V}_0 \tag{32}$$

Subtracting (31) from (32), and accounting for the linearity of each functional, yields the Galerkin orthogonality condition $a(\mathbf{u} - \mathbf{u}_h, \mathbf{v}_h) = 0$. If the true error of the FE solution is defined as $\mathbf{e} = \mathbf{u} - \mathbf{u}_h$, the Galerkin orthogonality reads as $a(\mathbf{e}, \mathbf{v}_h) = 0$, i.e. the error $\mathbf{e}$ is orthogonal to any test function from $\boldsymbol{V}_{0,h}$. The error satisfies the residual equation

$$a(\mathbf{e}, \mathbf{v}) = a(\mathbf{u}, \mathbf{v}) - a(\mathbf{u}_h, \mathbf{v}) = l(\mathbf{v}) - a(\mathbf{u}_h, \mathbf{v}), \quad \forall \mathbf{v} \in \boldsymbol{V}_0 \tag{33}$$

The above introduced true error $\mathbf{e}$ is a local measure. It is possible, in the case of singularities, that $\mathbf{e}$ is locally infinite, while the overall solution may well be acceptable. To measure the global error, the complementary energy norm of $\mathbf{e}$ may be used:

$$\|\mathbf{e}\|_{E^*}^2 = a^*(\mathbf{e}, \mathbf{e}) = \sum_e \|\mathbf{e}_e\|_{E^*}^2 \tag{34}$$

The element contribution is

$$\begin{aligned} \|\mathbf{e}_e\|_{E^*}^2 &= a^*(\mathbf{e}_e, \mathbf{e}_e) \\ &= \int_{\Omega_{h,e}} \left[\mathbf{m}\left(\mathbf{u}_e\right) - \mathbf{m}_h\left(\mathbf{u}_{h,e}\right)\right]^T \mathbf{C}_B^{-1} \left[\mathbf{m}\left(\mathbf{u}_e\right) - \mathbf{m}_h\left(\mathbf{u}_{h,e}\right)\right] \mathrm{d}\Omega \end{aligned} \tag{35}$$

In (35), $\mathbf{m}_h$ are the moments of the FE solution, and $\mathbf{m}$ are the exact moments.

We will briefly present computation of $\|\mathbf{e}_{e,h}\|_{E^*}^2$ by the recovery based technique. An approximation of exact moments $\mathbf{m}$ is recovered from the FE solution $\mathbf{m}_h\left(\mathbf{u}_{h,e}\right)$ by postprocessing. The recovered moments, denoted by $\mathbf{m}^*$, are obtained as

$$\mathbf{m}^*(\xi)|_{\Omega_{h,e}} = \sum_I \mathbf{m}_I^* N_I(\xi), \tag{36}$$

where I runs over the nodes of the element, ξ are isoparametric coordinates, N_I are standard Lagrange shape functions, and $\mathbf{m}_I^*$ are nodal parameters for interpolation. The local and global error indicators are computed as

$$\begin{aligned} \|\mathbf{e}_{h,e}\|^2_{E^*} &= \int\limits_{\Omega_{h,e}} (\mathbf{m}^* - \mathbf{m}_h)^T \mathbf{C}_B^{-1} (\mathbf{m}^* - \mathbf{m}_h)\, \mathrm{d}\Omega, \\ \|\mathbf{e}_h\|^2_{E^*} &= \textstyle\sum_e \|\mathbf{e}_{h,e}\|^2_{E^*} \end{aligned} \tag{37}$$

The superconvergent patch recovery (SPR) technique can be used to obtain $\mathbf{m}_I^*$ in (36). It takes advantage of the superconvergence property of the FE solution [15, 29]. Namely, the error in displacements/rotations is the smallest at the nodes of the element, whereas the error in rotation gradients (moments) is the smallest at the points inside the element usually coinciding with the Gauss points. At such points, the order of the convergence of the gradients is at least one order higher than what would be anticipated from the approximation polynomials. This is known as the superconvergence property. The values of moments at Gauss points ξ_{gp} (or other optimal sampling points), denoted as $\mathbf{m}_h(\xi_{gp})$, serve as the basis for the reconstruction of $\mathbf{m}_I^*$. Further details of the SPR are given e.g. in [29]. The above described procedure suits any DK element.

Global and local error indicators, $\|\mathbf{e}_h\|_{E^*}$, and $\|\mathbf{e}_{h,e}\|_{E^*} = \eta_e^* = \eta_e$, respectively, are computed for the DKT with the above described SPR procedure. The global "true error" $\|\mathbf{e}\|_{E^*}$ is computed by using the reference solution

$$\|\mathbf{e}\|^2_{E^*} = \textstyle\sum_e \displaystyle\int\limits_{\Omega_{h,e}} (\mathbf{m}^{ref} - \mathbf{m}_h)^T \mathbf{C}_B^{-1} (\mathbf{m}^{ref} - \mathbf{m}_h)\, \mathrm{d}\Omega \tag{38}$$

If available, the reference solution $\mathbf{m}^{ref}$ is the strong form solution, see e.g. [3, 4, 9, 21, 22, 24, 26]. Otherwise, the reference solution is obtained by a very fine mesh of Argyris plate elements [5], which use the 5th order polynomal to approximate the transverse displacement. The relative local and global error indicators may be defined as

$$\eta_{e,r}^* = \frac{\|\mathbf{e}_{h,e}\|_{E^*}}{\|\mathbf{n}_{h,e}\|_{E^*}}, \qquad \|\mathbf{n}_{h,e}\|^2_{E^*} = \int\limits_{\Omega_e} \mathbf{m}_h^T \mathbf{C}_B^{-1} \mathbf{m}_h \, \mathrm{d}s, \tag{39}$$

$$\eta_r^* = \frac{\|\mathbf{e}_h\|_{E^*}}{\|\mathbf{n}_h\|_{E^*}}, \qquad \|\mathbf{n}_h\|^2_{E^*} = \sum_e \|\mathbf{n}_{h,e}\|^2_{E^*} = \|\mathbf{n}_h\|^2 \tag{40}$$

The discretization error indicators can control an adaptive FE meshing, e.g. [18]. The goal is to generate a mesh where the local element error is approximately equal for every element to a prescribed value $\bar{\eta}_e$. It is possible to deduce a desired element size from the a-priori error estimate of the local element error $\bar{\eta}_e = C\bar{h}^p$ where the element size is denoted by $\bar{h}$, C is a constant and p the polynomial order of the FE interpolation. By knowing $\bar{\eta}_e$ and the local error indicator η_e^*, the desired element size $\bar{h}$ can be obtained as

$$\eta_e^* = Ch^p; \quad \bar{\eta}_e = C\bar{h}^p \quad \Rightarrow \quad \bar{h} = h(\bar{\eta}_e/\eta_e^*)^{1/p} \tag{41}$$

The software `gmsh` [12] is employed to generate meshes. It enables mesh generation according to the scalar field $\bar{h}$.

5 Examples

The above presented DK finite elements were implemented in the Fortran based finite element program FEAP [25]. They were also implemented in the C++ based finite element program AceFEM [16]. The computed FE stress resultants are not continuous across element edges, which makes the assessment of the results difficult. To produce readable plots, the nodal values are computed from the element integration point values using the standard averaging technique, see [23], Sect. 5.5. The shear forces are computed from the equilibrium equations (4). To obtain the derivatives of the moments, the bilinear interpolation of the moments on Ω_h^e is first constructed from the integration point values. The examples below are meant to be illustrative and academic; this is the reason that specific units are not provided.

5.1 Uniformly Loaded Clamped Circular Plate

A clamped circular plate of radius $R = 5$ and thickness $t = 0.01$ under uniform loading $f = 1$ is considered. The material has Young's modulus $E = 6.825 \times 10^9$ and Poisson's ratio $\nu = 0.3$. The closed form solution of this problem is available e.g. in [26]. The reference solution is given in Fig. 2, our solution with the DK plate elements is given in Figs. 3 and 4. The meshes used in the convergence analysis are shown in Fig. 6. Results are influenced by the mesh distortion (quadrilaterals) and mesh orientation (triangles). This influence is, however, noticeable only for the equilibrium shear forces. The comparison of the convergence, shown in Fig. 7, reveals that the convergence of the Argyris element [5] is not superior to that of the DK elements (Fig. 5).

5.2 Uniformly Loaded Hard Simply Supported Skew Plate

The analysis of Morley's $\alpha = 30°$ skew plate [20] under uniform loading $f = 1$ with thickness $t = 0.1$ and side length $a = 10$ (Fig. 8) is considered. The plate is hard simply supported on all sides. The Young's modulus is $E = 10.92 \times 10^7$ and Poisson's ratio is $\nu = 0.3$.

The most interesting feature of the solution concerns two singular points at the two obtuse corners of the plate, which strongly influence the quality of the computed results (e.g. see [20]). The strength of the singularity is $\lambda = 6/5 = 1.2$. In the vicinity of the corner, the moment resultants m_{xx}, m_{yy} vary as $r^{\lambda-2}$, while the moments

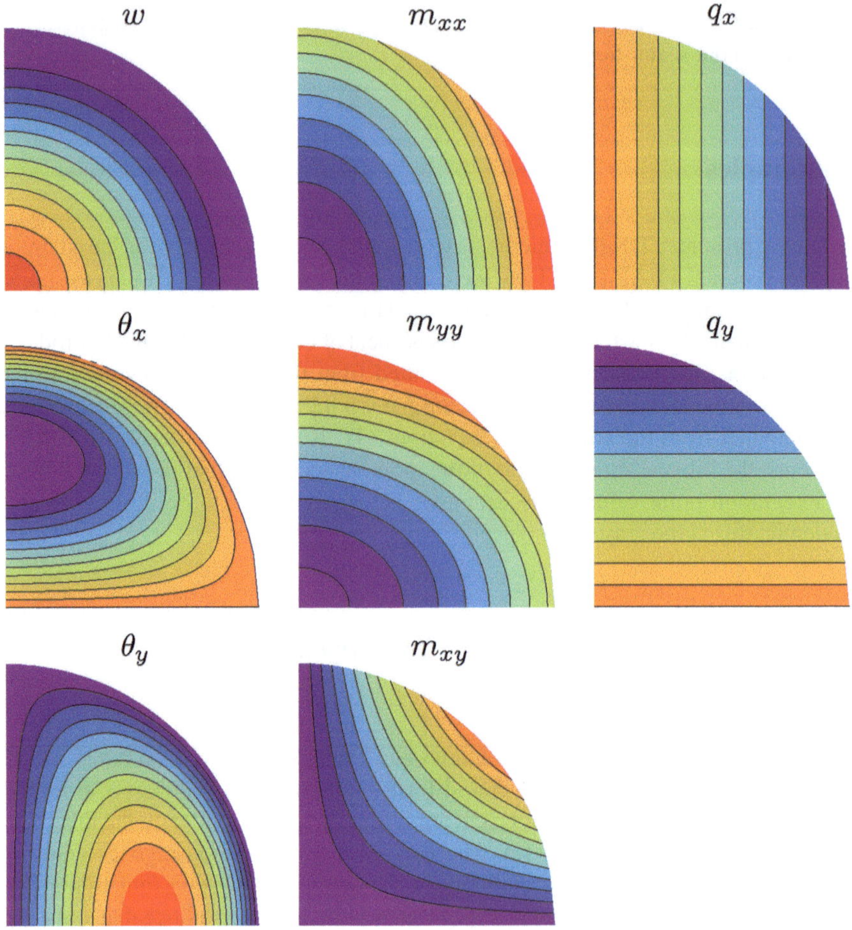

Fig. 2 Reference solution of the uniformly loaded clamped circular plate

resultant m_{xy} and shear forces q_x and q_y vary as $r^{\lambda-3}$, where r is the distance from the corner. As the analytic solution is not available, a finite element computation with Argyris element [5] on a fine mesh (the elements with side $h = 0.1$) is taken as the reference solution. The boundary conditions used for the hard simply supported edge at $y = const.$ are: $w_I = 0, \theta_{x,I} = 0$ for the DK elements and $w_I = 0, w_{I,x} = 0$ for the Argyris element. At the hard simply supported sloped edge $y/x = \tan\alpha$ the rotation θ_n is set to zero, which results in the boundary condition: $\theta_{x,I} - \tan\alpha\, \theta_{y,I} = 0$ for the DK elements and $w_{I,y} + \tan\alpha\, w_{I,x} = 0$ for the Argyris element.

The FE solution with the DK elements qualitatively matches the reference solution except at the obtuse corners. The quality of the FE solution is influenced by the mesh distortion (see Fig. 13). The performance of the DKT does not the reach the performance of the DKQ. The comparison of the converge rates shows that the

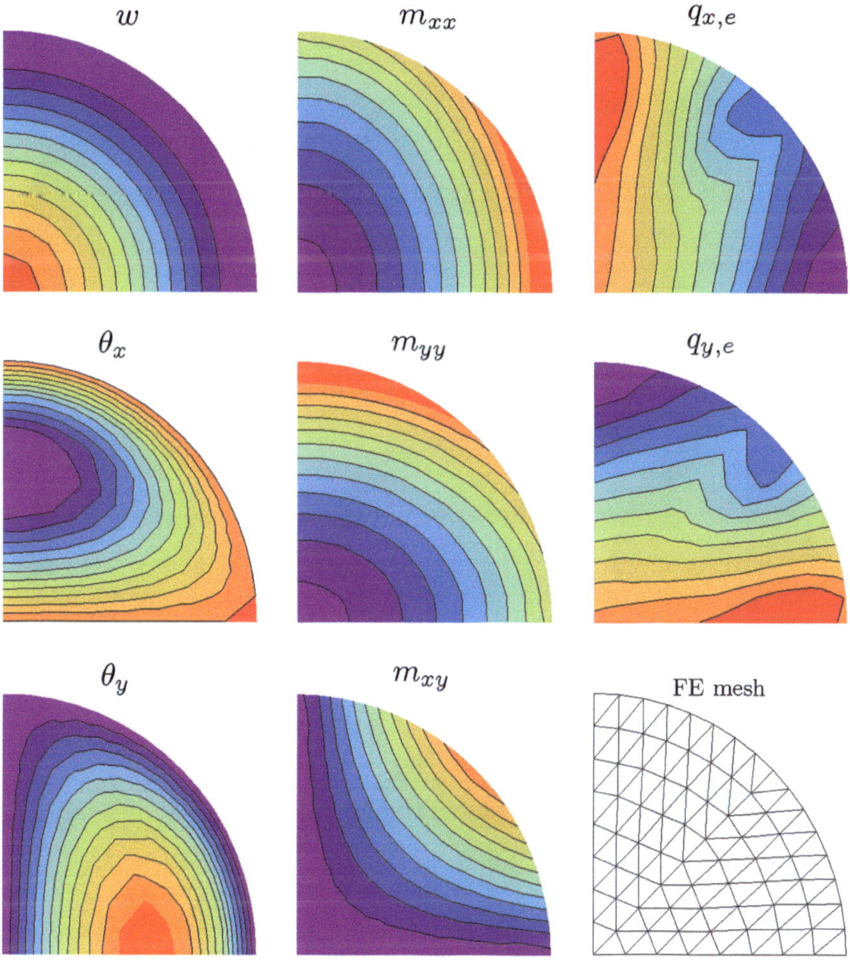

Fig. 3 FE solution of the uniformly loaded clamped circular plate with the DKT plate elements

superior accuracy of the Argyris element is lost. In Figs. 9, 10 and 11, the singularity can be observed at the obtuse corners in m_{xx}, m_{yy} and m_{xy} and in q_x and q_y. The contour plots are not very suitable to study the singularities due to the fixed finite values of contour lines. At the obtuse corners (singularities) the gradients of the solution are infinite and in the limit, the optimal size of the elements is zero. With the decreasing mesh size, the singularity is captured increasingly better. The singularity manifests in the convergence rates (see Fig. 13). The monotonic convergence is observed for the DK elements, while the convergence of the Argyris element is not monotonic (Fig. 12).

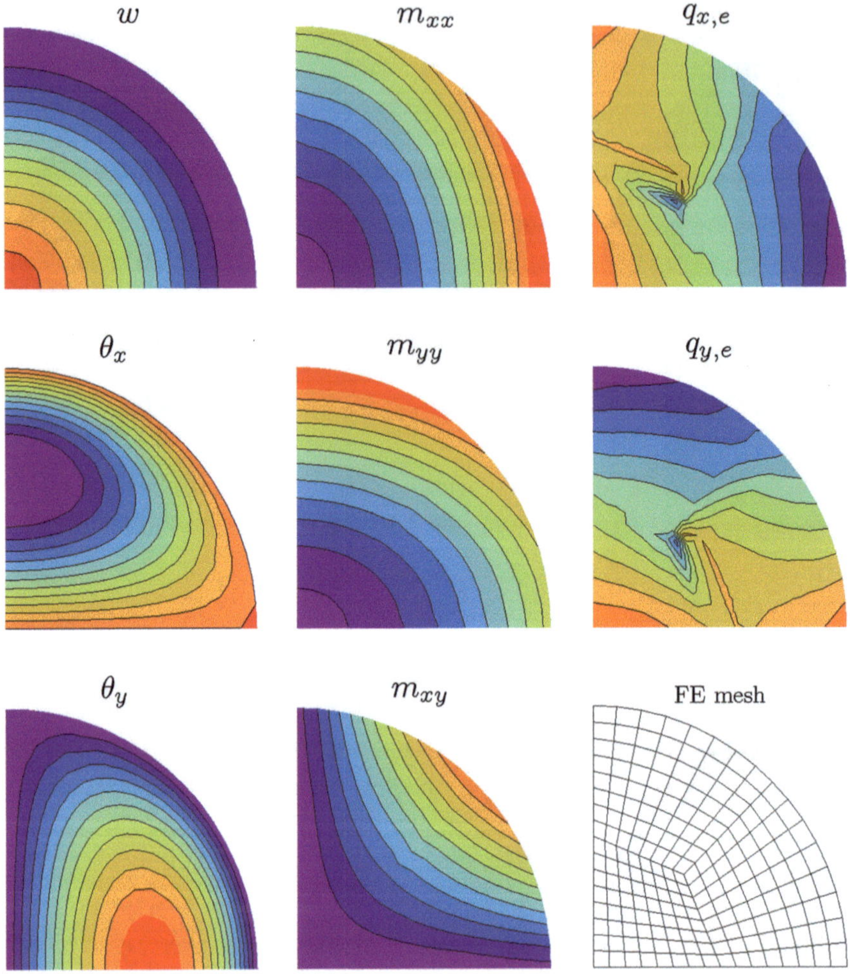

Fig. 4 FE solution of the uniformly loaded clamped circular plate with the DKQ plate elements

5.3 Uniformly Loaded L-Shaped Plate

The L-shaped plate with thickness $t = 0.01$, side length $a = 10$, simple supports on all sides, and uniform pressure loading $f = 1$ is considered. The material data is: $E = 10.92 \times 10^9$, and $\nu = 0.3$. The solution exhibits a singularity in stress resultant components at the obtuse corner. The singularity for m_{xx} and m_{yy} is governed by the term $r^{\lambda-2}$ and for m_{xy} by the term $r^{\lambda-3}$, where r is the distance from the singular point, and λ is the exponent which depends on the opening angle α, i.e. $\lambda = \pi/\alpha$ [19]. In our case, $\alpha = 3\pi/2$, thus $\lambda = 2/3$.

	w	θ_x	θ_y	m_{xx}	m_{yy}	m_{xy}	q_x	q_y
	0. ~ 0.00125	−0.0045 ~ −0.00413	0. ~ 0.000375	−2.3 ~ −1.92	−2.3 ~ −1.92	0. ~ 0.0917	−2.5 ~ −2.29	−2.5 ~ −2.29
	0.00125 ~ 0.0025	−0.00413 ~ −0.00375	0.000375 ~ 0.00075	−1.92 ~ −1.53	−1.92 ~ −1.53	0.0917 ~ 0.183	−2.29 ~ −2.08	−2.29 ~ −2.08
	0.0025 ~ 0.00375	−0.00375 ~ −0.00338	0.00075 ~ 0.00113	−1.53 ~ −1.15	−1.53 ~ −1.15	0.183 ~ 0.275	−2.08 ~ −1.88	−2.08 ~ −1.88
	0.00375 ~ 0.005	−0.00338 ~ −0.003	0.00113 ~ 0.0015	−1.15 ~ −0.767	−1.15 ~ −0.767	0.275 ~ 0.367	−1.88 ~ −1.67	−1.88 ~ −1.67
	0.005 ~ 0.00625	−0.003 ~ −0.00263	0.0015 ~ 0.00188	−0.767 ~ −0.383	−0.767 ~ −0.383	0.367 ~ 0.458	−1.67 ~ −1.46	−1.67 ~ −1.46
	0.00625 ~ 0.0075	−0.00263 ~ −0.00225	0.00188 ~ 0.00225	−0.383 ~ 0.	−0.383 ~ 0.	0.458 ~ 0.55	−1.46 ~ −1.25	−1.46 ~ −1.25
	0.0075 ~ 0.00875	−0.00225 ~ −0.00188	0.00225 ~ 0.00263	0. ~ 0.383	0. ~ 0.383	0.55 ~ 0.642	−1.25 ~ −1.04	−1.25 ~ −1.04
	0.00875 ~ 0.01	−0.00188 ~ −0.0015	0.00263 ~ 0.003	0.383 ~ 0.767	0.383 ~ 0.767	0.642 ~ 0.733	−1.04 ~ −0.833	−1.04 ~ −0.833
	0.01 ~ 0.0113	−0.0015 ~ −0.00113	0.003 ~ 0.00338	0.767 ~ 1.15	0.767 ~ 1.15	0.733 ~ 0.825	−0.833 ~ −0.625	−0.833 ~ −0.625
	0.0113 ~ 0.0125	−0.00113 ~ −0.00075	0.00338 ~ 0.00375	1.15 ~ 1.53	1.15 ~ 1.53	0.825 ~ 0.917	−0.625 ~ −0.417	−0.625 ~ −0.417
	0.0125 ~ 0.0138	−0.00075 ~ −0.000375	0.00375 ~ 0.00413	1.53 ~ 1.92	1.53 ~ 1.92	0.917 ~ 1.01	−0.417 ~ −0.208	−0.417 ~ −0.208
	0.0138 ~ 0.015	−0.000375 ~ 0.	0.00413 ~ 0.0045	1.92 ~ 2.3	1.92 ~ 2.3	1.01 ~ 1.1	−0.208 ~ 0.	−0.208 ~ 0.

Fig. 5 Legend for Figs. 2, 3 and 4

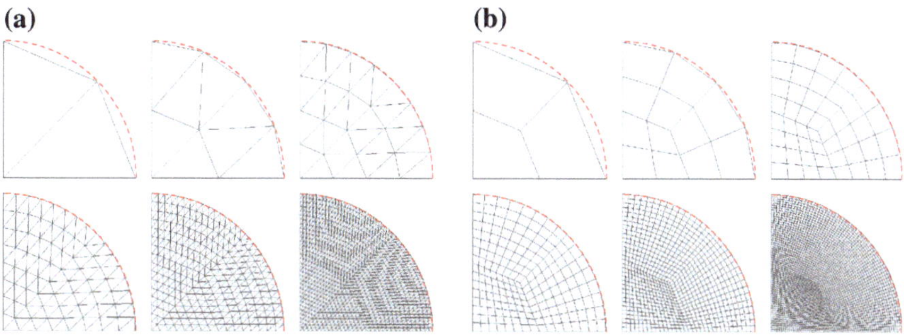

Fig. 6 The meshes used for the FE analysis of uniformly loaded clamped circular plate. **a** Sequence of meshes of triangular elements. **b** Sequence of meshes of quadrilateral elements

The first solution was obtained on the regular structured mesh shown in Fig. 14. The problem was further reanalyzed in six iterations.

(1) The most straightforward mesh refinement strategy was a uniform refinement of a structured mesh.

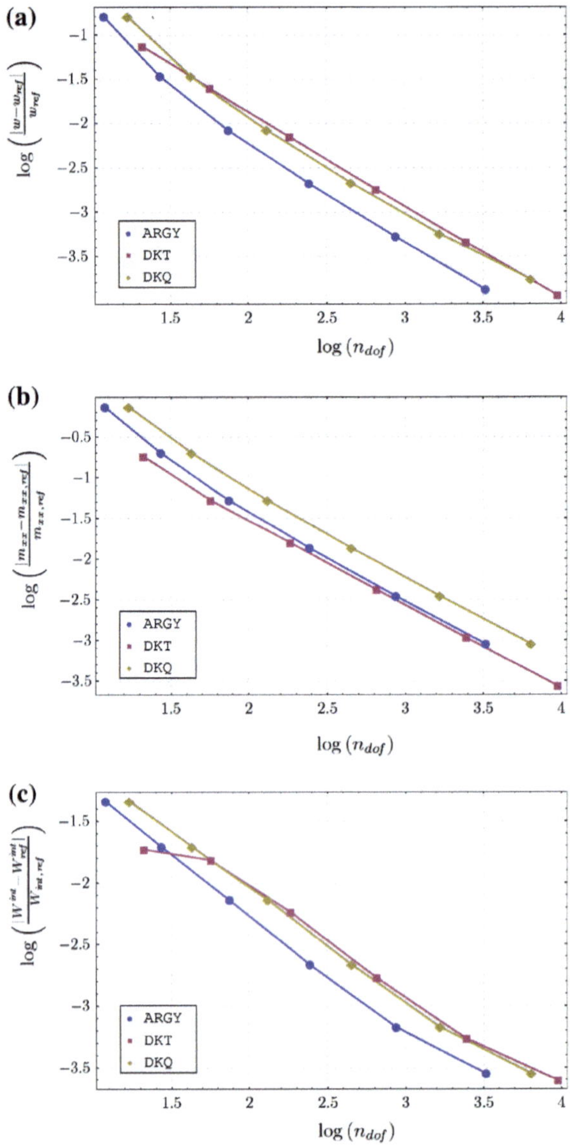

Fig. 7 Convergence of FE solutions for the uniformly loaded clamped circular plate. **a** Displacement w at *center*. **b** Moment resultant m_{xx} at *center*. **c** Energy norm

(2) Another mesh refinement strategy, that belongs to the same category, was a uniform refinement of an unstructured mesh. For that purpose, the initial structured mesh from Fig. 14 was replaced by an unstructured mesh with approximately the same number of elements.

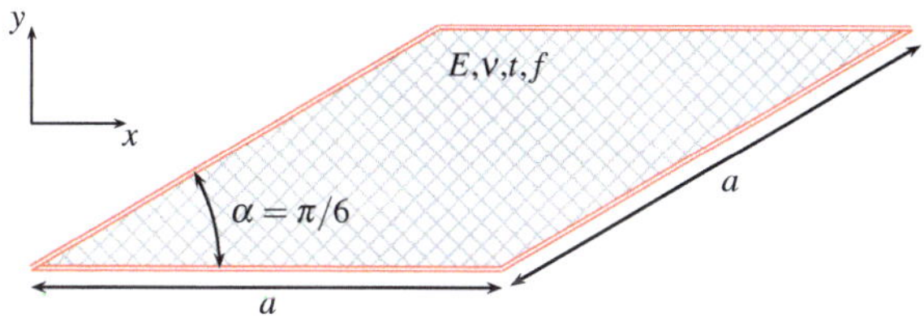

Fig. 8 Uniformly loaded simply supported skew plate

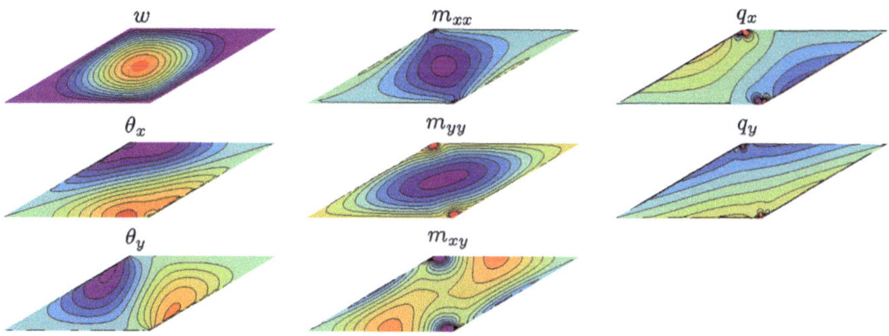

Fig. 9 Reference solution of the uniformly loaded simply supported skew plate

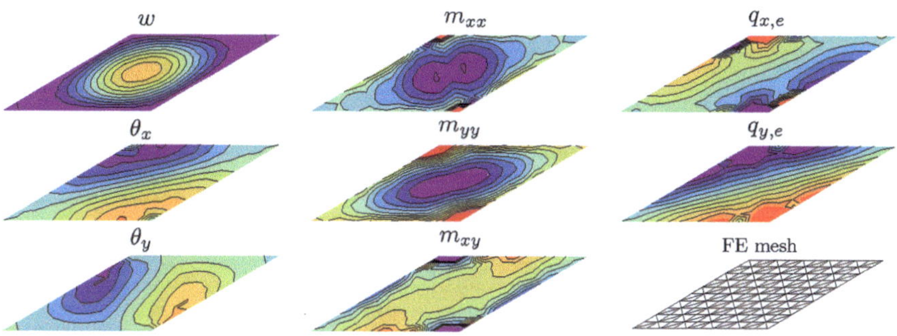

Fig. 10 FE solution of the uniformly loaded simply supported skew plate with the DKT

(3) The adaptive meshing based on above defined error indicator was the third strategy. The "true error" indicator was computed by using reference solution obtained with Argyris elements [5] with side $h = 0.1$. At each iteration, distribution of error indicator $\eta_e = \eta_e^*$ was computed. Based on the estimated new element-size density (41), a new mesh was generated with $p = 1$ and $\bar{\eta}_e = 0.05$.

Comparison of the convergence is shown in Fig. 15. The convergence of the adaptive meshing (3) is considerably quicker from uniform mesh refinements (1) and (2). Table 1 shows results after the third iteration. The comparison reveals that the distribution of the local error indicator is considerably more uniform for (3) than for (1), the later shown in the first row. This is one of the primary goals of adaptive meshing:

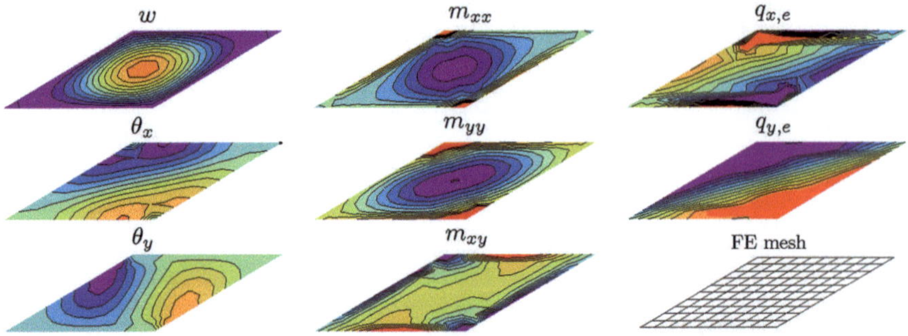

Fig. 11 FE solution of the uniformly loaded simply supported skew plate with the DKQ

	w	θ_x	θ_y	m_{xx}	m_{yy}	m_{xy}	q_x	q_y
	0. ∼ 0.00005	−0.0003 ∼ −0.00025	−0.000145 ∼ −0.000121	−1.5 ∼ −1.38	−2.2 ∼ −1.98	−0.5 ∼ −0.417	−1.5 ∼ −1.25	−3. ∼ −2.5
	0.00005 ∼ 0.0001	−0.00025 ∼ −0.0002	−0.000121 ∼ −0.0000967	−1.38 ∼ −1.25	−1.98 ∼ −1.75	−0.417 ∼ −0.333	−1.25 ∼ −1.	−2.5 ∼ −2.
	0.0001 ∼ 0.00015	−0.0002 ∼ −0.00015	−0.0000967 ∼ −0.0000725	−1.25 ∼ −1.13	−1.75 ∼ −1.53	−0.333 ∼ −0.25	−1. ∼ −0.75	−2. ∼ −1.5
	0.00015 ∼ 0.0002	−0.00015 ∼ −0.0001	−0.0000725 ∼ −0.0000483	−1.13 ∼ −1.	−1.53 ∼ −1.3	−0.25 ∼ −0.167	−0.75 ∼ −0.5	−1.5 ∼ −1.
	0.0002 ∼ 0.00025	−0.0001 ∼ −0.00005	−0.0000483 ∼ −0.0000242	−1. ∼ −0.875	−1.3 ∼ −1.08	−0.167 ∼ −0.0833	−0.5 ∼ −0.25	−1. ∼ −0.5
	0.00025 ∼ 0.0003	−0.00005 ∼ 0.	−0.0000242 ∼ 0.	−0.875 ∼ −0.75	−1.08 ∼ −0.85	−0.0833 ∼ 0.	−0.25 ∼ 0.	−0.5 ∼ 0.
	0.0003 ∼ 0.00035	0. ∼ 0.00005	0. ∼ 0.0000242	−0.75 ∼ −0.625	−0.85 ∼ −0.625	0. ∼ 0.0833	0. ∼ 0.25	0. ∼ 0.5
	0.00035 ∼ 0.0004	0.00005 ∼ 0.0001	0.0000242 ∼ 0.0000483	−0.625 ∼ −0.5	−0.625 ∼ −0.4	0.0833 ∼ 0.167	0.25 ∼ 0.5	0.5 ∼ 1.
	0.0004 ∼ 0.00045	0.0001 ∼ 0.00015	0.0000483 ∼ 0.0000725	−0.5 ∼ −0.375	−0.4 ∼ −0.175	0.167 ∼ 0.25	0.5 ∼ 0.75	1. ∼ 1.5
	0.00045 ∼ 0.0005	0.00015 ∼ 0.0002	0.0000725 ∼ 0.0000967	−0.375 ∼ −0.25	−0.175 ∼ 0.05	0.25 ∼ 0.333	0.75 ∼ 1.	1.5 ∼ 2.
	0.0005 ∼ 0.00055	0.0002 ∼ 0.00025	0.0000967 ∼ 0.000121	−0.25 ∼ −0.125	0.05 ∼ 0.275	0.333 ∼ 0.417	1. ∼ 1.25	2. ∼ 2.5
	0.00055 ∼ 0.0006	0.00025 ∼ 0.0003	0.000121 ∼ 0.000145	−0.125 ∼ 0.	0.275 ∼ 0.5	0.417 ∼ 0.5	1.25 ∼ 1.5	2.5 ∼ 3.

Fig. 12 Legend valid for Figs. 9, 10 and 11

to make the local error indicator approximately equal for all elements of the mesh. Another observation can be made: although the average of local error indicator is smaller for the uniformly refined mesh (see histogram in the first row of Table 1), the total energy norm of the error is still larger (see Fig. 15).

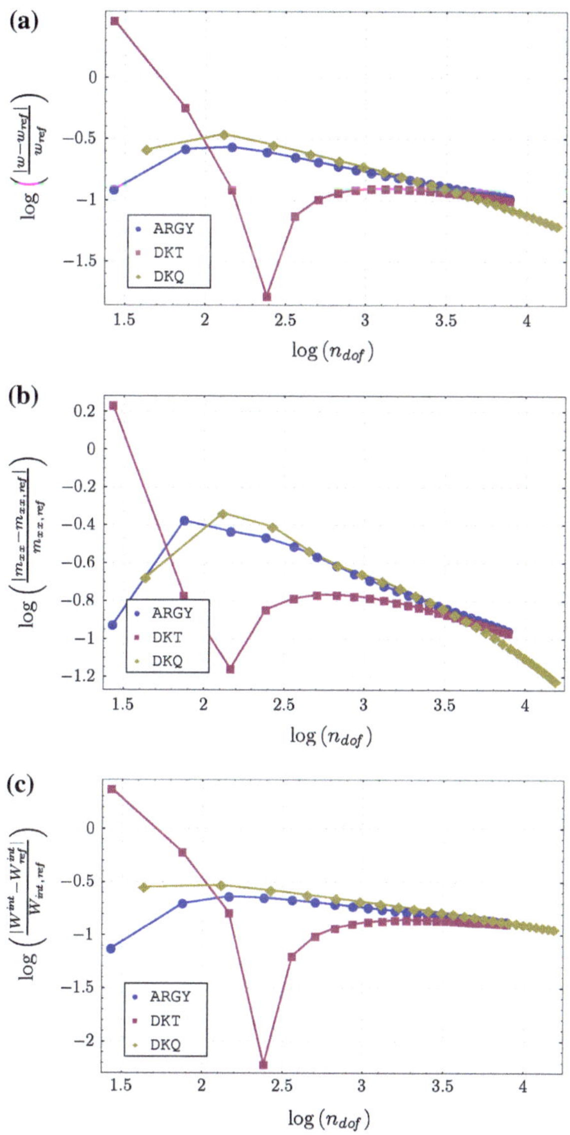

Fig. 13 Comparison of the convergence of FE solutions for the uniformly loaded simply supported skew plate. **a** Displacement w at *center*. **b** Moment m_{xx} at *center*. **c** Energy norm

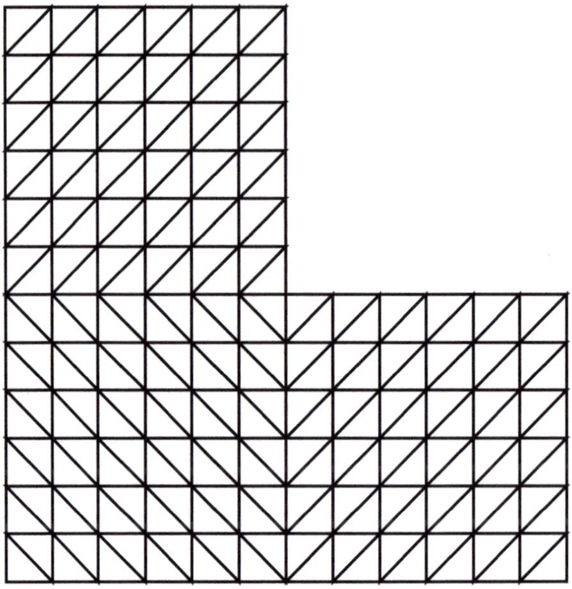

Fig. 14 L-shaped plate under uniform loading—initial mesh

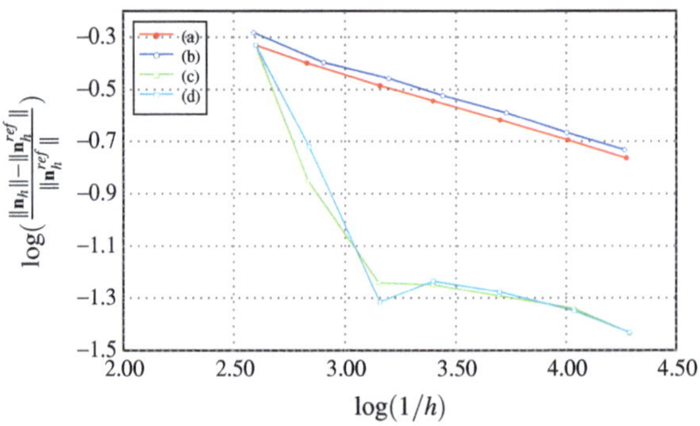

Fig. 15 L-shaped plate under uniform loading. Convergence for: **a** Uniform mesh refinement of structured mesh, **b** Uniform mesh refinement of unstructured mesh, **c** Adaptive meshing based on true error, **d** Adaptive meshing based on SPR

Table 1 Adaptive meshing of L-shaped plate

	Mesh after 3rd refinement step	Local element error η_e 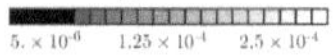	Histogram of local element error η_e
Uniformly refined mesh 1716 elements, η_r: 18.1 %	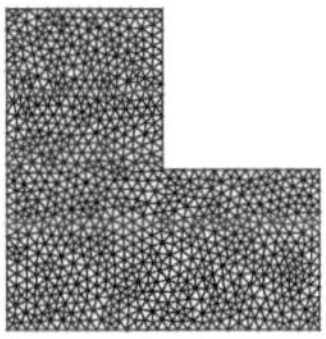		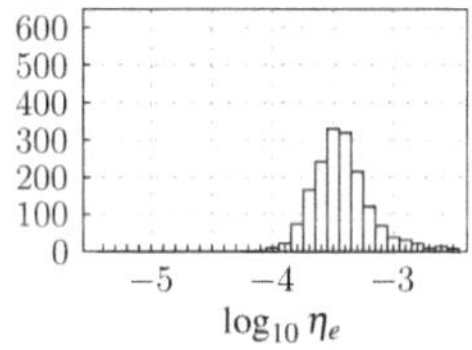
Mesh based on true error 1559 elements, η_r: 5.8 %	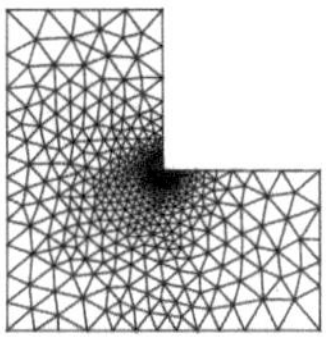		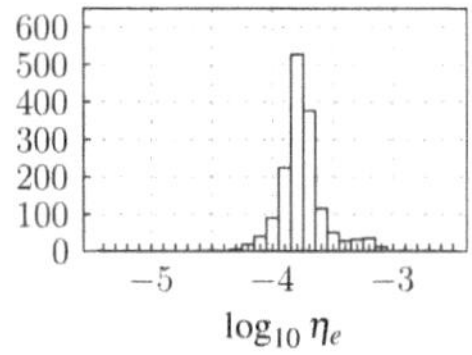
Mesh based on SPR 1557 elements, η_r: 5.6 %	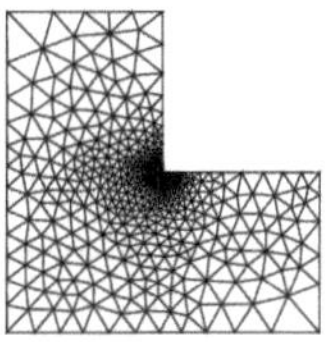	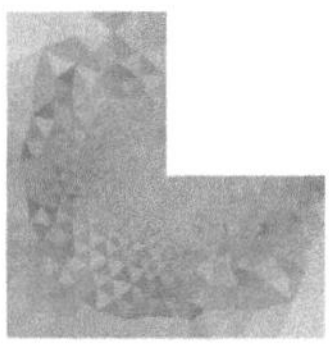	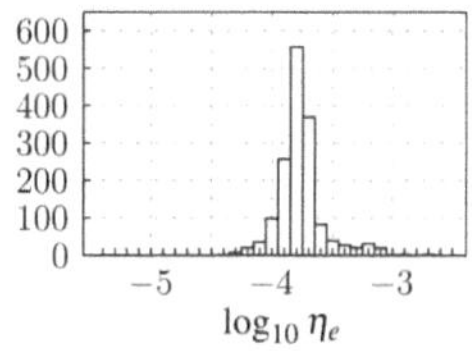

6 Conclusions

The theory, the computer code implementation and the discretization error computation for the Discrete-Kirchhoff plate finite elements have been presented and discussed. Those elements take the best of the Reisner–Mindlin and the Kirchhoff plate models and provide higher-order approximation of plate curvatures even for small number of nodes. As such they are very attractive candidates for solving numerous problems in mechanics, engineering and biology. They can be extended to the corresponding thin shell finite elements, however, this topic has not been addressed in the present work. An extension of the here presented work, showing how the modeling error can be computed for the plate bending problems, is provided in [7].

References

1. Ainsworth, M., Oden, J.T.: A unified approach to a posteriori error estimation using element residual methods. Numer. Math. **65**(1), 23–50 (1993)
2. Ainsworth, M, Oden, J.T.: A Posteriori Error Estimation in Finite Element Analysis. Wiley, New York, (2000)
3. Altenbach, H., Altenbach, J., Naumenko, K.: Ebene Flächentragwerke. Springer, Berlin (1998)
4. Altenbach, H., Eremeyev, V.: Direct approach based analysis of plates composed of functionally graded materials. Arch. Appl. Mech. **78**(5), 775–794 (2008)
5. Argyris, J.H., Fried, I., Scharpf, D.W.: The TUBA family of plate elements for the matrix displacement method. Aeronaut. J. Royal Aeronaut. Soc. **72**, 701–709 (1968)
6. Batoz, J.L., Bathe, K.J., Ho, L.W.: A study of three-node triangular plate bending elements. Int. J. Num. Meth. Eng. **15**, 1771–1812 (1980)
7. Bohinc, U., Ibrahimbegovic, A., Brank, B.: Model adaptivity for finite element analysis of thin or thick plates based on equilibrated boundary stress resultants. Eng. Comput. **26**(1/2), 69–99 (2009)
8. Bohinc, U., Brank, B., Ibrahimbegović, A.: Discretization error for the discrete Kirchhoff plate finite element approximation. Comput. Methods Appl. Mech. Eng. **269**(1), 415–436 (2014)
9. Brank, B.: On boundary layer in the Mindlin plate model: Levy plates. Thin-walled Struct. **46**(5), 451–465 (2008)
10. Clough, R.W, Tocher, J.L.: Finite element stiffness matrices for analysis of plate bending. Proceedings of Conference on Matrix Methods in Structural Mechanics, WPAFB, Ohio, pp. 66–80 (1965)
11. Felippa, C.A, Clough, R.: A refined quadrilateral element for analysis of plate bending. Proceedings of Conference on Matrix Methods in Structural Mechanics, WPAFB, Ohio, pp. 23–69 (1965)
12. Geuzaine, C., Remacle, J.F.: Gmsh: a three-dimensional finite element mesh generator with built-in pre- and post-processing facilities. Int. J. Num. Meth. Eng. **79**(11), 1309–1331 (2009)
13. Hughes, T.J.R.: The Finite Element Method: Linear Static and Dynamic Finite Element Analysis. Dover publications, Minelo (2000)
14. Ibrahimbegovic, A.: Quadrilateral finite elements for analysis of thick and thin plates. Comput. Methods Appl. Mech. Eng. **110**, 195–209 (1993)
15. Ibrahimbegovic, A.: Nonlinear Solid Mechanics: Theoretical Formulations and Finite Element Solution Methods. Springer, Berlin (2009)
16. Korelc, J.: AceGen and AceFEM user manuals (2006)
17. Ladevèze, P., Leguillon, D.: Error estimate procedure in the finite element method and applications. SIAM J. Numer. Anal. **20**(3), 485–509 (1983)
18. Lee, C.K., Hobbs, R.E.: Automatic adaptive refinement for plate bending problems using reissner-mindlin plate bending elements. Int. J. Num. Meth. Eng. **41**, 1–63 (1998)
19. Lee, K.H., Lim, G.T., Wang, C.M.: Thick Levy plates re-visited. Int. J. Solids Struct. **39**, 127–144 (2002)
20. Morley, L.S.D.: Skew Plates and Structures. Pergamon Press, Oxford (1963)
21. Naumenko, K., Altenbach, J., Altenbach, H., Naumenko, V.: Closed and approximate analytical solutions for rectangular mindlin plates. Acta Mech. **147**, 153–172 (2001)
22. Reddy, J., Wang, C.: An overview of the relationships between solutions of classical and shear deformation plate theories. Compos. Sci. Technol. **60**, 2327–2335 (2000)
23. Taylor, R.: FEAP-A finite element analysis program, programmer manual. University of California, Berkeley, http://www.ce.berkeley.edu/rltedn (2013)
24. Taylor, R.L, Govindjee, S.: Solution of clamped rectangular plate problems. Report UCB/SEMM-2002/09 (2002)
25. Taylor, R.L.: FEAP-a finite element analysis program-Version 7.3. University of California, Berkeley (2000)
26. Timoshenko, S., Woinowsky-Krieger, S.: Theory of Plates and Shells. McGraw-Hill, New York (1959)

27. Yunus, S.M., Pawlak, T.P., Wheeler, M.J.: Application of the Zienkiewicz-Zhu error estimator for plate and shell analysis. Int. J. Num. Meth. Eng. **29**(6), 1281–1298 (1990)
28. Zienkiewicz, O.C., Zhu, J.Z.: The superconvergent patch recovery and a posteriori error estimates. Part 1: the recovery technique. Int. J. Numer. Meth. Eng. **33**, 1331–1364 (1992)
29. Zienkiewicz, O.C., Taylor, R.L.: Finite Element Method. Elsevier, London (2000)

Shell Theory-Based Estimation of Local Elastic Characteristics of Biological Cells

Elizaveta S. Drozd, Gennadi I. Mikhasev, Marina G. Botogova, Sergei A. Chizhik and Maria E. Mychko

Abstract The shell theory-based approach for the elastic analysis of biological cells is proposed. This approach introduces the estimation of the local Young's modulus of the single cell on the basis of the shell theory and data of the atomic force microscopy (AFM). This method is applicable to evaluate the elastic properties of the cell membrane which stretched under action of the atomic force microscope indenter (AFMI). The cell is represented by a thin shallow spherical shell experiencing the concentrated outward force. The influence of cytoskeleton on the cell deformation is disregarded. Taking into account microscopic sizes of the cell and the indenter tip, the internal nanoscale parameter is introduced into the constitutive equations. On the basis of the experimental data and developed shell model we give a rough estimate of local Young's modulus for the red blood cell.

1 Introduction

Mechanical properties are fundamental characteristics of cells and tissues. They define a number of cytophysiological and cytopathological processes. The cell mechanical parameters may be used as certain markers of various pathologies

E.S. Drozd (✉) · S.A. Chizhik · M.E. Mychko
Heat and Mass Transfer Institute of National Academy of Sciences of Belarus, Brovki St. 4, 220072 Minsk, Belarus
e-mail: drozd.elizaveta@gmail.com

S.A. Chizhik
e-mail: chizhik_sa@tut.by

M.E. Mychko
e-mail: akumananoda@gmail.com

G.I. Mikhasev · M.G. Botogova
Belarusian State University, Nezavisimosti Ave 4, 220030 Minsk, Belarus
e-mail: mikhasev@bsu.by

M.G. Botogova
e-mail: botogova@tut.by

H. Altenbach and G.I. Mikhasev (eds.), *Shell and Membrane Theories in Mechanics and Biology*, Advanced Structured Materials 45, DOI 10.1007/978-3-319-02535-3_7

[23, 32]. The study of elastic properties allows us to obtain new knowledge on the biological cells and is also of clinical interest.

There are several methods to evaluate the elastic properties of the biological cells: micropipette aspiration [15, 18, 27, 31, 33], filtering [3, 34], fluctuations of the cell's shape [4, 42], the cell deformation (created by shear stress) by ektocytometer, optical channel or microplates [7, 18, 20], and micromanipulation (optical and magnetic tweezers, magnetic twisting cytometry) [5, 22, 24, 43]. However, the most modern and accurate approach developed during the last decade is the one based on the technology of the atomic force microscopy (AFM) including the force spectroscopy [10, 11, 28, 35, 37–40]. This method allows to determine the value of the local modulus as a result of the indenter impact (AFM probe) on the test material. It should be however noted that the study of biological objects by using any approach is very complicated procedure. It includes the following obligatory steps: preparing samples by a special way, optimizing the measurement parameters and reliable interpreting the experimental results. Analysis of the available literature on the estimation of the cell mechanical properties displays a significant scatter in experimental data obtained by using the AFM. First of all, this fact can be explained by the lack of uniformity in measurements and hence makes the quantitative analysis and comparison of the results to be rather complicated [21, 41]. Secondly, when studying the surface layer of a single cell (with thickness varying from several nanometers to some hundred nanometers) with the AFM methods, we deal with a composite material, the mechanical properties of which are mainly conditioned by the properties of the cortical actin cytoskeleton, two- or three-dimensional network from natural biopolymers (proteins) [41]. And the cytoskeleton structure reorganization results in changing the mechanical properties of cells. In our opinion, the satisfactory results may be reached by developing the realistic continuum mechanical model describing a response of the biological cell on the microscopic indenter.

There are different mechanical models, including the shell ones, which are applied for the mechanical characterization of living cells (e.g., see the survey paper by [25]). Within the shell model, cells are often assumed to have a spherical shape. For instance, [44] developed the spherical shell model to describe motion of the bacteria (cocci) in an ultrasonic field. Recently, [36] studied aspiration of spherical cells represented by a hyperelastic isotropic momentless spherical shell, and the simplified spherical shell model for estimation of the local Young's modulus of erythrocyte on the basis of the AFM data was proposed by [12].

The basic goal of this study is to create the reliable shell theory-based model describing a deformation of the biological cell when subjected to the concentrated adhesion forces. Taking into account microscopic sizes of the cell and the indenter tip, we aim to derive an equation for the shell deflection including the nanoscale effects. Using the experimental data of the nanoindentation, it is possible to obtain the force curve by recording the cantilever deflection while the tip is in contact with the biological body. Among all conditions of the cell-indenter interactions, the regime of adhesion between the indenter tip and cell will be considered. Under this type of interaction (when the action of normal force is directed outwards), the cell membrane is only stretched. In this case the influence of the cytoskeleton on the strain-stress state of the cell represented by an elastic shell will be negligible and may be disregarded [12].

2 Methods

2.1 Force Spectroscopy Method

The procedure of force spectroscopy constitutes a standard mode of the AFM. The method consists in the realization of the contact deformation of a specimen using the probe and in quantifying the relationship between probe interaction force and distance [6]. We record the cantilever deflection as the sample moves up, reaches the tip and is retracted. The force curve is obtained by monitoring the movement of the reflected laser beam from the rear of the cantilever. The force curves are the relation between the bend of the cantilever and the position of the probe. Knowing the displacement of the sample in the vertical direction and the amplitude of cantilever bending, one can calculate both the resulting deformation of the sample and the total external force that has been applied to it. The force curve is applied to the calculation of Young's modulus (Fig. 1).

Figure 1 shows the different stages of the indentation process. At the beginning of the force curve recording, the tip is distant from the sample and approaches, but does not contact it (a). Since there is no contact the cantilever deflection is constant. As soon as the tip actually touches the sample, the cantilever moves upward (b). Further, as the tip indents the sample, the cantilever arm moves downward (c). The sample is then retracted displaying a reverse behavior to action (c), as indicated by the curve (d) which shows the deflection during retraction. However, when adhesion between sample and tip occurs, the tip will adhere to the sample beyond the point of contact (e), until it finally breaks free again and the deflection returns to zero (f). It should be noted that the approach curve exactly retraces the pathway to point (e) provided there is no piezohysteresis.

We are able to quantify the local elastic properties of a living cell using the force spectroscopy mode. The force curve is obtained by recording the cantilever

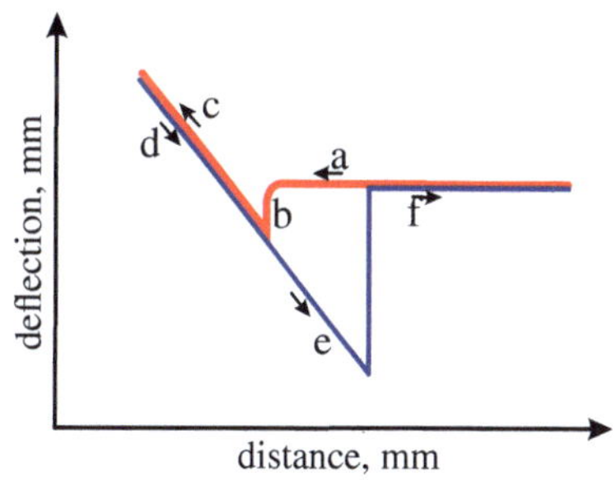

Fig. 1 Regime of static force spectroscopy: typical force curve

deflection as the tip is brought into contact at the fixed point and then retracted. The Young's modulus may be calculated using the Hertz model describing the elastic deformation of two bodies in contact under load [17]. When applying the Hertz model the indented sample is assumed to be extremely thick in comparison to the indentation depth. In this case the elastic modulus may be calculated as described earlier [9, 23]. Alternatively, the shell theory will be developed to describe the regime of adhesion of a cell with the indenter tip; in this case the cell thickness will be assumed to be very thin with respect to the characteristic size of a cell.

2.2 Shell Theory-Based Model

As mentioned above, a thickness of the biological cell varies in the interval from several to hundreds nanometers, and its characteristic size is a value of about several millimeters. For instance, for the red blood cell the average values of these parameters are $h \approx 10\,\text{nm}$ and $R \approx 2\,\mu\text{m}$ respectively. So, when ignoring the internal cytoskeleton, a cell may be represented by a thin-walled elastic structures. Here, we consider the case of adhesion between the cell and indenter. Note that the radius of indenter (about 40 nm) is much less than the characteristic size of the cell. Then, at small magnitudes of the adhesion forces acting from the indenter tip, in a vicinity of the contact point the cell may be modeled by a shallow spherical shell.

Let R be the radius of the spherical shell representing the cell in a vicinity of the contact point, E the local Young's modulus, and ν Poisson ratio of the material. The midsurface is referred to the orthogonal rectangular coordinate system $\overline{x} = R_c x$, $\overline{y} = R_c y$, where x, y are dimensionless coordinates. The shell is assumed to be under action of the concentrated normal outward force Z.

Taking into account microscopic sizes of the cell and a nanoscopic radius of the indenter tip, we aim in passing to study the influence of the nanoscale effect on deformations of the microscale object. With that end in view, we apply the nonlocal version of elasticity pioneered by [13, 14]. Let σ_{ij} and $\sigma_{ij}^{(m)}$ be microscopic and macroscopic stresses in the shell, respectively. According to the nonlocal elasticity theory, these stresses are linked as follows

$$\Xi \sigma_{ij} = \sigma_{ij}^{(m)}, \tag{1}$$

where Ξ is the appropriate linear differential operator which takes into account the effect of the elastic nonlocality. For the two-dimensional strain-stress state it is written as [14]

$$\Xi = 1 - \left(\frac{e_0 a}{R_c}\right)^2 \Delta, \tag{2}$$

where Δ is the Laplace operator in the dimensionless coordinate system x, y, a parameter e_0 is the material dimensionless constant of nonlocality, and a is the internal characteristic length of the material. For instance, [13] gives the value $e_0 = 0.39$, and for discreet nanoscale structures such as carbon nanotubes, the parameter $a = 0.142$ nm is chosen to be the length of the C-C bond. As concerns biological cells, the parameter $e_0 a$ remains indefinite. So, studying the mechanical behaviors of protein microtubules, [16] varied the small scale parameter $e_0 a$ from 0 to 70 nm.

To describe small deformations of the spherical shallow shell under action of the normal concentrated force we apply the theory of [30] which takes into account the transverse shears. According to this theory, the governing equation, with Eqs. (1) and (2) in mind, may be rewritten as follows

$$\left(\frac{1}{R_c^2}\Delta+\frac{2}{R^2}\right)\left[D\left(\frac{1}{R_c^2}\Delta+\frac{1+\nu}{R^2}\right)w-\frac{1}{R}\left(1-\frac{h^2}{5(1-\nu)R_c^2}\Delta\right)\Phi\right]$$

$$=\left[1-\frac{(2-\nu)h^2\Delta}{10(1-\nu)R_c^2}\right]\Xi Z(x,y), \tag{3}$$

$$\left(\frac{1}{R_c^2}\Delta+\frac{2}{R^2}\right)\left[\frac{1}{Eh}\left(\frac{1}{R_c^2}\Delta+\frac{1-\nu}{R^2}\right)\Phi+\frac{w}{R}\right]=\frac{-\nu}{2ER_c^2}\Delta\Xi Z(x,y),$$

where $D = Eh^3/[12(1-\nu^2)]$ is the flexural rigidity of the shell, w is the normal deflexion, and Φ is the stress function.

As opposed to equations for macro-scale shells [26, 30], the modified Eq. (3) contain the additional operator Ξ (the derivation of similar equations for nano-scale shells may be found in the paper by [29]) influenced by the nonlocal parameter e_0.

To eliminate the influence of boundary conditions, we assume that the shell is infinite in all directions. Then the concentrated force $Z(x, y) = P\delta(x, y)$ applied in the point $x = 0$, $y = 0$ may be presented by the Fourier integral

$$P(x,y)=\frac{P}{\pi^2}\int_0^\infty\int_0^\infty \cos\alpha x\cos\beta y\,\mathrm{d}\alpha\,\mathrm{d}\beta, \tag{4}$$

where $\delta(x, y)$ is the delta function.

The unknown functions from Eq. (3) may be also presented as

$$w(x,y)=\int_0^\infty\int_0^\infty w_{\alpha\beta}(\alpha,\beta)\cos\alpha x\cos\beta y\,\mathrm{d}\alpha\,\mathrm{d}\beta,$$

$$\Phi(x,y)=\int_0^\infty\int_0^\infty \Phi_{\alpha\beta}(\alpha,\beta)\cos\alpha x\cos\beta y\,\mathrm{d}\alpha\,\mathrm{d}\beta. \tag{5}$$

Substituting Eqs. (4) and (5) into Eq. (3) , and performing the Fourier inversion, one obtains

$$w = \frac{PR_c^2}{\pi^2 D} \int_0^\infty \int_0^\infty K_1 \left[1 + \frac{(e_o a)^2}{R_c^2}(\alpha^2 + \beta^2)\right] \frac{\cos\alpha x \cos\beta y \mathrm{d}\alpha \mathrm{d}\beta}{(\alpha^2 + \beta^2 - 2k_R)F(\alpha, \beta)}, \tag{6}$$

where

$$\begin{aligned} K_1 &= \Big[(1 + (\eta - \varepsilon)(\alpha^2 + \beta^2))(\alpha^2 + \beta^2 - (1 - \nu)k_R) \\ &\qquad - \bar{\varepsilon}(1 + \eta(\alpha^2 + \beta^2))(\alpha^2 + \beta^2)\Big] \\ F(\alpha, \beta) &= \left[\alpha^2 + \beta^2 - k_R)^2 + 4\kappa^4(1 + \eta(\alpha^2 + \beta^2)\right], \\ 4\kappa^4 &= \frac{EhR_c^4}{R^2 D} = \frac{12(1 - \nu^2)R_c^4}{R^2 h^2}, \\ \bar{\varepsilon} &= \frac{\nu h}{2R}, \quad \varepsilon = \frac{\nu h^2}{10(1 - \nu)R_c^2}, \quad \eta = \frac{h^2}{5(1 - \nu)R_c^2} \end{aligned} \tag{7}$$

The characteristic size may be chosen as follows

$$R_c = \frac{\sqrt{Rh}}{\sqrt[4]{12(1 - \nu^2)}}. \tag{8}$$

Then, proceeding to the polar coordinate system by equations

$$\alpha = \gamma \cos\varphi, \quad \beta = \gamma \sin\varphi, \quad x = r\cos\theta, \quad y = r\sin\theta, \tag{9}$$

one gets

$$w = \frac{PR_c^2}{2\pi D} \int_0^\infty K_2(\gamma) \left[1 + \frac{(e_o a)^2}{R_c^2}\gamma^2\right] \frac{J_0(\gamma r)\gamma \mathrm{d}\gamma}{(\gamma^2 - 2k_R)F(\gamma)}, \tag{10}$$

where $J_0(\gamma r)$ is the zeroth-order Bessel function of the first kind, and

$$\begin{aligned} K_2 &= \left[(1 + (\eta - \varepsilon)\gamma^2) - (\gamma^2 - (1 - \nu)k_R) - \bar{\varepsilon}(1 + \eta\gamma^2)\gamma^2\right], \\ F(\gamma) &= (\gamma^2 - k_R)^2 + \eta\gamma^2 + 1, \quad k_R = \frac{1}{\sqrt{12(1 - \nu^2)}} \frac{h}{R}. \end{aligned} \tag{11}$$

Since $h^2/R_c \ll 1$ and $k_R \ll 1$, then the term $(1 + k_R^2 + 2\eta k_R)^{-1}$ can be neglected, and

$$\frac{1}{F(\gamma)} \approx \left[1 - \frac{\gamma^2}{1 + \gamma^4}(\eta - 2k_R)\right] \frac{1}{1 + \gamma^4}. \tag{12}$$

Performing simple transformations of functions under the integral (see details in the paper by [26]), one gets the final equations for the normal displacement of the shell subjected to action of the concentrated normal force P:

$$w = w^* + w^{**}, \tag{13}$$

where

$$\begin{aligned} w^* &= \frac{PR_c^2}{2\pi D}\left\{-\mathrm{kei}(r) - k_R\left[(1+\nu)\left(\frac{\pi}{2}Y_0(r\sqrt{2k_R}) + \mathrm{ker}(r)\right) + \frac{1}{2}r\,\mathrm{ker}'(r)\right]\right. \\ &\left.+\left[(\eta-\varepsilon)\,\mathrm{ker}(r) + \frac{\eta}{4}r\,\mathrm{ker}'(r)\right]\right\} \end{aligned} \tag{14}$$

is the term obtained earlier for a macro-scale shell [26] and

$$w^{**} = \frac{P(e_o a)^2}{2\pi D}\left\{\mathrm{ker}(r) + \dot{\mathrm{kei}}(r)\left[\eta(2-\bar{\varepsilon}) - \varepsilon - k_R(3+\nu)\right]\right\} \tag{15}$$

is the new summand taking into account the nano-scale effect. In Eqs. (14) and (15), $\mathrm{kei}(x)$ and $\mathrm{ker}(x)$ are the Kelvin functions.

Although Eqs. (14) and (15) have singularity in the point $r = 0$, they can be applied for estimation of the maximum displacement of the shell for the case when the force P is distributed over the surface of a small circle of the radius c. When taking into account properties of the Bessel and Kelvin functions, one has

$$w_0 = w_0^* + w_0^{**}, \tag{16}$$

where

$$\begin{aligned} w_0^* &\approx \frac{PR\sqrt{12(1-\nu^2)}}{Eh^2\pi}\left[\frac{1}{2c^2} + \frac{1}{2c}\mathrm{ker}'(c) - \frac{1}{2}k_R(1+\nu)\ln\sqrt{2k_R} - \frac{k_R}{4}\right] \\ &+\frac{3P}{5\pi Eh}(1+\nu)(2-\nu)\left[\mathrm{ker}(c) + \frac{c\,\mathrm{ker}'(c)}{2(2-\nu)}\right], \end{aligned} \tag{17}$$

$$w_0^{**} \approx \frac{P(e_0 a)^2}{2\pi D}\left\{\mathrm{ker}(c) - \left[\frac{1}{c^2} + \frac{1}{c}\mathrm{ker}'(c)\right]\left[\eta(2-\bar{\varepsilon}) - \varepsilon - k_R(3+\nu)\right]\right\}. \tag{18}$$

Equation (17) have been derived by [26], and term (18) taking into account the scale effect is the new one.

3 Determination of Contact Area

Let c^* be a radius of the contact area between the cell and tip. This contact is provided by the forces of adhesion under withdrawal of the tip out of the cell. It should be noted that a parameter c^* is hardly measurable one during an experiment. When the indenter being taken out the sample, the contact area decreases so that it is very difficult to fix and quantify a radius c^* in the moment of detachment of the tip.

Another available way to determine the contact area is based on the theory of [19]. Let $\varDelta\varGamma$ be the surface energy of the sample. This energy is acquired as result of the surface forces appearing in the sample due to cohesion between the sample and tip. According to this theory the contact area is defined by the following equation

$$c^* = \sqrt[3]{R_t K \left(P + 3\pi R_t \varDelta\varGamma + \sqrt{(3\pi R_t \varDelta\varGamma)^2 + 6\pi R_t \varDelta\varGamma P}\right)}, \tag{19}$$

where

$$K = \frac{3}{4}\left(\frac{(1-\nu^2)}{E} + \frac{(1-\nu_t^2)}{E_t}\right), \tag{20}$$

R_t is the tip radius, E_t, ν_t are the Young's modulus and Poisson's ratio of the tip. Since $E \ll E_t$, it is assumed $K \approx 3(1-\nu^2)/E$ in what follows.

If the surface forces are neglected ($\varDelta\varGamma = 0$), then Eq. (19) is transformed into the Hertz formula

$$c^* = \sqrt[3]{P R_t K}. \tag{21}$$

When the sample is stretched, then the force $P = -P_s$ ($P_s > 0$) is negative. Increasing the force magnitude P_s results in decrease of the contact area (radius c^*). Detachment of the indenter tip from the cell takes place when the magnitude of the stretching force reaches the value

$$P_s = \frac{3}{2}\pi R_t \varDelta\varGamma. \tag{22}$$

Then the surface energy of the cell is as follows;

$$\varDelta\varGamma = \frac{2P_s}{3\pi R_t}. \tag{23}$$

Substituting Eq. (23) into Eq. (19) results in the following equation for the contact area radius

$$c^* = \sqrt[3]{\frac{3P_s R_t (1 - \nu^2)}{4E}}, \tag{24}$$

which will be used in our subsequent calculations.

4 Estimation of Local Young's Modulus

Let $w_{\max}$ be the maximum deflection of the sample in the moment of its detachment from the cantilever. The parameters $w_{\max}$, P_s are assumed to be known ones; they are quantified during an experiment. Returning to the shell theory, we assume that $w_{\max} = w_0$, where w_0 is defined by (16)–(18); the function w_0 depends on the external force P_s, the local Young's modulus E, thickness h and the dimensionless radius c of the contact area. Taking into account Eq. (24), one obtains the following equations

$$w_{\max} = w_0[P_s, E, h, c(P_s, E)], \quad c = c^*/R_c \tag{25}$$

with respect to the modulus E.

We have performed series of the AFM probes and computations for the red blood cell at the following parameters: $R_t = 40$ nm, the shell radius $R = 2$ μm (the total radius of the red blood cell in the point of indentation [8], Poisson ratio $\nu = 0,5$ [1, 2], $e_0 a = 0$. The shell thickness h was initially taken to be equal to 10 nm, which corresponds to the average thickness of the cell membrane (without its cytoskeleton). Solving Eq. (25) at different values of parameters P_s, $w_{\max}$ (the quantified maximum displacements $w_{\max}$ were varied in the interval from 10–100 nm), we compared the obtained values of E with data found by the Hertz model.

Figure 2 shows the diagrams for average values of Young's modulus (in relative units) at different deformations of the cell.

The diagrams marked by blue were constructed by using both the Hertz model and data of nanoindentations of the sample (at $w_{\exp} > 0$), and the diagrams in red correspond to data found from the shell model and experimental data by stretching the sample (at $w_{\exp} < 0$). All magnitudes shown in this figure as well as in others were found by averaging data of a whole number of indentations and adhesive tensions of the sample, then the average values were normalized by the magnitude $E = 338.5$ kPa. As result of linearity of the shell model, the shell-based values E_{sh} do not depend on the displacement $w_{\exp}$. However, the Hertz model demonstrates very strong dependence on $w_{\exp}$: increasing the sample deformation results in decrease of the local modulus E_H. It may be seen that both models give close results for very small deformations. Considerable gap in results at large values of $|w_{\exp}|$ can be explained by influence of both the cell internal components and compressive strains

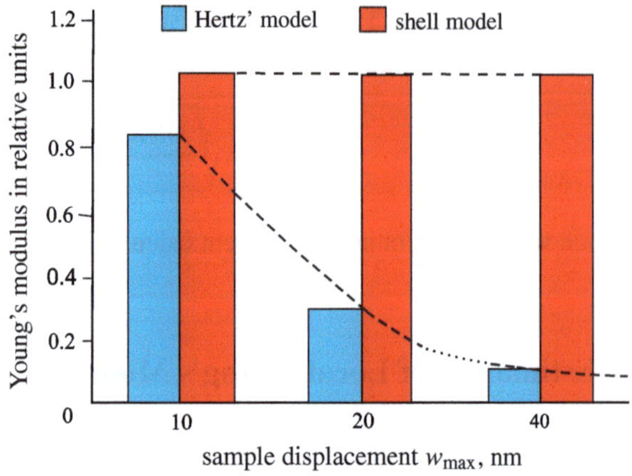

Fig. 2 Local Young's modulus in relative units versus the sample displacement w_{max} found by the shell and Hertz models

(when using Hertz model). As concerns the shell model, it is linear and can not be used for prediction of large deformations. In addition, applying the shell model for the stretched cell we ignored the cell cytoskeleton. It should be also noted that a value of E found at large displacements of the sample can not be considered as local Young's modulus of a cell; most probably it might be interpreted as the effective stiffness of all structure of the microscopic biological object.

It is obvious that the mechanical properties of a cell is influenced by its internal components (cytoskeleton). The cytoskeleton consists of a dense three-dimensional network of filaments. Among them it is possible to distinguish at least three types: microtubules, microfilaments and intermediate filaments. They provide mechanical stability of the surface layer of cytoplasm and create conditions that allows for the cell to change its shape and move. Filaments have sufficient resistance to bending in large scales and remarkable resiliency in small scales [16]. The thickness of microtubules is approximately 24 nm, the thickness of microfilaments (f-actin) is about 5–8 nm and the intermediate filaments diameter is 8–10 nm [8]. The thickness of a cell membrane is stated to vary from 3 to 11 nm. Thus, the overall thickness of a cell membrane and neighboring elements of the cytoskeleton is in the range about 10–40 nm. To take into account the influence of the cytoskeleton on the elastic properties of a cell, we introduced into the shell model the effective thickness which was varied in the range mentioned above. Numerical calculations by Eq. (25) performed at $P_s = 0.1$ nN, $w_{exp} = 10$ nm revealed that increasing the shell thickness h from 10 nm to 40 nm leads to near tenfold decrease of the local Young's modulus (from 338–34 kPa).

The next question that we are interested in: is there such equivalent shell thickness at which Young's modulus calculated by using both models are close? Coincidence of moduli would determine the degree of participation of the cellular component in the processes of compression and stretching. The results of this evaluation are shown in Fig. 3. It may be seen that the values of the local elastic modulus estimated on the basis of the shell and Hertz models are close to each other when the maximum sample deformation w_{exp} is near the value of the shell thickness (at $h/w_{exp} \approx 1.15$).

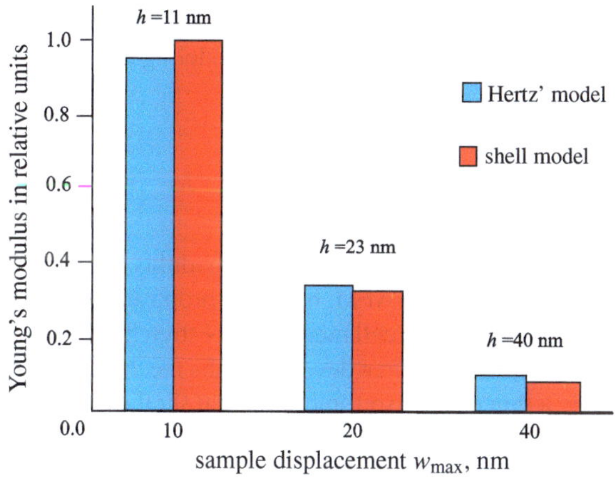

Fig. 3 Local Young's modulus in relative units vs. the sample displacement w_{max} found by the shell and Hertz models at different cell thicknesses $t_1 = 11, 23, 40$ nm

Finally, taking into account the microscopic sizes of the cell and indenter, we attempted to estimate the effect of the internal scale on local properties of the sample. According to the nonlocal theory of elasticity of [13, 14], the principle parameter characterizing the nonlocal effect is the scale coefficient ae_0, where a and e_0 were introduced above. The satisfactory estimation of this parameter for biological micro- and nano-objects is still unsolved problem. For instance, studying behavior of the protein microtubule, [16] varied the small scale parameter ae_0 from 0 to 70 nm. Following [16], we have also performed the series of accurate calculations at different values of ae_0. It was found that impact of the parameter ae_0 on the local modulus E is negligibly small: at $ae_0 = 70$ nm, the additional term (18) gives correction (for E) not exceeding 0.01 %. Very weak influence of the scale parameter ae_0 on the assessment of the modulus E might be explained by large diameter of the indenter (about 40 nm) with respect to the cell size. As shown in the paper by [29], the scale parameter ae_0 should be taken into consideration for prediction of mechanical behavior which is characterized by high variability of deformations along even if one direction on the surface of a nanoscale shell object. In our case, decreasing the area of application of the external force Z would result in increasing variability of the cell deflection in a vicinity of the contact point between the tip and cell. However, utilization of more thin indenter tip for indentation of biological cells generates the following serious problems:

(a) the microscope should be thoroughly standardized;
(b) all probes should be done with very high accuracy;
(c) outcome scatter for different probes becomes more significant;
(d) one rises the risk of the membrane penetration.

In our opinion, further improvement of the shell model for prediction of mechanical properties of biological cells might be done by introducing the cell components into model. As the first step of this development, the cytoskeleton structure could be represented by an elastic foundation for a shell modeling a cell.

5 Conclusions

The approach for estimation of the local Young's modulus of a biological cell, based on the shell theory and data of the AFM, has been proposed. The developed method is applicable for the case when the cell membrane is stretched by the adhesion forces between the AFMI and sample. When developing the mechanical model, the cell has been represented by a thin elastic isotropic spherical shell with the radius equaled to the total one of the sample in the point of indentation. The differential equations written in terms of the normal displacements and stress function and taking into account the shear deformations and nano-scale effect have been considered as governing ones. In the developed model the internal components of a cell (cytoskeleton) are disregarded. The solution of these equations with the external concentrated force has been constructed using the Fourier transformation. The found solution has been modified for the case of the adhesion forces applied to the small circular area. The radius of the contact area has been defined by the theory of Johnson-Kendall-Roberts. Substitution of this radius into the obtained solution of the governing equations has allowed to derive the transcendental equation with respect to unknown Young's modulus.

The series of the probes for the red blood cell corresponding to conditions of adhesion and indentation as well has been made. The averaged data (the force acting from the AFMI and the maximum displacement of the sample) of these probes have been introduced into the mathematical model. Alternatively, the Hertz theory has been applied to estimate the local mechanical properties of the cell when subjected to the indentation forces.

Comparative analysis of data obtained by using the two approaches allows to conclude:

- both models give close results only for very small deformations of the cell (about 10 nm) and at the effective cell thickness having the order of the maximum deflexion of the sample;
- introducing the nano-scale parameter into the mechanical model does not give some noticeable correction for the local Young's modulus found at the generally accepted size of the indenter tip (about 40 nm);
- more accurate estimations of the local mechanical properties of a cell might be done by incorporation of the cell components into the mechanical model; influence of the cytoskeleton could be taking into account by introducing the nonhomogeneous elastic foundation into the shell model.

Acknowledgments The research leading to these results has received funding from the People Programme (Marie Curie Actions) of the European Union's Seventh Framework Programme FP7/2007–2013/under REA grant agreement PIRSES-GA-2013-610547-TAMER.

References

1. A-Hassan, E., Heinz, W., Antonik, M., D'Costa, N., Nageswaran, S., Schoenenberger, C., Hoh, J.: Relative microelastic mapping of living cells by atomic force microscopy. Biophys. J. **74**(3), 1564–1578 (1998)
2. Alcaraz, J., Buscemi, L., Grabulosa, M., Trepat, X., Fabry, B., Farré, R., Navajas, D.: Microrheology of human lung epithelial cells measured by atomic force microscopy. Biophys. J. **84**(3), 2071–2079 (2003)
3. Baerlocher, G., Schlappritzi, E., Straub, P., Reinhart, W.: Erythrocyte deformability has no influence on the rate of erythrophagocytosis in vitro by autologous human monocytes/macrophages. Br. J. Haematol. **86**(3), 629–634 (1994)
4. Baumann, M.: Cell ageing for 1 day alters both membrane elasticity and viscosity. Pflugers Arch. **445**(5), 551–555 (2003)
5. Berrios, J., Schroeder, M., Hubmayr, R.: Mechanical properties of alveolar epithelial cells in culture. J. Appl. Physiol. **91**(1), 65–73 (2001)
6. Burnham, N., Colton, R.: Measuring the nanomechanical properties and surface forces of materials using an atomic force microscope. J. Vac. Sci. Technol. **A7**, 2906–2913 (1989)
7. Caille, N., Thoumine, O., Tardy, Y., Meister, J.J.: Contribution of the nucleus to the mechanical properties of endothelial cells. J. Biomech. **35**(2), 177–187 (2002)
8. Cherenkevich, S., Martinovic, K.A.G.N.: Biological Membranes. Belarusian State University, Minsk (2009)
9. Chizhik, S., Huang, Z., Gorbunov, V., Myshkin, N., Tsukruk, V.: Micromechanical properties of elastic polymeric materials as probed by scanning force microscopy. Langmuir **14**, 2606–2609 (1998)
10. Costa, K., Sim, A., Yin, F.: Non-hertzian approach to analyzing mechanical properties of endothelial cells probed by atomic force microscopy. J. Biomech. Eng **128**(2), 176–184 (2006)
11. Drozd, E, Chizhik, S.: Estimation of the elastic modulus for highly flexible materials by the method of indentation and separation of a spherical indenter at nanoscale (in Russ.). Natl. Acad. Sci. Reports **54**, 117–122 (2010)
12. Drozd, E., Mikhasev, G., Chizhik, S.: Evaluation of the local elasticity modulus of biological cells on the basis of the shells theory. Ser. Biomech. **27**(3–4), 17–22 (2012)
13. Eringen, A.: On differential equations of nonlocal elasticity and solutions of screw dislocations and surface waves. J. Appl. Phys. **54**, 4703–4710 (1983)
14. Eringen, A.: Nonlocal Continuum Field Theories. Springer, New York (2002)
15. Ewans, E., Waugh, R., Melnik, L.: Elastic area compressibility modulus of red cell membrane. Biophys. J. **16**(6), 585–595 (1976)
16. Gao, Y., Lei, F.M.: Small scale effects on the mechanical behaviors of protein microtubules based on the nonlocal elasticity theory. Biochem. Biophys. Res. Commun. **387**, 467–471 (2009)
17. Hertz, H.: Ueber die Berührung fester elastischer Körper. J für die reine und angewandte Mathematik **92**, 156–171 (1881)
18. Hochmuth, R., Mohandas, N., Blackshear, P.: Measurement of the elastic modulus for red cell membrane using a fluid mechanical technique. Biophys. J. **13**(8), 747–762 (1973)
19. Johnson, K., Kendall, K., Roberts, A.: Surface energy and the contact of elastic solids. Proc. R. Soc. **A324**, 301–313 (1971)
20. Kaneta, T., Makihara, J., Imasaka, T.: An optical channel: a technique for the evaluation of biological cell elasticity. Anal. Chem. **73**(24), 5791–5795 (2001)
21. Kuznetsova, T., Starodubtseva, M., Yegorenkov, N., Chizhik, S., Zhdanov, R.: Atomic force microscopy probing of cell elasticity. Micron **38**(8), 824–833 (2007)
22. Laurent, V., Fodil, R., Cañadas, P., Féréol, S., Louis, B., Planus, E., Isabey, D.: Partitioning of cortical and deep cytoskeleton responses from transient magnetic bead twisting. Annal. Biomed. Eng. **31**(10), 1263–1278 (2003)
23. Lee, G., Lim, C.: Biomechanics approaches to studying human diseases. Trends Biotechnol. **25**(3), 111–118 (2007)

24. Li, C., Liu, K.: Nanomechanical characterization of red blood cells using optical tweezers. J. Mater. Sci. - Mater. Med. **19**(4), 1529–1535 (2008)
25. Lim, C., Zhoua, E., Quek, S.: Mechanical models for living cells - a review. J. Biomech. **39**(2), 195–216 (2006)
26. Lukasiewicz, S.: Local Loads in Plates and Shells. Polish Scentific Publishers, Warszawa (1979)
27. Maggakis-Kelemen, C., Biselli, M., Artmann, G.: Determination of the elastic shear modulus of cultured human red blood cells. Biomed. Tech. **47**(1), 106–109 (2002)
28. Mahaffy, R., Park, S., Gerde, E., Käs, J., Shih, C.: Quantitative analysis of the viscoelastic properties of thin regions of fibroblasts using atomic force microscopy. Biophys. J. **86**(3), 1777–1793 (2004)
29. Mikhasev, G.: On localized modes of free vibrations of single-walled carbone nanotubes embedded in nonhomogeneous elastic medium. Z Angew. Math. Mech. **94**(1–2), 130–141 (2014)
30. Naghdi, P.: On the theory of thin elastic shells. Quart. Appl. Math. **14**, 369–380 (1957)
31. Nash, G., Wyard, S.: Erythrocyte membrane elasticity during in vivo ageing. Biochim. et. Biophys. Acta. **643**(2), 175–269 (1981)
32. Radmacher, M.: Studying the mechanics of cellular processes by atomic force microscopy. Methods Cell Biol. **83**, 91–189 (2007)
33. Rand, R., Burton, A.: Mechanical properties of the cell membrane. I. Membrane stiffness and intracellular pressure. Biophys. J. **4**(2), 115–135 (1964)
34. Rasia, R., Valverde, J., Rosasco, M.: Blood preservation. bacteriological, immunohematological, hematological and hemorrheological studies. Sangre **43**(1), 71–76 (1998)
35. Rosenbluth, M., Lam, W., Fletcher, D.: Force microscopy of nonadherent cells: a comparison of leukemia cell deformability. Biophys. J. **90**(8), 2994–3003 (2006)
36. Rudenko, O., Kolesnikov, A.: Aspiration of a nonlinear elastic spherical membrane. Int. J. Eng. Sci. **80**, 62–73 (2014)
37. Scheffer, L., Bitler, A., Ben-Jacob, E., Korenstein, R.: Atomic force pulling: probing the local elasticity of the cell membran. Europ. Biophys. J. **30**(2), 83–90 (2001)
38. Sirghi, L., Ponti, J., Broggi, F., Rossi, F.: Probing elasticity and adhesion of live cells by atomic force microscopy indentation. Europ. Biophys. J. **37**(6), 935–945 (2008)
39. Smith, B., Tolloczko, B., Martin, J., Grütter, P.: Probing the viscoelastic behavior of cultured airway smooth muscle cells with atomic force microscopy: stiffening induced by contractile agonist. Biophys. J. **88**(4), 2994–3007 (2005)
40. Starodubtseva, M., Chizhik, S., Yegorenkov, N., Nikitina, I., Drozd, E.: Formatex Research Center. Badajoz, chap Study of the mechanical properties of single cells as biocomposites by atomic force microscopy, Microscopy: Science, Technology, Applications and Education Vol. 3, 470–477 (2010)
41. Starodubtseva, M.N.: Mechanical properties of cells and ageing. Ageing Res. Rev. **10**(1), 16–25 (2011)
42. Tishler, R., Carlson, F.: A study of the dynamic properties of the human red blood cell membrane using quasi-elastic light-scattering spectroscopy. Biophys. J. **65**(6), 2586–2600 (1993)
43. Tseng, Y., Kole, T., Wirtz, D.: Micromechanical mapping of live cells by multiple-particle-tracking microrheology. Biophys. J. **83**(6), 3162–3176 (2002)
44. Zinin, P.V., Allen, J.S.: Deformation of biological cells in the acoustic field of an oscillating bubble. Phys. Rev. E **79**, 021 (2009)

On the Direct Approach in the Theory of Second Gradient Plates

Victor A. Eremeyev and Holm Altenbach

Abstract The aim of the paper is to formulate the two-dimensional governing equations in the theory of elastic second gradient plates, that is plates which constitutive equations include second gradients of strain and/or stress measures. Here we use so-called direct approach to modeling of plates and shells. According to the approach a plate is considered as a material deformed surface and all equations are written as for two-dimensional continuum. Here we use the six-parameter theory of shells.Within the framework of the six-parameter theory the kinematics of a shell is described by two independent fields of translations and rotations. We introduce the linear constitutive equations of six-parameter second gradient plates. Considering a deflection of a plate we discuss peculiarities of these models.

1 Introduction

Recently the interest to nonlocal models of thin-walled structures grows with respect of investigations such nano-structures as nanofilms, fullerenes, nanotubes, etc., see, for example, [5, 14, 16, 17, 19, 22–24, 26–28] and references therein. In fact, at the micro- and nanoscales the nonlocal interactions are more important than at the macroscale. In particular, nowadays it is clear that many observed unusual properties of nanostructured materials relate with surface phenomena. After Mindlin [20] the

V.A. Eremeyev (✉) · H. Altenbach
Lehrstuhl für Technische Mechanik, Institut für Mechanik, Fakultät für Maschinenbau, Otto-von-Guericke-Universität Magdeburg, Universitätsplatz 2, 39106 Magdeburg, Germany
e-mail: victor.eremeyev@ovgu.de; eremeyev.victor@gmail.com

H. Altenbach
e-mail: holm.altenbach@ovgu.de

V.A. Eremeyev
Southern Scientific Center of Russian Academy of Science, South Federal University, Rostov-on-Don, Russia

H. Altenbach and G.I. Mikhasev (eds.), *Shell and Membrane Theories in Mechanics and Biology*, Advanced Structured Materials 45, DOI 10.1007/978-3-319-02535-3_8

second gradient elasticity is widely used for description of surface tension in solids together with Gurtin-Murdoch model of the surface elasticity [13] and others types of boundary reinforcements, see [9, 15, 29]. The enhancements of the theory of plates and shells taking into account surface stresses are discussed in [2, 3] among others.

In this paper we discuss the direct approach to formulate the gradient-type linear theory of plates and shells considering an extension of the non-linear micropolar shell theory. This variant of the shell theory is called six-parameter theory of shells [4, 6, 10, 11, 18] and the reference therein. Within the micropolar shell theory the deformations of the shell are described by two kinematically independent fields of translations and rotations. Each point of the micropolar shell base surface has six degrees of freedom as in the rigid body dynamics. The surface stress and couple stress tensors are used including the drilling moment. The advantages of the six-parameter shell model consist of correct description of kinematics of multifolded shells, of interaction of a shell with a rigid body, etc.

The paper is organized as follows. In Sect. 2 we recall the basic equations of the isotropic micropolat shell theory. The gradient-type constitutive equations are presented in Sect. 3. We use the so-called direct approach. We introduce the constitutive equations for isotropic plates and shells considered as deformable material surfaces that is as for 2D elastic continua. These constitutive equations express linear dependence of two stress tensors on two conjugated strain measures considering the second gradients of strains and/or stress tensors. Finally we discuss the difference between models considering the deflection of a plate in Sect. 4.

2 Governing Equations

The equilibrium of any part of a micropolar shell is represented by the balance of momentum and balance of angular momentum. As a result the equilibrium equations take the form [11]

$$\boldsymbol{\nabla} \cdot \mathbf{T} + \mathbf{f} = \mathbf{0}, \qquad \boldsymbol{\nabla} \cdot \mathbf{M} + \mathbf{T}_\times + \mathbf{c} = \mathbf{0}, \tag{1}$$

where $\boldsymbol{\nabla}$ is the surface nabla-operator, $\mathbf{T}$ and $\mathbf{M}$ are the surface stress and couple stress tensors, respectively, $\mathbf{f}$ and $\mathbf{c}$ are the external surface forces and couples, and $\mathbf{T}_\times$ denotes the vectorial invariant of the second-order tensor $\mathbf{T}$.

Within the framework of the linear micropolar shell theory the surface stress resultants have the following form:

$$\mathbf{T} = T^{\alpha\beta}\mathbf{i}_\alpha \otimes \mathbf{i}_\beta + T^{\alpha 3}\mathbf{i}_\alpha \otimes \mathbf{n}, \qquad \mathbf{M} = M^{\alpha\beta}\mathbf{i}_\alpha \otimes \mathbf{i}_\beta + M_{\alpha 3}\mathbf{i}_\alpha \otimes \mathbf{n} \qquad (\alpha, \beta = 1, 2),$$

where $\mathbf{i}_1, \mathbf{i}_2$ are the base vectors in the tangent space and $\mathbf{n}$ is the unit normal vector to the shell base surface, see Fig. 1.

Fig. 1 Micropolar shell

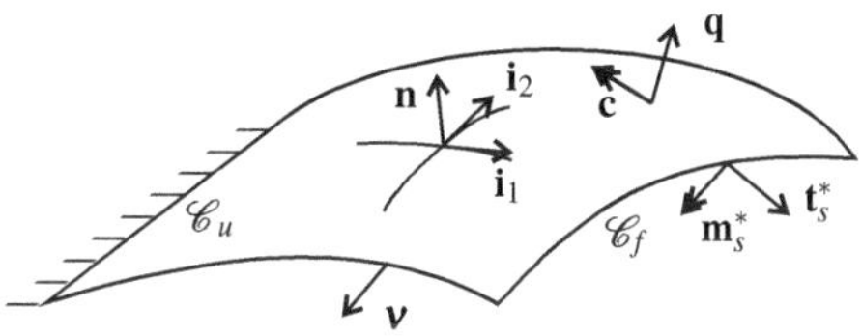

In Cartesian coordinates Eq. (1) takes the following form

$$\begin{array}{ll} T_{11,1}+T_{21,2}+q_1=0, & M_{11,1}+M_{21,2}+T_{23}+c_1 \qquad =0,\\ T_{12,1}+T_{22,2}+q_2=0, & M_{12,1}+M_{22,2}-T_{13}+c_2 \qquad =0,\\ T_{13,1}+T_{23,2}+q_3=0, & M_{13,1}+M_{23,2}+T_{12}-T_{21}+c_3=0. \end{array} \tag{2}$$

The kinematics of the shell is described by two surface fields. The first one is the vector of translations **u** while the second field is the rotation vector **w**. The nonsymmetric linear strain measures are

$$\mathbf{e}=\nabla\mathbf{u}+\mathbf{A}\times\mathbf{w}, \qquad \mathbf{k}=\nabla\mathbf{w} \qquad (\mathbf{A}\equiv\mathbf{I}-\mathbf{n}\otimes\mathbf{n}),$$

where **I** is the unit tensor.

The static and kinematic boundary conditions take the form

$$\boldsymbol{\nu}\cdot\mathrm{T}=\mathbf{t}_s^*, \quad \boldsymbol{\nu}\cdot\mathrm{M}=\mathbf{m}_s^* \quad \text{along } \mathscr{C}_f, \qquad \mathbf{v}=\mathbf{v}^0, \quad \mathbf{w}=\mathbf{w}^0 \ \text{ along } \mathscr{C}_u. \tag{3}$$

For the isotropic shell the surface strain energy density is given by

$$\begin{aligned} 2W &= \alpha_1\mathrm{tr}^2\mathrm{e}_{\|}+\alpha_2\mathrm{tr}\,\mathrm{e}_{\|}^2+\alpha_3\mathrm{tr}\left(\mathrm{e}_{\|}\cdot\mathrm{e}_{\|}^T\right)+\alpha_4\mathbf{n}\cdot\mathrm{e}^T\cdot\mathrm{e}\cdot\mathbf{n}\\ &\quad+\beta_1\mathrm{tr}^2\mathrm{k}_{\|}+\beta_2\mathrm{tr}\,\mathrm{k}_{\|}^2+\beta_3\mathrm{tr}\left(\mathrm{k}_{\|}\cdot\mathrm{k}_{\|}^T\right)+\beta_4\mathbf{n}\cdot\mathrm{k}^T\cdot\mathrm{k}\cdot\mathbf{n}. \end{aligned} \tag{4}$$

Here $\mathrm{e}_{\|}=\mathrm{e}\cdot\mathrm{A}$, $\mathrm{k}_{\|}=\mathrm{k}\cdot\mathrm{A}$, and α_i, β_i are the elastic parameters, $i=1,2,3,4$, which satisfy the inequalities

$$\begin{array}{l} 2\alpha_1+\alpha_2+\alpha_3>0,\ \ \alpha_2+\alpha_3>0,\ \ \alpha_3-\alpha_2>0,\ \ \alpha_4>0,\\ 2\beta_1+\beta_2+\beta_3>0,\ \ \beta_2+\beta_3>0,\ \ \beta_3-\beta_2>0,\ \ \beta_4>0. \end{array} \tag{5}$$

The corresponding surface stress and couple stress tensors are given by

$$\begin{aligned} \mathrm{T}\equiv\frac{\partial W}{\partial \mathrm{e}} &= \alpha_1\mathrm{A}\mathrm{tr}\,\mathrm{e}_{\|}+\alpha_2\mathrm{e}_{\|}^T+\alpha_3\mathrm{e}_{\|}+\alpha_4\mathrm{e}\cdot\mathbf{n}\otimes\mathbf{n},\\ \mathrm{M}\equiv\frac{\partial W}{\partial \mathrm{k}} &= \beta_1\mathrm{A}\mathrm{tr}\,\mathrm{k}_{\|}+\beta_2\mathrm{k}_{\|}^T+\beta_3\mathrm{k}_{\|}+\beta_4\mathrm{k}\cdot\mathbf{n}\otimes\mathbf{n}. \end{aligned} \tag{6}$$

In what follows we discuss the extension of the constitutive equations (6) considering second gradients of strains and/or stress measures.

3 Constitutive Equations of Gradient Theory of Micropolar Plates and Shells

The 3D linear gradient elasticity is based on constitutive equations of the following form, see [1, 12, 21] for details,

$$\boldsymbol{\sigma} - \ell^2 \tilde{\Delta} \boldsymbol{\sigma} = \mathrm{C} : \boldsymbol{\varepsilon}, \tag{7}$$

with $2\boldsymbol{\varepsilon} = \tilde{\boldsymbol{\nabla}} \mathbf{u} + \tilde{\boldsymbol{\nabla}} \mathbf{u}^T$, $\tilde{\Delta} = \tilde{\boldsymbol{\nabla}} \cdot \tilde{\boldsymbol{\nabla}}$ or

$$\boldsymbol{\sigma} = \mathrm{C} : (\boldsymbol{\varepsilon} - \ell^2 \tilde{\Delta} \boldsymbol{\varepsilon}), \tag{8}$$

where C is the stiffness tensor, $\tilde{\boldsymbol{\nabla}}$ is the 3D nabla-operator, $\boldsymbol{\sigma}$ is the stress tensor, and ℓ is the length-scale parameter. More general constitutive relations are possible, for example, such as

$$\boldsymbol{\sigma} - \ell_1^2 \tilde{\Delta} \boldsymbol{\sigma} = \mathrm{C} : (\boldsymbol{\varepsilon} - \ell_2^2 \tilde{\Delta} \boldsymbol{\varepsilon}) \tag{9}$$

with two length-scale parameters ℓ_1 and ℓ_2, see Aifantis [1].

Ru and Aifantis [25] proved the theorem which gives the simple method for solution of boundary-value problems based on constitutive Eq. (8). The method consists of solution of two problems

$$\tilde{\boldsymbol{\nabla}} \cdot \boldsymbol{\sigma}_l + \mathbf{f} = \mathbf{0}, \tag{10}$$

where $\boldsymbol{\sigma}_l = \mathrm{C} : \boldsymbol{\varepsilon}_l$, $2\boldsymbol{\varepsilon}_l = \tilde{\boldsymbol{\nabla}} \mathbf{u}_l + \tilde{\boldsymbol{\nabla}} \mathbf{u}_l^T$ and

$$\mathbf{u} - \ell^2 \tilde{\Delta} \mathbf{u} = \mathbf{u}_l. \tag{11}$$

For the linear problems the Ru–Aifantis theorem [25] gives a simple method for solution of BVPs for constitutive Eq. (8) which is based on two relatively simple problems. Using the Ru–Aifantis theorem and the through-the-thickness integration procedure described in [6, 18], we obtain the following constitutive equations of the linear isotropic shell:

$$\begin{aligned} \mathbf{T} &= (1 - \ell^2 \Delta)[\alpha_1 \mathbf{A} \mathrm{tr} \mathbf{e}_{\|} + \alpha_2 \mathbf{e}_{\|}^T + \alpha_3 \mathbf{e}_{\|} + \alpha_4 \mathbf{e} \cdot \mathbf{n} \otimes \mathbf{n}], \\ \mathbf{M} &= (1 - \ell^2 \Delta)[\beta_1 \mathbf{A} \mathrm{tr} \mathbf{k}_{\|} + \beta_2 \mathbf{k}_{\|}^T + \beta_3 \mathbf{k}_{\|} + \beta_4 \mathbf{k} \cdot \mathbf{n} \otimes \mathbf{n}], \end{aligned} \tag{12}$$

where

$$\Delta = \boldsymbol{\nabla} \cdot \boldsymbol{\nabla}, \quad \mathbf{e}_{\|} = \mathbf{e} \cdot \mathbf{A}, \quad \mathbf{k}_{\|} = \mathbf{k} \cdot \mathbf{A},$$

α_k and β_k, $k = 1, 2, 3, 4$, are elastic moduli, and ℓ is the length-scale parameter as in [25]. We call (12) the Aifantis-type constitutive equations of shells.

Let us note that direct transition of the Ru–Aifantis theorem to the formulation of 2D resultant equations requires the proper consideration of the boundary conditions on the plate or shell faces. As a result, the 2D elastic parameters may depend on type of the boundary conditions and can be expressed via 3D elastic parameters in more complex way than in the case of Cauchy continuum. In what follows we apply the direct method to formulation of the constitutive equations of gradient type plates and shells. In other words we formulate the constitutive equations as for 2D elastic continua described by two strain measures and two conjugates stress tensors considering the second gradients of strains and/or stress tensors. The corresponding 2D material properties are considered as independent on its 3D counterparts. Extending the model (12) we introduce the following constitutive models.

3.1 Aifantis-Type Equations

The stress resultant tensor **T** is a linear function with respect to $(1-\ell_s^2\Delta)\mathbf{e}$ and the couple stress tensor **M** is a linear function with respect to $(1-\ell_b^2\Delta)\mathbf{k}$

$$\begin{aligned} \mathbf{T} &= (1-\ell_s^2\Delta)[\alpha_1\mathbf{A}\mathrm{tr}\mathbf{e}_\parallel + \alpha_2\mathbf{e}_\parallel^T + \alpha_3\mathbf{e}_\parallel + \alpha_4\mathbf{e}\cdot\mathbf{n}\otimes\mathbf{n}], \\ \mathbf{M} &= (1-\ell_b^2\Delta)[\beta_1\mathbf{A}\mathrm{tr}\mathbf{k}_\parallel + \beta_2\mathbf{k}_\parallel^T + \beta_3\mathbf{k}_\parallel + \beta_4\mathbf{k}\cdot\mathbf{n}\otimes\mathbf{n}]. \end{aligned} \tag{13}$$

3.2 Eringen-Type Equations

Now the gradient is introduced on the left hand side of Eq. (6)

$$\begin{aligned} (1-\ell_s^2\Delta)\mathbf{T} &= \alpha_1\mathbf{A}\mathrm{tr}\mathbf{e}_\parallel + \alpha_2\mathbf{e}_\parallel^T + \alpha_3\mathbf{e}_\parallel + \alpha_4\mathbf{e}\cdot\mathbf{n}\otimes\mathbf{n}, \\ (1-\ell_b^2\Delta)\mathbf{M} &= \beta_1\mathbf{A}\mathrm{tr}\mathbf{k}_\parallel + \beta_2\mathbf{k}_\parallel^T + \beta_3\mathbf{k}_\parallel + \beta_4\mathbf{k}\cdot\mathbf{n}\otimes\mathbf{n}. \end{aligned} \tag{14}$$

Here are introduced two length-scale parameters ℓ_s and ℓ_b. Obviously, using (13) and (14), one can introduce more general equations of strain gradient shell theory. Indeed, combining (13) and (14), we propose the following constitutive equations.

3.3 Generalized Equations

Now the gradient is introduced on both sides of Eq. (6)

$$\begin{aligned} (1-\ell_{1s}^2\Delta)\mathbf{T} &= (1-\ell_{2s}^2\Delta)[\alpha_1\mathbf{A}\mathrm{tr}\mathbf{e}_\parallel + \alpha_2\mathbf{e}_\parallel^T + \alpha_3\mathbf{e}_\parallel + \alpha_4\mathbf{e}\cdot\mathbf{n}\otimes\mathbf{n}], \\ (1-\ell_{1b}^2\Delta)\mathbf{M} &= (1-\ell_{2b}^2\Delta)[\beta_1\mathbf{A}\mathrm{tr}\mathbf{k}_\parallel + \beta_2\mathbf{k}_\parallel^T + \beta_3\mathbf{k}_\parallel + \beta_4\mathbf{k}\cdot\mathbf{n}\otimes\mathbf{n}] \end{aligned} \tag{15}$$

with four length-scale parameters $\ell_{1s}, \ell_{2s}, \ell_{1b}$ and ℓ_{2b}. Obviously, that in addition to the form of differential equations for deflection the crucial point is the corresponding boundary conditions discussed in the case of 3D gradient elasticity, for example, in Aifantis [1], Eringen [12], Mindlin [21], dell'Isola and Seppecher [7], dell'Isola et al. [8].

4 Example

To illustrate the difference between the models we consider bending of a square plate under constant transverse load. For the sake of simplicity let us assume that $\mathbf{f} = q\mathbf{n}$, $\mathbf{c} = \mathbf{0}$, and $\ell_s = \ell_b$. The translation and rotation fields are assumed to be given by

$$\mathbf{u} = u(x, y)\mathbf{n}, \quad \mathbf{w} = w_1(x, y)\mathbf{i}_1 + w_2(x, y)\mathbf{i}_2.$$

Hence, the equilibrium Eq. (2) reduce to three scalar equations

$$M_{11,1}+M_{21,2}+T_{23} = 0, \quad M_{12,1}+M_{22,2}-T_{13}-0, \quad T_{13,1}+T_{23,2}+q = 0. \tag{16}$$

Equation (16) can be transformed as follows

$$M_{12,11} + M_{22,21} - M_{11,12} - M_{21,22} + q = 0. \tag{17}$$

Finally, (17) takes the following form:

1. Non-gradient theory ($\ell = 0$)

$$D\Delta\Delta u = q - \frac{1}{\alpha}\Delta q, \qquad D = \frac{Eh^3}{12(1-\nu^2)}. \tag{18}$$

2. Aifantis-type gradient theory

$$(1 - \ell^2\Delta)D\Delta\Delta u = q - \frac{1}{\alpha}\Delta q. \tag{19}$$

3. Eringen-type gradient theory

$$D\Delta\Delta u = (1 - \ell^2\Delta)\left(q - \frac{1}{\alpha}\Delta q\right). \tag{20}$$

Here α is the parameter related with the transverse shear stiffness. Analysis of solutions of (18)–(20) present the difference between the models.

5 Conclusions

We discussed here the theory of the second-gradient micropolar or 6-parameter plates. As a result,

- the gradient theory of linear isotropic micropolar plates and shells is presented;
- Ru-Aifantis' theorem is used as a special case to derive the form of constitutive equations;
- Eringen-, Aifantis-type and general 2D constitutive equations are proposed for isotropic second-grade plates and shells. These equations represent linear dependence of surface stress and couple stress tensors on conjugated strain measures taking into account the presence of second gradients of strains and stresses.

Acknowledgments The first author was supported by the RFBR with the grant No. 12-01-00038

References

1. Aifantis, E.: Update on a class of gradient theories. Mech. Mater. **35**(3–6), 259–280 (2003)
2. Altenbach, H., Eremeyev, V.A.: On the shell theory on the nanoscale with surface stresses. Int. J. Eng. Sci. **49**(12), 1294–1301 (2011)
3. Altenbach, H., Eremeyev, V.A., Morozov, N.F.: Surface viscoelasticity and effective properties of thin-walled structures at the nanoscale. Int. J. Eng. Sci. **59**, 83–89 (2012)
4. Altenbach, J., Altenbach, H., Eremeyev, V.A.: On generalized Cosserat-type theories of plates and shells. A short review and bibliography. Arch. Appl. Mech. **80**(1), 73–92 (2010)
5. Challamel, N., Ameur, M.: Out-of-plane buckling of microstructured beams: Gradient elasticity approach. J. Eng. Mech. **139**(8), 1036–1046 (2013)
6. Chróścielewski, J., Makowski, J., Pietraszkiewicz, W.: Statics and Dynamics of Multyfolded Shells. Wydawnictwo IPPT PAN, Warszawa, Nonlinear Theory and Finite Elelement Method (in Polish) (2004)
7. dell'Isola, F., Seppecher, P.: The relationship between edge contact forces, double forces and interstitial working allowed by the principle of virtual power. Comptes rendus de l'Académie des sciences Série 2 **321**(8), 303–308 (1995)
8. dell'Isola, F., Seppecher, P., Madeo, A.: How contact interactions may depend on the shape of Cauchy cuts in Nth gradient continua: approach "à la D'Alembert. ZAMP **63**, 1119–1141 (2012)
9. Duan, H.L., Wang, J., Karihaloo, B.L.: Theory of elasticity at the nanoscale. Advances in Applied Mechanics, pp. 1–68. Elsevier, San Diego (2008)
10. Eremeyev, V.A., Zubov, L.M.: Mechanics of Elastic Shells (in Russian). Nauka, Moscow (2008)
11. Eremeyev, V.A., Lebedev, L.P., Altenbach, H.: Foundations of Micropolar Mechanics. SpringerBriefs in Applied Sciences and Technologies. Springer, Heidelberg (2013)
12. Eringen, A.C.: Nonlocal Continuum Field Theories. Springer, New York (2002)
13. Gurtin, M.E., Murdoch, A.I.: A continuum theory of elastic material surfaces. Arch. Ration. Mech. Anal. **57**(4), 291–323 (1975)
14. Ieşan, D.: Deformation of thin chiral plates in strain gradient elasticity. European J. Mech, A. Solids **44**, 212–221 (2014)

15. Javili, A., McBride. A., Steinmann, P.: Thermomechanics of solids with lower-dimensional energetics: On the importance of surface, interface, and curve structures at the nanoscale. A unifying review. Appl. Mech. Rev. **65**:010, 802-1–31 (2012)
16. Lazopoulos, K.: On bending of strain gradient elastic micro-plates. Mech. Res. Commun. **36**(7), 777–783 (2009)
17. Lazopoulos, K., Lazopoulos, A.: Nonlinear strain gradient elastic thin shallow shells. Eur. J. Mech. A. Solids **30**(3), 286–292 (2011)
18. Libai, A., Simmonds, J.G.: The Nonlinear Theory of Elastic Shells, 2nd edn. Cambridge University Press, Cambridge (1998)
19. Mikhasev, G.: On localized modes of free vibrations of single-walled carbon nanotubes embedded in nonhomogeneous elastic medium. ZAMM **94**(1–2), 130–141 (2014)
20. Mindlin, R.D.: Second gradient of strain and surface-tension in linear elasticity. Int. J. Solids Struct. **1**(4), 417–438 (1965)
21. Mindlin, R.D., Eshel, N.N.: On first strain-gradient theories in linear elasticity. Int. J. Solids Struct. **4**(1), 109–124 (1968)
22. Ramezani, S.: A shear deformation micro-plate model based on the most general form of strain gradient elasticity. Int. J. Mech. Sci. **57**(1), 34–42 (2012)
23. Ramezani, S.: Nonlinear vibration analysis of micro-plates based on strain gradient elasticity theory. Nonlinear Dyn. **73**(3), 1399–1421 (2013)
24. Reddy, J., Srinivasa, A., Arbind, A., Khodabakhshi, P.: On gradient elasticity and discrete peridynamics with applications to beams and plates. Adv. Mater. Res. **745**, 145–154 (2013)
25. Ru, C.Q., Aifantis, E.C.: A simple approach to solve boundary-value problems in gradient elasticity. Acta Mechanica **101**(1–4), 59–68 (1993)
26. Serpilli, M., Krasucki, F., Geymonat, G.: An asymptotic strain gradient Reissner-Mindlin plate model. Meccanica **48**(8), 2007–2018 (2013)
27. Tsiatas, G.: A new Kirchhoff plate model based on a modified couple stress theory. Int. J. Solids Struct. **46**(13), 2757–2764 (2009)
28. Wang, B., Zhou, S., Zhao, J., Chen, X.: A size-dependent Kirchhoff micro-plate model based on strain gradient elasticity theory. Eur. J. Mech. A. Solids **30**(4), 517–524 (2011)
29. Wang, J., Huang, Z., Duan, H., Yu, S., Feng, X., Wang, G., Zhang, W., Wang, T.: Surface stress effect in mechanics of nanostructured materials. Acta Mech. Solida Sin. **24**, 52–82 (2011)

A Shell Theory for Carbon Nanotube of Arbitrary Chirality

Antonino Favata and Paolo Podio-Guidugli

Abstract This write-up summarizes the developments in Bajaj et al. [1], Favata and Podio-Guidugli [8, 9], along the lines of the presentation recently given by one of us in the occasion of the international conference 'SMT in MB'. A characterization is given of the mechanical response of the linearly elastic, anisotropic shell we propose to associate to a single-wall carbon nanotube of arbitrary chirality.

1 Introduction

When carbon nanotubes (CNTs) are employed as components of a mechanical nanodevice, they are modeled as elastic beams or shells and their mechanical response is characterized in terms of an as-small-as-possible number of stiffness and inertia parameters. To define and evaluate these parameters in terms of the interaction forces keeping C atoms together in a given cylindrical shape is the common goal of all modelers; a way to achieve it is to to describe the mechanical behavior of CNTs by a *bottom-up method*, which bridges between three different scales:

- the microscopic scale of *molecular mechanics*;
- a mesoscopic scale, at which concepts from *discrete structure mechanics* apply; and
- the macroscopic scale of *continuous structure mechanics*.

(see, e.g., Li and Chou [10], Odegard [12, 13], Shen and Li [15], Wan and Delale [16], and the literature cited therein).

A. Favata (✉)
Department of Civil, Environmental and Mechanical Engineering,
University of Trento, via Mesiano 77, 38123 Trento, Italy
e-mail: antonino.favata@unitn.it

P. Podio-Guidugli
Accademia Nazionale dei Lincei and Department of Mathematics,
University of Rome Tor Vergata, via della Ricerca Scientifica 1, 00133 Rome, Italy
e-mail: ppg@uniroma2.it

H. Altenbach and G.I. Mikhasev (eds.), *Shell and Membrane Theories in Mechanics and Biology*, Advanced Structured Materials 45, DOI 10.1007/978-3-319-02535-3_9

As to what model type to pick at the macroscopic scale, whether beam-like or shell-like, a conclusive evidence is still wanted, for a number of reasons. One of these reasons, perhaps not the least important, was the lack of a fit shell theory.

The mechanics of chiral SWCNTs is not an easy subject. Within the framework of discrete structure mechanics, stick-and-spiral models have been used in Chang and Gao [5] and Shen and Li [15] to determine how the elastic properties of chiral SWCNTs depend on size; closed expressions for chirality and size dependencies of elastic properties of SWCNTs are given in Chang and Gao [5], Chang et al. [6]. The ideas of Chang and Gao [5], Chang et al. [6] are exploited in Chang [4] to adapt Donnell's theory of linearly elastic shells to chiral SWCNTs. Moreover, this time in the context of continuum structure mechanics, chirality-dependent properties are investigated in Ru [14], where a shell model is proposed for SWCNTs of arbitrary chiral angle, whose linearly elastic response is deduced from the orthotropic response to plane-stress states by a procedure involving a small-angle rotation of the coordinate system.

In this paper, we give an abridged exposition of the linearly elastic shell model we developed in a series of papers [1, 8, 9]. In our opinion, our model captures fairly well the main peculiarities of the linear mechanical response of CNTs, *whatever their chirality*; needless to say, nonlinear effects such as those investigated in Zhang et al. [19] are out of the model's reach.

2 Geometry and Classification

Chirality is a geometrical character that influences heavily the mechanical, electrical, and thermal, properties of CNTs, especially when their radius is small; a variety of chirality-dependent mechanical phenomenologies is described in Cao and Chen [2], Chang [3, 4], Chang et al. [7], Liang and Upmanyu [11], Wang et al. [17] and Warner et al. [18].

In imagination, a single-wall carbon nanotube (SWCNT) can be obtained by rolling up into a cylindrical shape a *graphene*—that is, a monolayer flat sheet of graphite—visualized as a two-dimensional Bravais lattice with hexagonal unit cell. There are many ways to roll a graphene up, sorted by introducing a geometrical object, the *chiral vector*:

$$\boldsymbol{\chi} = n\mathbf{a}_1 + m\mathbf{a}_2, \quad n \geq m, \tag{1}$$

where n, m are integers, and $\mathbf{a}_1, \mathbf{a}_2$ are *lattice vectors*, such as those at a mutual angle of $\pi/3$ radians shown in Fig. 1. Once a pair of lattice vectors is chosen, the ordered pair (n, m) specifies the *chirality* of the SWCNT to be, whose axis and minimal length are identified by the *axial vector*

$$\boldsymbol{\tau} = t_1\mathbf{a}_1 + t_2\mathbf{a}_2, \tag{2}$$

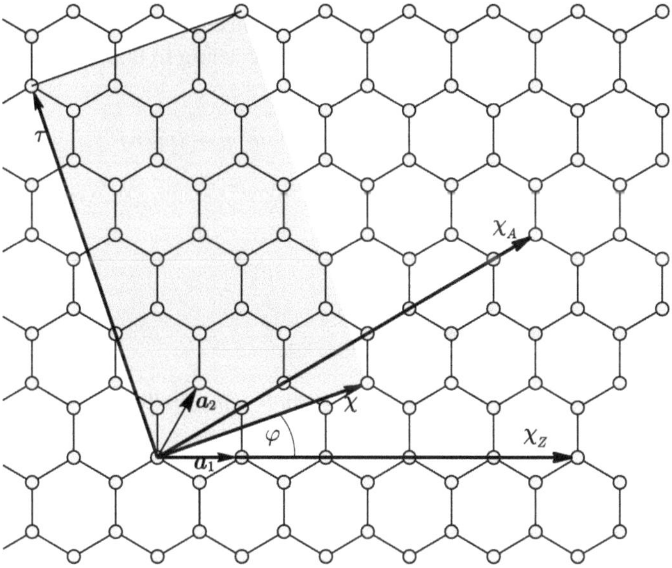

Fig. 1 The graphene part involved in rolling up a (2, 1)-chiral CNT

where t_1 and t_2 are integers such that

$$t_1 = \frac{n + 2m}{d_R}, \qquad t_2 = -\frac{2n + m}{d_R}, \tag{3}$$

with

$$d_R := \gcd(2n + m, n + 2m).$$

It is not difficult to check that $\boldsymbol{\chi} \cdot \boldsymbol{\tau} = 0$, and that the *chiral angle*

$$\varphi =: \arccos \frac{\boldsymbol{\chi} \cdot \mathbf{a}_1}{|\boldsymbol{\chi}|\,|\mathbf{a}_1|}$$

depends on (n, m) as follows:

$$\tan \varphi = \sqrt{3}\,\frac{m}{2n + m}. \tag{4}$$

When $n > m > 0$, the SWCNT under examination is termed *chiral*. The $(n, 0)$- and (n, n)-nanotubes, at times referred to collectively as *achiral*, are termed, respectively, *zigzag* and *armchair*; in Fig. 1, their chiral vectors are denoted by, respectively, $\boldsymbol{\chi}_Z \equiv \mathbf{a}_1$ ($\varphi_Z = 0$ radians) and $\boldsymbol{\chi}_A$ ($\varphi_A = \pi/6$ radians). The *nominal radius* ρ_0 of a (n,m)-SWCNT is defined to be the radius of the cylinder on which the centers of the

C atoms are placed after an ideal rolling-up operation entailing no energy expenditure for the inevitable distortion of the C–C bonds; according to this definition,

$$\rho_0 = \widehat{\rho}_0(n, m) := \frac{\sqrt{3}}{2\pi} n \sqrt{1 + m/n + (m/n)^2}\, s, \tag{5}$$

where s is the length of the C–C bond.

3 Achiral Carbon NanoShells

Our point of departure is the theory of *carbon nanoshells* (CNSs) developed in Bajaj et al. [1], Favata and Podio-Guidugli [8], that is, a macroscopic theory that models achiral SWCNT as linearly elastic shells, whose geometric and constitutive parameters are induced from the available microscopic information. We now give an abridged exposition of that theory, in preparation of a similar exposition of its extension to cover chiral CNSs, to be given in the next section.

3.1 Macroscopic Mechanics

In the majority of papers where CNTs are regarded macroscopically as shells, textbook theories induced from classic three-dimensional *isotropic* elasticity are used; consequently, the shell response is characterized in terms of two elastic moduli. We think it better to view a SWCNT, whatever its chirality, as an *orthotropic* cylindrical shell whose midsurface has a tangent plane coinciding with the orthotropy plane, so that the shell geometry agrees point-wise with the geometry intrinsic to the chosen type of material response (see Fig. 2, where the three little cylinders suggest what probes one should cut out of the shell body in order to determine its material moduli). Moreover, we accept the so-called *Kirchhoff-Love Ansatz*, i.e., we assume that a shell of the type under study cannot change its thickness and that its fibers orthogonal to the referential mid surface remain orthogonal to it after any admissible deformation.

Under our assumptions, the mechanical response of the material comprising the cylindrical shells we treat can be expressed in terms of five elastic moduli: $E_1, E_2, \nu_{12}, \nu_{21}$, and G. The first two are Young-like moduli, the second two are Poisson-like, the fifth is associated with shearing with respect to the $(\mathbf{e}_1, \mathbf{e}_2)$ pair of directions; the first four are not independent, because it turns out that, mandatorily,

$$\frac{E_1}{E_2} = \frac{\nu_{12}}{\nu_{21}}\,. \tag{6}$$

The relative *elasticity tensor*—that is, the fourth-order tensor $\mathbb{C}$ whose linear action on strain $\mathbf{E}$ delivers the accompanying stress $\mathbf{S} = \mathbb{C}[\mathbf{E}]$ – is:

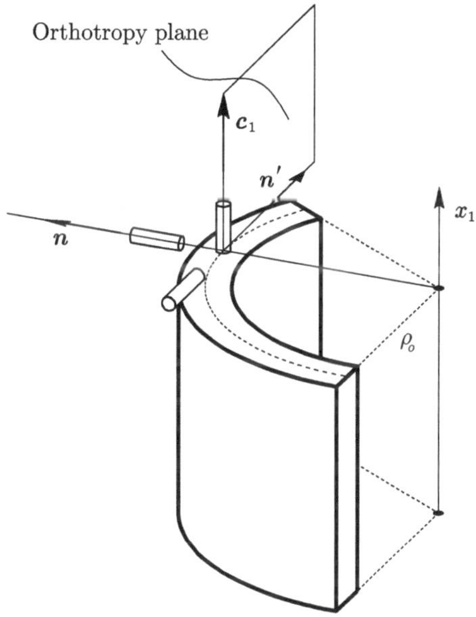

Fig. 2 The geometrical elements relevant to describe the material response of CNTs modeled as shells

$$\mathbb{C} = \Delta^{-1}(E_1\mathbf{W}_1 \otimes \mathbf{W}_1 + E_2\mathbf{W}_2 \otimes \mathbf{W}_2 + 2G\,\Delta\,\mathbf{W}_3 \otimes \mathbf{W}_3 + E_1\nu_{21}(\mathbf{W}_1 \otimes \mathbf{W}_2 + \mathbf{W}_2 \otimes \mathbf{W}_1), \tag{7}$$

where

$$\Delta := 1 - \nu_{12}\,\nu_{21}, \quad \mathbf{W}_\alpha = \mathbf{e}_\alpha \otimes \mathbf{e}_\alpha \ (\alpha \text{ not summed}), \quad \sqrt{2}\,\mathbf{W}_3 = \mathbf{e}_1 \otimes \mathbf{e}_2 + \mathbf{e}_2 \otimes \mathbf{e}_1.$$

We refer the reader to Favata and Podio-Guidugli [8] for the construction of the linearly elastic theory of Kirchhoff-Love shells associated with a given $\mathbb{C}$. As anticipated, we designate such shells by CNSs when the elastic moduli in (7)—that is, the information needed for our modeling of a SWCNT at the macroscopic scale of continuous structure mechanics (CSM)—can be deduced from the available nanoscopic chemical-physical information. In the next three subsections, we build the indispensable bridge between the nanoscopic and macroscopic viewpoints by way of an intermediate passage through a *meta-nanoscopic* viewpoint, in a manner that we now summarize.

At the nanoscopic scale, the bonding and non-bonding interatomic energies are evaluated firstly in the fashion of molecular mechanics (MM). Next, a mechanical caricature of a SWCNT as an orderly arrangement of pin-jointed sticks and (axial and spiral) springs is drawn, within the framework of discrete structure mechanics (DSM). The viewpoints of MM and DSM are connected by equating the energies per bond, for each of which an approximate quadratic expression is accepted: two types of bonding energies—hence two purely constitutive parameters, the stiffness constants of the two types of springs—enter the expression for the DSM energy of a bond unit; they are taken equal to the two corresponding nanoscopic stiffness parameters, regarded as a known input from MM. This simple measure opens the way to the posing and solving, within the framework of DSM, of a number of equilibrium problems for SWCNTs.

3.2 Nanoscopic and Meta-Nanoscopic Mechanics

At nanometer scale, atomic interactions are modeled by either quantum mechanics (QM) or MM; in both cases, one strives to capture how and to what extent the system's energy varies with changes in atomic positions.

QM can describe rigorously the electronic structure of a material complex, but its computational cost quickly becomes prohibitive as the number of atoms involved increases. MM is based on the *Born–Oppenheimer approximation* for the Hamiltonian of a collection of heavy and light particles, a cornerstone of computer simulations. Consistent with this simplifying assumption, the total energy U of such a material complex is expressed as a sum of two terms:

$$U = U_b + U_{nb},$$

where U_b and U_{nb} denote, respectively, the collections of *bonding* and *non-bonding* energies. In principle, the latter term accounts for van der Waals and Coulomb interactions; however, when modeling SWCNT, the contribution of Coulomb interactions are usually neglected, because nonbonding interactions involve essentially neutral carbon atoms. Within the same modeling context, the former term consists of four parts:

$$U_b = U_\rho + U_\theta + U_\omega + U_\tau,$$

where U_ρ, U_θ, U_ω and U_τ denote the energies associated, respectively, with *stretching* (of a covalent bond), *angle variation* (between two covalent bonds), *torsion* (around bonds), and the so-called *improper torsion* (see e.g. [12]). The third and fourth contributions are considered negligible when compared with U_ρ and U_θ, whose harmonic approximations are:

$$U_\rho = \frac{1}{2}\sum k_\rho(\rho - \rho^{\text{ref}})^2,$$
$$U_\theta = \frac{1}{2}\sum k_\theta(\theta - \theta^{\text{ref}})^2.$$

Thus, the only bond-stiffness constants of importance are k_ρ and k_θ; they can be obtained by *ab initio* QM evaluations or fitted to experiments.

We view a meta-nanoscopic mechanical model of a SWCNT as consisting of pin-jointed rigid sticks and linearly elastic springs. There are two types of springs:

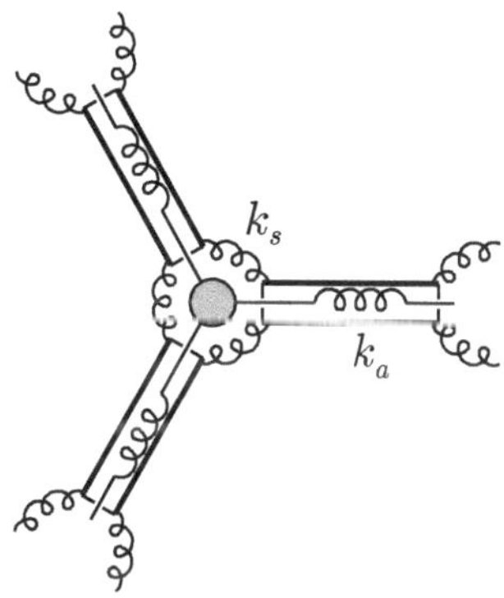

Fig. 3 A stick-and-spring caricature of an achiral basic unit

1. *axial*, sensitive to changes in distance of the two C atoms sitting at the ends of the coaxial stick, of stiffness k_a;
2. *spiral*, sensitive to changes in angle of the two sticks they are attached to, of stiffness k_a. Our mechanical model of an achiral basic unit looks as in Fig. 3.

The bridge between the nanoscopic and meta-nanoscopic scales is achieved by setting $k_a = k_\rho$ and $k_s = k_\vartheta$.

3.3 From Meta-Nanoscopic to Macroscopic Mechanics

To take the step from discrete to continuous structure mechanics, a DSM model of SWCNT has to be assimilated to a suitable model taken from the library of CSM. The main difficulty resides in the number and nature of the parameters needed to specify a CSM model: no matter what rod or shell theory one chooses, the relative stiffness notion consists of a list of more than two combinations of both *constitutive* and *geometric* parameters, the latter reflecting theory-specific concepts of *thinness* and *slenderness*. Now, while there is a natural nanoscopic notion of slenderness for SWCNTs, to assess their thinness is a controversial issue, because a preliminary estimate of their *effective thickness* is required; and, no doubt both slenderness and thinness of a SWCNT depend on its chirality.

No matter whether we model a SWCNT as a shell or as a stick-and-spring complex, we may regard it as a *cylindrical probe*. Two parameters characterize the mechanical response of a probe in an axial traction experiment, and one in torsion; their verbal definitions are, respectively,

$$s_{Ax} := \frac{\text{axial load}}{\text{axial deformation}}, \qquad \nu_{Ax} := -\frac{\text{radial deformation}}{\text{axial deformation}}, \tag{8}$$

and

$$s_{To} = \frac{\text{axial torque}}{\text{axial twist}}; \tag{9}$$

s_{Ax} and s_{To} are stiffness measures, ν_{Ax} is a Poisson-like modulus of transverse contraction. Two additional parameters are needed to characterize the response of a SWCNT to radial loading, namely, the specific stiffness and Poisson-like moduli

$$s_{Rd} := \frac{\text{radial load}}{\text{radial deformation}}, \quad \nu_{Rd} := -\frac{\text{axial deformation}}{\text{radial deformation}}.$$

Mathematical expressions for these five parameters within the frameworks of both continuous and discrete structure mechanics are given in Bajaj et al. [1], for both achiral types of SWCNTs. Here is how.

The CSM expressions involve the five elastic moduli $E_1, E_2, \nu_{12}, \nu_{21}$, and G, as well as two geometric parameters, ε and ρ_o, that is, the shell's thickness and model-surface radius. The DSM expressions depend on the nanoscopic spring-stiffness moduli k_a and k_s, as well as on s, the C–C bond length, and ρ_0, the radius of the cylinder on which the C atoms arrange themselves. Our simple method consists in equating as many as needed of the continuous and discrete expressions (there are more than seven mutually independent such expressions, of which we need only six, in view of (6); anyone of those that are not used can serve to assess the suitability of an obtained six-parameter list). At the end of the day, each of the seven macroscopic parameters can be expressed as a function of the two nanoscopic parameters and of the chirality index n, what makes our shell theory 'nanoscopically informed'.

4 Chiral Carbon NanoShells

When it comes to chiral SWCNTs, the local geometry of the material and the global geometry of the associated cylindrical shell cease to agree, because the orthotropy axes do not coincide anymore with the chiral and roll-up axes. However, the *anisotropic* response of a (n, m)-chiral SWCNT can be induced, alternatively, from the orthotropic response of the corresponding $(n, 0)$-zigzag or (n, n)-armchair SWCNTs.

Let the *tensor product* $\boxtimes$ of two second-order tensors $\mathbf{A}, \mathbf{B}$ be the fourth-order tensor defined as follows by its linear action on the collection of second-order tensors:

$$\mathbf{A} \boxtimes \mathbf{B}[\mathbf{C}] := \mathbf{A}\mathbf{C}\mathbf{B}^T, \quad \text{for each 2nd order tensor } \mathbf{C}.$$

In particular, for $\mathbf{Q}$ an orthogonal tensor, the tensor product

$$\mathbf{Q} \boxtimes \mathbf{Q} =: \mathbb{Q}$$

defines the fourth-order tensor that delivers the *orthogonal conjugate* with respect to $\mathbf{Q}$ of a given second-order tensor $\mathbf{C}$:

$$\mathbb{Q}\mathbf{C} = \mathbf{Q}\mathbf{C}\mathbf{Q}^T.$$

Now, let $\mathbf{Q}$ be a rotation of ψ radians about an axis parallel to $\mathbf{e}_3$:

$$\mathbf{Q} = \widehat{\mathbf{Q}}(\psi, \mathbf{e}_3) := \cos\psi\,(\mathbf{e}_1 \otimes \mathbf{e}_1 + \mathbf{e}_2 \otimes \mathbf{e}_2) - \sin\psi\,(\mathbf{e}_1 \otimes \mathbf{e}_2 - \mathbf{e}_2 \otimes \mathbf{e}_1) + \mathbf{e}_3 \otimes \mathbf{e}_3. \tag{10}$$

For such a rotation, consider the fourth-order tensor

$$\widetilde{\mathbb{C}} := \mathbb{Q}^T \mathbb{C} \mathbb{Q}. \tag{11}$$

Note that the Cartesian components of $\widetilde{\mathbb{C}}$ with respect to the orthonormal frame $(\mathbf{e}_1, \mathbf{e}_2, \mathbf{e}_3)$, namely,

$$\widetilde{\mathbb{C}}_{ijhk} = Q_{li} Q_{mj} Q_{nh} Q_{pk} \mathbb{C}_{lmnp}, \tag{12}$$

are identical to the Cartesian components with the same indices of $\mathbb{C}$ with respect to the orthonormal frame

$$(\widetilde{\mathbf{e}}_1, \widetilde{\mathbf{e}}_2, \mathbf{e}_3), \quad \widetilde{\mathbf{e}}_\alpha = \mathbf{Q}\mathbf{e}_\alpha, \quad (\alpha = 1, 2);$$

for example,

$$\widetilde{\mathbb{C}}_{1111} = \mathbb{C}_{1111} \cos^4\psi + \mathbb{C}_{2222} \sin^4\psi + 2(\mathbb{C}_{1122} + \mathbb{C}_{1212}) \sin^2\psi \cos^2\psi.$$

Chirality enters (11) in two ways: because $\mathbb{C}$ is the elasticity tensor of a $(n, 0)$-zigzag SWCNT, whose representation is given in (7); and because we compose the mapping $\widehat{\mathbf{Q}}(\cdot, \mathbf{e}_3)$ introduced in (10) with the function

$$\psi = \widehat{\psi}(n, m) := \begin{cases} 0, & \text{if } m = 0, \\ \dfrac{\pi}{3} + \varphi(n, m) & \text{if } m \in (0, n], \end{cases} \tag{13}$$

that is to say, in view of (4),

$$\widehat{\psi}(n, m) = \begin{cases} 0, & \text{if } m = 0, \\ \arctan\left(\sqrt{3}\,\dfrac{n + m}{n - m}\right) & \text{if } m \in (0, n]. \end{cases} \tag{14}$$

For

$$\widetilde{\mathbf{Q}}(n, m) := \widehat{\mathbf{Q}}(\widehat{\psi}(n, m), \mathbf{e}_3)$$

the rotation mapping associated to a given (n, m)-chiral SWCNT by way of the composition operation we just mentioned, we get:

- for $m = 0$, ($\psi = \varphi = 0$; $\widetilde{\mathbf{Q}}(n, 0) = \mathbf{I}$, with $\mathbf{I}$ the 2nd order identity tensor; and) $\widetilde{\mathbb{C}} = \mathbb{C}$;

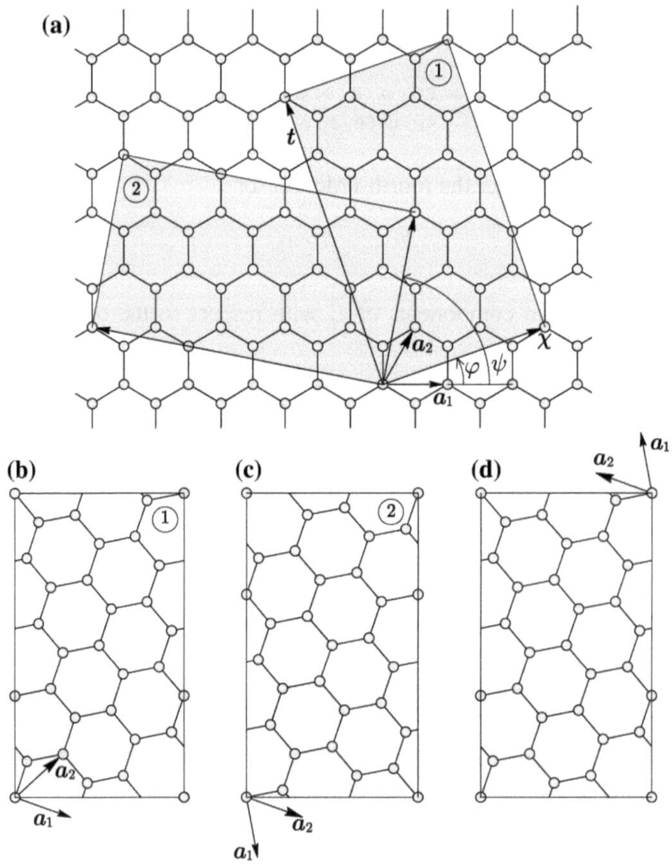

Fig. 4 Visualizing the action of a rotation by ψ about the axis $\mathbf{e}_3$. **a** The graphene portion corresponding to a (2,1)-CNT marked ①, and its rotation of an angle $\pi/3$, marked ②. **b** Rotation of ①. **c** Rotation of ②, so as to have roll-up axes parallel to ① in **b**, **d** rotation by π of ②, making evident that its atomic arrangement is the same as that of ①

- for $n = m$, ($\psi = \pi/2$, $\varphi = \pi/6$, $\widetilde{\mathbf{Q}}(n,n) = \mathbf{e}_2 \otimes \mathbf{e}_1 - \mathbf{e}_1 \otimes \mathbf{e}_2 + \mathbf{e}_3 \otimes \mathbf{e}_3$, and) $\widetilde{\mathbb{C}}$ becomes the elasticity tensor of a $(n,n) =$ armchair SWCNT, for which, according to (12),

$$\widetilde{\mathbb{C}}_{1111} = \mathbb{C}_{2222}, \quad \widetilde{\mathbb{C}}_{2222} = \mathbb{C}_{1111}, \text{ etc.};$$

- for $m \in (0,n)$, ($\psi \in (0,\pi/2)$, $\varphi \in (0,\pi/6)$, and) $\widetilde{\mathbb{C}}$ captures the elastic response for the generic (n,m) chirality.

Figure 4 is meant to help visualizing the action of a rotation by ψ about the axis $\mathbf{e}_3$.

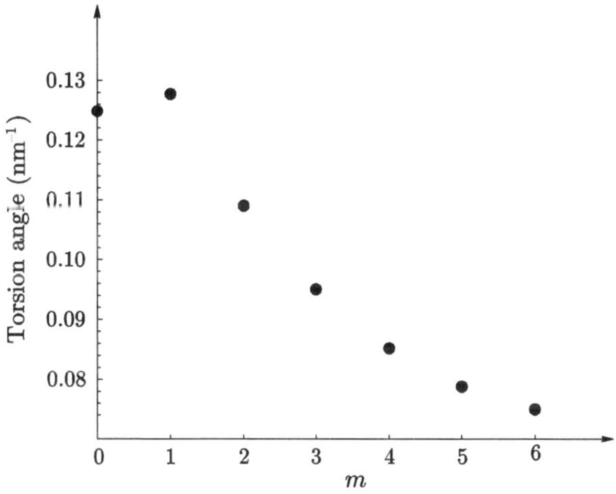

Fig. 5 Torsion angle for $(6, m)$-CNSs

In Fig. 4a, the graphene portion corresponding to a (2, 1)-SWCNT, whose chiral angle is φ, is marked ①; its rotation by an angle $\pi/3$ gives portion ②; the angle ψ is shown. In Fig. 4b, c ① and ② have been rotated so as to have parallel roll-up axes; in Fig. 4d, ② has been further rotated by π radians, so as to make evident the fact that its atomic arrangement is the same as that of ①. But, such a final rotation belongs to the symmetry group of an orthotropic material. Thus, the SWCNTs obtained by rolling up the graphene portions ① and ② depicted in Fig. 4a have the same mechanical response.

5 Some Test Equilibrium Problems

Equilibrium problems for CNSs of arbitrary chirality can be formulated in full generality, as shown in Favata and Podio-Guidugli [9]. Surprisingly enough, a number of *axisymmetric* equilibrium problems can be solved explicitly; some qualitative results related to the torsion problem are shown in the following figures.

Figures 5 and 6 permit to visualize the chirality dependence of, respectively, *torsion angle* and *torsion stiffness*; we see a coherent dependence on chirality of these two quantities, with maximal torsion rigidity achieved by armchair CNSs.

The axial strain accompanying the twist induced by a given torque T—an effect that must be null for both zigzag and armchair CNSs, as physical intuition suggests—is exemplified in Fig. 7; because of the problem symmetries, one would expect the effect to be maximal for intermediate chiralities, and in fact we see that, for n even, a maximum occurs at $m = n/2$.

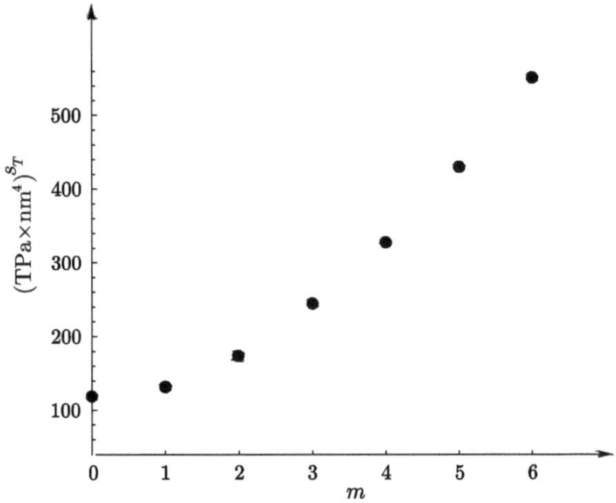

Fig. 6 Torsion stiffness of (6, m)-CNSs

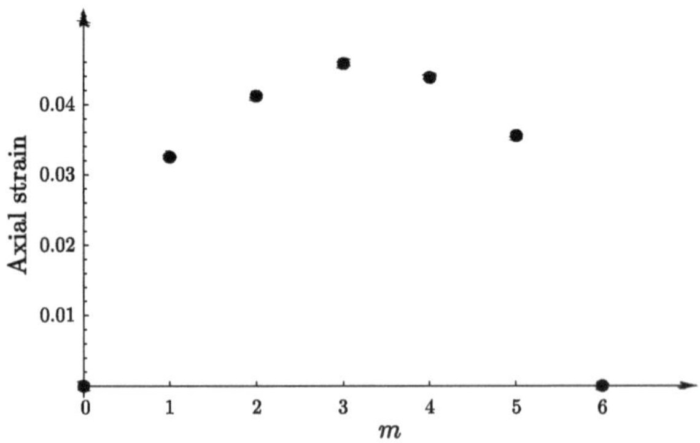

Fig. 7 Axial strain of twisted (6, m)-CNSs

6 Conclusions

We have presented a continuum theory of SWCNTs of arbitrary chirality, modeled as anisotropic linearly elastic shells.

Within our theory, a number of equilibrium problems in terms of displacements can be solved *explicitly and in closed form*–to our knowledge, for the first time. Our model results from a bottom-up sequence of modeling steps, from the atomistic to the continuum scale, at the end of which the model's macroscopic parameters are determined in terms of the available nanoscopic chemical-physical information.

In particular, we have presented our formulation of the torsion problem, when balanced end torques are applied. Our solution accounts both qualitatively and quantitatively for the expected *coupling of torsional and extensional effects* that chirality brings about. Were the relevant experimental results available, it would not be difficult to derive the solutions to other boundary-value problems where analogous couplings are in order, such as the axial traction or extension problems.

References

1. Bajaj, C., Favata, A., Podio-Guidugli, P.: On a nanoscopically-informed shell theory of single-wall carbon nanotubes. Eur. J. Mech. A/Sol. **42**, 137–157 (2013)
2. Cao, G., Chen, X.: The effects of chirality and boundary conditions on the mechanical properties of single-walled carbon nanotubes. Int. J. Sol. Struc. **44**, 5447–5465 (2007)
3. Chang, T.: Torsional behavior of chiral singlewalled carbon nanotubes is loading direction dependent. Appl. Phys. Lett. **90**(201), 910 (2007)
4. Chang, T.: A molecular based anisotropic shell model for single-walled carbon nanotubes. J. Mech. Phys. Sol. **58**, 1422–1433 (2010)
5. Chang, T., Gao, H.: Size-dependent elastic properties of a single nanotube via a molecular model. J. Mech. Phys. Solids. **51**, 1059–1074 (2003)
6. Chang, T., Geng, J., Guo, X.: Chirality- and size-dependent elastic properties of single-walled carbon nanotubes. Appl. Phys. Lett. **87**(251), 929 (2005)
7. Chang, T., Geng, J., Guo, X.: Prediction of chirality- and size-dependent elastic properties of single-walled carbon nanotubes via a molecular mechanics model. Proc. R. Soc. A. **462**, 2523–2540 (2006)
8. Favata, A., Podio-Guidugli, P.: A new CNT-oriented shell theory. Eur. J. Mech. A/Sol. **35**, 75–96 (2012)
9. Favata, A., Podio-Guidugli, P.: A shell theory for chiral single-wall carbon nanotubes. Eur. J. Mech. A/Sol. **45**, 198–210 (2014)
10. Li, C., Chou, T.W.: A structural mechanics approach for the analysis of carbon nanotubes. Int. J. Solids Struct **40**(10), 2487–2499 (2003)
11. Liang, H., Upmanyu, M.: Axial strain-induced torsion in singlewalled carbon nanotubes. Phys. Rev. Lett. **96**(165), 501 (2003)
12. Odegard, G.: Modeling of carbon nanotube/polymer composites. In: Schulz, M., Kelkar, A., Sundaresan, M. (eds.) Nanoengineering of Structural, Functional and Smart Materials. CRC Press, Boca Raton (2005)
13. Odegard, G.: Equivalent-continuum modeling of nanostructured materials. In: Rieth, M., Schommers, W. (eds.) Handbook of Theoretical and Computational Nanotechnology. American Scientific Publishers, Valencia (2006)
14. Ru, C.: Chirality-dependent mechanical behavior of carbon nanotubes based on an anisotropic elastic shell model. Math. Mech. Sol. **14**, 88–101 (2009)
15. Shen, L., Li, J.: Transversely isotropic properties of single-walled carbon nanotube. Phys. Rev. B **69**(045), 414 (2004)
16. Wan, H., Delale, F.: A structural mechanics approach for predicting the mechanical properties of carbon nanotubes. Meccanica **45**, 43–51 (2010)
17. Wang, L., Zheng, Q., Liu, J., Jiang, Q.: Size dependence of the thin-wall models for carbon nanotubes. Phys. Rev. Lett. **95**(105), 501 (2005)
18. Warner, J., Young, N., Kirkland, A., Briggs, G.: Resolving strain in carbon nanotubes at the atomic level. Nat. Mater. **10**, 958–962 (2011)
19. Zhang, Y., Tan, V., Wang, C.: Effect of chirality on buckling behavior of singlewalled carbon nanotubes. J. Appl. Phys. **100**(074), 304 (2006)

Finite Axisymmetric Deformation of an Inflatable Anisotropic Toroidal Membrane

Sergei B. Filippov and Peter E. Tovstik

Abstract Finite axisymmetric deformation of a thin toroidal shell under action of internal pressure is studied. The shell is reinforced by two systems of threads located along parallels and meridians and is considered as anisotropic membrane. The nonlinear theory of membranes is used. To find membrane deformations and displacements the system of ordinary differential equations of the fourth order is delivered. The method of asymptotic integration in the case when the meridian radius is much smaller than the parallel one is elaborated. Asymptotic and numerical results are compared.

1 Introduction

Textile composites and pneumatic structures became increasingly popular in recent years for a variety of applications in civil engineering, architecture, aerospace engineering, hydraulic engineering, etc. [10]. Inflatable membrane structures are extremely light and elegant. Typical examples of inflatable toroidal shells include tires, pneumatic jacks and inflatable pools.

The membrane theory is the particular case of the nonlinear Kirchhoff-Love shell theory presented in Libai [8] and Pietraszkiewicz [11]. The simulation and the design of inflatable membrane structures is normally performed with membrane finite elements [10]. For solution of one-dimensional problems the methods of numerical or asymptotic ODEs integration may be applied.

Deformation of cylindrical and spherical membranes and tent structures is studied in Oñate and Kröplin [10] by means of FEM. In Kolesnikov and Zubov [5] the pure bending deformation of the circular cylindrical membrane are analyzed. The problem

S.B. Filippov (✉) · P.E. Tovstik
Department of Theoretical and Applied Mechanics, St. Petersburg State University, Universitetsky pr. 28, Peterhof 198504, St. Petersburg, Russia
e-mail: s_b_filippov@mail.ru

P.E. Tovstik
e-mail: peter.tovstik@mail.ru

H. Altenbach and G.I. Mikhasev (eds.), *Shell and Membrane Theories in Mechanics and Biology*, Advanced Structured Materials 45, DOI 10.1007/978-3-319-02535-3_10

was reduced to the system of nonlinear ordinary differential equations which was solved numerically.

A toroidal membrane of circular cross section, composed of homogeneous incompressible elastic material and inflated by uniform internal pressure have been considered in Kydoniefs and Spencer [6]. The approximate solutions are based on the assumption that the ratio of radius of the circle which generates the torus to its over-all radius is small. Some numerical results are obtained for the special case of Mooney materials.

Axisymmetric deformation of a pressurized toroidal membrane composed of an isotropic elastic material was studied in Li and Steigmann [7]. The equilibrium equations are reduced to a first order system of ordinary differential equations and solved numerically. In its unstressed state the torus has the form of a cylinder. Wrinkling of the membrane is taken approximately into account by using a relaxed strain energy derived from Ogden's three-term strain energy function. The paper of Jacques and Potier-Ferry [4] gives a overview of wrinkling with applications and extensive bibliographies.

In Tamadapu and DasGupta [13] the finite inflation of a hyperelastic toroidal membrane under internal pressure was studied. The problem is discretized and solved up through a sequence of approximations based on the Ritz expansion of the field variables combined with a potential energy density perturbation and Newton-Raphson method. The stability of the inflated configurations in terms of impending wrinkling of the membrane has been investigated on the principal stretch parameter plane for both isotropic and transversely isotropic material cases.

The dynamics of in-plane surface deformation modes of an inflated toroidal membrane has been studied in Tamadapu and DasGupta [12]. The effect of geometric and material properties on the modal dynamics is analyzed.

In Eriksson and Nordmark [1] the evaluation of quasi-static equilibrium solutions for inflatable space membrane structures is discussed. Analytical instability results are shown for a spherical and derived for a cylindrical case. These are compared to numerical simulations based on a flat linearly interpolated triangular space membrane element.

In this paper the results concisely presented in Filippov [2, 3] are developed. The axisymmetric finite deformation under internal pressure of an anisotropic toroidal membrane is studied. In its unstressed state the membrane has the form of a cylinder. Wrinkling of the membrane is taken into account. As examples, the incompressible Neo-Hookean and linear-elastic materials are considered. The stability of the equilibrium states is not discussed.

2 Main Assumptions

We suppose that the toroidal membrane is made of a cylindrical textile composite pipe which contains two systems of threads located along parallels and meridians. The lengths of non-deformed threads are equal accordingly L and l.

We assume that the distance between the adjacent fibers is small in comparison with the membrane sizes. After averaging we get the 2D medium. We consider this medium as an anisotropic membrane. Connecting end faces of the pipe, we obtain a toroidal membrane.

In the presence of the internal pressure q all fibers oriented along membrane meridians are stretched, but some fibers oriented along parallels are compressed if $q < q_*$, where q_* is the minimal value of pressure at which all parallels are stretched. The membrane can not hold the compression stresses. Therefore in case $q < q_*$ the part of its surface is covered by folds. If the internal pressure q increases then the area covered by folds decreases. At $q \geq q_*$ the whole membrane will be stretched.

3 Basic Equations

For the toroidal membrane the following geometrical relations are valid:

$$\begin{aligned} \lambda_1 &= \frac{ds}{ds_0}, \qquad \lambda_2 = \frac{r}{R}, \qquad R = \frac{L}{2\pi}, \\ \frac{dr}{ds} &= -\sin\theta, \qquad \frac{d\hat{z}}{ds} = \cos\theta, \\ \frac{1}{R_1} &= \frac{d\theta}{ds}, \qquad \frac{1}{R_2} = \frac{\cos\theta}{r}, \end{aligned} \tag{1}$$

where $s_0 \in [0, l]$ and $s(s_0)$ are lengths of the meridian arch before and after deformation, $r(s_0)$ is the radius of a parallel, $\hat{z}(s_0)$ is the height of a parallel above a point O (see Fig. 1), $\lambda_1(s_0)$ and $\lambda_2(s_0)$ are the stretch ratios of meridians and parallels, R_1 and R_2 are the radii of the surface curvature, θ is the angle between tangent to a meridian and the vertical direction.

The equilibrium equations in the projection to the tangent to the meridian and the normal to the membrane are

$$\frac{d(rT_1)}{ds} + T_2 \sin\theta = 0, \qquad \frac{T_1}{R_1} + \frac{T_2}{R_2} = q, \tag{2}$$

where q is the internal pressure, T_1 and T_2 are stress resultants.

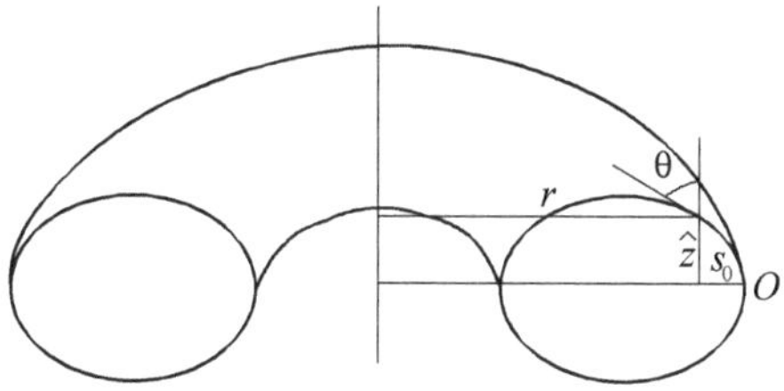

Fig. 1 Toroidal membrane

Assume that the shell and threads are made of incompressible homogeneous isotropic materials with elastic potentials

$$G_i \Phi_i(\lambda_1, \lambda_2, \lambda_3), \quad i = 1, 2, 3, \tag{3}$$

where G_i are constants of the material. For the threads oriented along meridians and parallels the indexes $i = 1$ and $i = 2$ are used accordingly, and index $i = 3$ is used for the shell material. In case of small deformations G_i is the shear modulus.

When the material is incompressible, equality $\lambda_1 \lambda_2 \lambda_3 = 1$ is valid. Besides, for the threads oriented along meridians $\lambda_2 = \lambda_3$. Therefore the potential (3) becomes

$$G_1 \Phi_1 \left(\lambda_1, \lambda_1^{-1/2}, \lambda_1^{-1/2}\right) = G_1 \Psi_1(\lambda_1)$$

For the threads oriented along parallels $\lambda_1 = \lambda_3$ and

$$G_2 \Phi_2 \left(\lambda_2^{-1/2}, \lambda_2, \lambda_2^{-1/2}\right) = G_2 \Psi_2(\lambda_2)$$

The shell potential has the form

$$G_3 \Phi_3 \left(\lambda_1, \lambda_2, (\lambda_1 \lambda_2)^{-1}\right) = G_3 \Psi_3(\lambda_1, \lambda_2)$$

The stress resultants in Eq. (2) are the sums of the tensile thread forces and the stresses arising as result of the shell deformation:

$$\begin{aligned} T_1 &= \frac{G_1 N_1 S_1}{\lambda_2} \frac{d\Psi_1}{d\lambda_1} + \frac{G_3 h_0}{\lambda_2} \frac{\partial \Psi_3}{\partial \lambda_1}, \\ T_2 &= \frac{G_2 N_2 S_2}{\lambda_1} \frac{d\Psi_2}{d\lambda_2} + \frac{G_3 h_0}{\lambda_1} \frac{\partial \Psi_3}{\partial \lambda_2}, \end{aligned} \tag{4}$$

where N_1, N_2 and S_1, S_2 are numbers of threads on unit of length in the cross-sectional direction and the cross-section areas of threads in the state before deformation for meridians and parallels, correspondingly, h_0 is the thickness of the shell before deformation. If Eq. (4) gives $T_2 < 0$ then according to membrane hypotheses it is necessary to put $T_2 = 0$ in system (2).

4 Dimensionless Equations

Let's introduce dimensionless variables by formulas

$$\begin{aligned} &z = \frac{\hat{z}}{R}, \quad s_0 = \rho\varphi, \quad \rho = \frac{l}{2\pi}, \quad \mu = \frac{\rho}{R}, \quad Q = \frac{qR}{G_0 h_0}, \\ &t_1 = \frac{T_1 \lambda_2}{G_0 h_0}, \quad t_2 = \frac{T_2 \lambda_1}{G_0 h_0}, \quad g_i = \frac{G_i N_i S_i}{G_0 h_0}, \quad i = 1, 2, \quad g_3 = \frac{G_3}{G_0}, \end{aligned} \tag{5}$$

where G_0 is an arbitrary constant, $\mu < 0.5$, and the parameters g_i characterize the relative stiffness of threads in comparison with the shell stiffness. By an appropriate choice G_0 any of values g_i can be made equal 1.

Using formulas (1), (2) and (5) we get the following equations for the unknown variables $\theta(\varphi)$, $t_1(\varphi)$, $\lambda_2(\varphi)$ and $z(\varphi)$:

$$\begin{aligned}
\frac{d\theta}{d\varphi} &= \frac{\mu}{t_1}(\lambda_1\lambda_2 Q - \hat{t}_2\cos\theta),\\
\frac{dt_1}{d\varphi} &= -\mu\hat{t}_2\sin\theta,\\
\frac{d\lambda_2}{d\varphi} &= -\mu\lambda_1\sin\theta,\\
\frac{dz}{d\varphi} &= \mu\lambda_1\cos\theta,\\
\hat{t}_2 &= \max\{t_2, 0\}, \quad 0 \le \varphi \le 2\pi.
\end{aligned} \tag{6}$$

As an example consider the potential

$$\Phi(\lambda_1, \lambda_2, \lambda_3) = \frac{1}{2}(\lambda_1^2 + \lambda_2^2 + \lambda_3^2 - 3)$$

corresponding to Neo-Hookean material. Assume that $\Phi_1 = \Phi_2 = \Phi_3 = \Phi$. Then for the dimensionless variables (5) constitutive relations (4) take the form

$$\begin{aligned}
t_1 &= g_3\left(\lambda_1 - \frac{1}{\lambda_1^3\lambda_2^2}\right) + g_1\left(\lambda_1 - \frac{1}{\lambda_1^2}\right),\\
t_2 &= g_3\left(\lambda_2 - \frac{1}{\lambda_1^2\lambda_2^3}\right) + g_2\left(\lambda_2 - \frac{1}{\lambda_2^2}\right)
\end{aligned} \tag{7}$$

If the shell stiffness is very small in comparison with stiffness of strings and strings are composed of linear-elastic material then

$$\begin{aligned}
T_1 &= \frac{E_1N_1S_1}{\lambda_2}\varepsilon_1,\\
T_2 &= \frac{E_2N_2S_2}{\lambda_1}\varepsilon_2,\\
\varepsilon_k &= \lambda_k - 1, \qquad k = 1, 2,
\end{aligned} \tag{8}$$

where E_1 and E_2 are Young's modulus for the threads oriented along meridians and parallels. Then the equations for the unknown variables $\theta(\varphi)$, $\lambda_1(\varphi)$, $\lambda_2(\varphi)$ and $z(\varphi)$ have the following form:

$$
\begin{aligned}
\frac{d\theta}{d\varphi} &= \frac{\mu(\lambda_1\lambda_2 Q - \eta\hat{\varepsilon}_2\cos\theta)}{\lambda_1 - 1}, \\
\frac{d\lambda_1}{d\varphi} &= -\mu\eta\hat{\varepsilon}_2\sin\theta, \\
\frac{d\lambda_2}{d\varphi} &= -\mu\lambda_1\sin\theta, \\
\frac{dz}{d\varphi} &= \mu\lambda_1\cos\theta, \\
\hat{\varepsilon}_2 &= \max\{\varepsilon_2, 0\} = \max\{\lambda_2 - 1, 0\}, \qquad 0 \le \varphi \le 2\pi,
\end{aligned}
\tag{9}
$$

where

$$
\eta = \frac{E_2 N_2 S_2}{E_1 N_1 S_1}
$$

If we take

$$
G_0 = E_1 N_1 S_1 / h_0, \quad g_3 = 0, \quad G_k = E_k/3, \quad \lambda_k - \lambda_k^{-2} = 3\varepsilon_k, \quad k = 1, 2
$$

then $g_1 = 1/3$, $g_2 = \eta/3$, $t_1 = \varepsilon_1$, $\hat{t}_2 = \eta\hat{\varepsilon}_2$. Substituting these expression for t_1 and $\hat{t}_2$ in Eq. (6) we get Eq. (9).

5 Numerical Solution

Owing to the problem symmetry relative to the plane $z = 0$, to find its periodic solution it is enough to construct the solution of equations (6) and (7) satisfying the boundary conditions

$$
\theta(0) = 0, \quad z(0) = 0, \quad \theta(\pi) = \pi, \quad z(\pi) = 0. \tag{10}
$$

To solve the boundary value problem (6, 7, 10) we use the initial-value or shooting procedure, presented in [9]. We set values of λ_2^0 and t_1^0 and obtain the numerical solution of equations (6, 7) satisfying the boundary conditions

$$
\theta(0) = 0, \quad t_1(0) = t_1^0, \quad \lambda_2(0) = \lambda_2^0, \quad z(0) = 0
$$

Then we find λ_2^0 and t_1^0 by solving numerically equations $\theta(\pi) = \pi$, $z(\pi) = 0$.

Consider three cases:

$$
(1)\ g_1 = g_2 = 1,\ g_3 = 0; \quad (2)\ g_1 = g_2 = g_3 = 1/2; \quad (3)\ g_1 = g_2 = 0,\ g_3 = 1. \tag{11}
$$

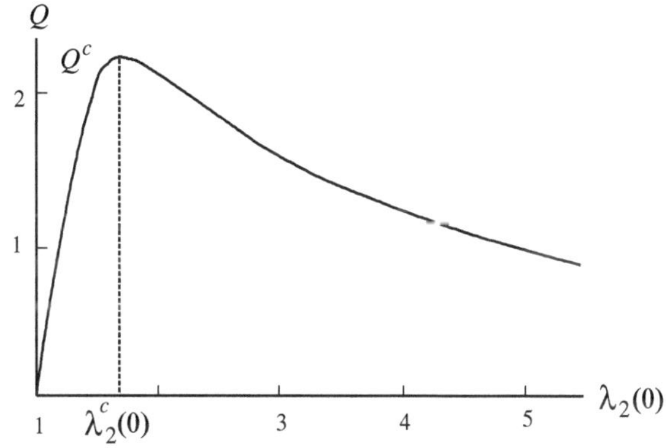

Fig. 2 Force-deflection curve in case $g_1 = g_2 = 0$, $g_3 = 1$

Table 1 Deformation of membrane versus the pressure Q for $g_1 = g_2 = g_3 = 1/2$

	Pre-critical deformations				Post-critical deformations			
Q	$t_1(0)$	$\lambda_2(0)$	$\lambda_2(\pi)$	$t_2(\pi)$	$t_1(0)$	$\lambda_2(0)$	$\lambda_2(\pi)$	$t_2(\pi)$
0.4	0.100	1.092	0.513	−3.636	14.62	12.20	5.207	2.189
0.7	0.196	1.148	0.551	−2.944	9.342	6.959	2.977	2.920
1.0	0.308	1.202	0.588	−2.394	5.778	4.852	2.086	1.962
1.5	0.547	1.304	0.652	−1.629	3.683	3.180	1.394	1.108
2.0	0.936	1.468	0.741	−0.862	2.406	2.257	1.038	0.441
2.2	1.318	1.645	0.823	−0.361	1.780	1.883	0.912	0.042
2.223	1.509	1.741	0.862	−0.170	–	–	–	–

In the first case only the threads stiffnesses is taken into account, in the second case it is assumed that the threads stiffnesses are equal to the shell stiffness, and in the third case the threads are absent.

The numerical analysis shows that for all three cases and also for the linear-elastic material the force-deflection curve Q versus $\lambda_2(0)$ has a peak point $(Q^c, \lambda_2^c(0))$, where Q^c is the critical pressure. The force-deflection curve for the case 3) and $\mu = 0.3$ is shown in Fig. 2. Pre-critical and post-critical equilibrium states correspond to $\lambda_2(0) < \lambda_2^c(0)$ and $\lambda_2(0) > \lambda_2^c(0)$.

Table 1 lists values of the stress resultants $t_1(0)$, $t_2(\pi)$ and the stretch ratios of outer and inner parallels $\lambda_2(0)$, $\lambda_2(\pi)$ for $\mu = 0.3$, $g_1 = g_2 = g_3 = 1/2$ and various values of the dimensionless pressure Q. Left and right parts of Table 1 contain the results for pre-critical and post-critical equilibrium states. In the considered case $\lambda_2^c(0) = 1.741$, $Q^c = 2.223$.

6 Approximate Solution for Completely Stretched Membrane

Consider the pre-critical equilibrium state. Assume that $t_2 \geq 0$ for $\varphi \in [0, \pi]$. If the parameter μ is small the functions t_1, t_2, $\lambda_1 - 1$, $\lambda_2 - 1$ and z are also small. Therefore we seek the approximate solution in the following form:

$$t_1 = \mu\alpha, \quad t_2 = \mu\delta, \quad \lambda_1 = 1 + \mu\beta, \quad \lambda_2 = 1 + \mu\gamma, \quad z = \mu\zeta. \tag{12}$$

After introducing the new variables $\alpha, \beta, \gamma, \delta$ and ζ, Eqs. (6) and (7) and boundary conditions (10) take the form

$$\frac{d\alpha}{d\varphi} = -\mu\delta\sin\theta, \quad \frac{d\theta}{d\varphi} = \frac{1}{\alpha}[(1+\mu\beta)(1+\mu\gamma)Q - \mu\delta\cos\theta],$$
$$\frac{d\zeta}{d\varphi} = (1+\mu\beta)\cos\theta, \quad \frac{d\gamma}{d\varphi} = -(1+\mu\beta)\sin\theta, \tag{13}$$

$$\mu\alpha = (g_1+g_3)(1+\mu\beta) - \frac{g_1}{(1+\mu\beta)^2} - \frac{g_3}{(1+\mu\beta)^3(1+\mu\gamma)^2},$$
$$\mu\delta = (g_2+g_3)(1+\mu\gamma) - \frac{g_2}{(1+\mu\gamma)^2} - \frac{g_3}{(1+\mu\gamma)^3(1+\mu\beta)^2}, \tag{14}$$

$$\theta(0) = \zeta(0) = \zeta(\pi) = 0, \quad \theta(\pi) = \pi \tag{15}$$

Let us substitute into Eqs. (13)–(15) the asymptotic expansions

$$\alpha = \alpha_0 + \mu\alpha_1, \quad \beta = \beta_0 + \mu\beta_1, \quad \gamma = \gamma_0 + \mu\gamma_1,$$
$$\delta = \delta_0 + \mu\delta_1, \quad \theta = \theta_0 + \mu\theta_1, \quad \zeta = \zeta_0 + \mu\zeta_1. \tag{16}$$

In the zeroth approximation we get

$$\alpha_0 = Q, \quad \theta_0 = \varphi, \quad \zeta_0 = \sin\varphi, \quad \gamma_0 = \cos\varphi + a_0,$$
$$\beta_0 = A_1 - A_2\gamma_0, \quad \delta_0 = A_3\gamma_0 + A_4, \tag{17}$$

where a_0 is a arbitrary constant which can be found from the construction of the first approximation,

$$A_1 = \frac{Q}{4g_3 + 3g_1}, \quad A_2 = \frac{2g_3A_1}{Q}, \quad A_3 = 4g_3 + 3g_2 - 2g_3A_2, \quad A_4 = QA_2$$

It follows from formulas (5), (12) and (17) that in the zeroth approximation the cross-section of the membrane is the circle of the radius ρ. The distance between the center of this circle and the torus center is $R + a_0\rho$.

The solutions of equations of the first approximation

$$\frac{d\alpha_1}{d\varphi} = -\delta_0 \sin\varphi, \quad \frac{d\theta_1}{d\varphi} = \beta_0 + \gamma_0 - \frac{\alpha_1}{Q} - \frac{\delta_0}{Q}\cos\varphi,$$
$$\frac{d\zeta_1}{d\varphi} = \beta_0 \cos\varphi - \theta_1 \sin\varphi, \quad \frac{d\gamma_1}{d\varphi} = -\beta_0 \sin\varphi - \theta_1 \cos\varphi \tag{18}$$

satisfy the boundary conditions

$$\theta_1(0) = \zeta_1(0) = \theta_1(\pi) = \zeta_1(\pi) = 0 \tag{19}$$

The first Eq. (18) has the solution

$$\alpha_1 = (A_3 a_0 + A_4)\cos\varphi + \frac{A_3}{2}\cos^2\varphi + a_1 Q, \tag{20}$$

where a_1 is an arbitrary constant. Substitution expressions (20) into the second equation (18) and taking first condition (19) into account leads to

$$\theta_1 = (A - 3c)\varphi + B\sin\varphi - \frac{c}{2}\sin 2\varphi,$$

where

$$A = A_1 + a_0(1 - A_2) - a_1, \quad B = 1 - A_2 - \frac{2(a_0 A_3 + A_4)}{Q}, \quad c = \frac{A_3}{4Q}$$

Equality $\theta_1(\pi) = 0$ holds if

$$A = 3c \tag{21}$$

The solution of the third Eq. (18) satisfying the boundary condition $\zeta_1(0) = 0$ has the form

$$\zeta_1 = -\frac{1}{2}(B + A_2)\varphi + (A_1 - a_0 A_2)\sin\varphi + \frac{B - A_2}{4}\sin 2\varphi + c\sin^3\varphi \tag{22}$$

Taking into account the condition $\zeta_1(\pi) = 0$ we get $B + A_2 = 0$. Hence

$$a_0 = \frac{Q - 2A_4}{2A_3}$$

Substitution of the expression for a_0 in formula (21) allows one to find a_1

$$a_1 = A_1 + \frac{(1 - A_2)(Q - 2A_4)}{2A_3} - 3c$$

Table 2 Values of Q_* for three cases

Case	Q_*		
	$\mu = 0.1$	$\mu = 0.01$	Asymptotics
$g_1 = g_2 = 1,\ g_3 = 0$	4.12	5.71	6.00
$g_1 = g_2 = g_3 = 1/2$	4.59	6.15	6.43
$g_1 = g_2 = 9,\ g_3 = 1$	4.71	5.78	6.00

The function γ_1 contains the constant term a_2 which can be found from the construction of the second approximation.

The condition $t_2 \geq 0$ at $\varphi \in [0, \pi]$ is necessary for correctness of the obtained solution. After substituting into the inequality $t_2 \geq 0$ the approximate expression $t_2 \simeq \mu\delta_0$, we get $Q \geq Q_0$, where $Q_0 = 2A_3$. The value Q_* for which by means of the asymptotic method the approximate expression Q_0 is obtained, represents a characteristic value of dimensionless pressure Q. In case of $Q < Q_*$ the part of the membrane is covered by folds, and when $Q > Q_*$ the membrane is completely stretched.

Table 2 lists the values of the dimensionless characteristic pressure Q_* for three cases (11). The second and third columns contain the values calculated by numerical solution of equations (6, 7) for $\mu = 0.1$ and $\mu = 0.01$. The last column contains the values $Q_0 = 2A_3$ found by the asymptotic approach. The error of the asymptotic formula $Q_* \simeq Q_0$ decreases with the parameter μ.

In the case $g_3 = 0$ and linear-elastic material we may seek the approximate solution of equations (9) with boundary conditions (10) in the form

$$\lambda_1 = 1 + \mu\beta_0 + \mu^2\beta_1 + \mu^3\beta_2,\ \lambda_2 = 1 + \mu\gamma_0 + \mu^2\gamma_1 + \mu^3\gamma_2,$$
$$z = \mu\zeta_0 + \mu^2\zeta_1 + \mu^3\zeta_2,\ \theta = \theta_0 + \mu\theta_1 + \mu^2\theta_2,\ Q_* = Q_0 + \mu Q_1. \quad (23)$$

In the considered case

$$\lambda_k - \lambda_k^{-2} = 3\varepsilon_k + O(\mu^2), \quad k = 1, 2,$$

therefore equations, boundary conditions and solutions in the zeroth and in the first approximations coincide with (17)–(23) if we choose

$$g_1 = 1/3, \quad g_2 = \eta/3, \quad A_1 = Q, \quad A_3 = \eta, \quad A_2 = A_4 = 0.$$

In particular,

$$Q_0 = 2\eta.$$

Hence the approximate value Q_0 of the dimensionless characteristic pressure Q_* is proportional to the non-dimensional stiffness η of the threads oriented along parallels.

The asymptotic formula $Q_0 = 2\eta$ for $\eta = 1$ gives $Q_* \simeq Q_0 = 2$. Solving numerically Eq. (9) with boundary conditions (10) for $\eta = 1$ and $\mu = 0.1$ we get $Q_* = 1.67$. The error of asymptotic result is 20 %, therefore it makes sense to construct the second approximation.

Substituting the asymptotic expansions (23) into (9), (10) we get

$$\lambda_2 = 1 + \mu\left(\frac{Q}{2\eta} + \cos\varphi\right) + \mu^2\left(\frac{Q^2}{\eta} + Q\cos\varphi - \frac{\eta}{4Q}\cos^2\varphi\right) \quad (24)$$

It follows from the condition $t_2 = \eta\varepsilon_2 = \eta(\lambda_2 - 1) \geq 0$ at $\varphi \in [0, \pi]$ and formula (24) that

$$f(\varphi) = -\mu\eta^2\cos^3\varphi + 4qQ\eta(1 + \mu Q)\cos\varphi + 2Q^2(1 + 2\mu Q) \geq 0, \quad \varphi \in [0, \pi]$$

The last inequality is valid if

$$\min_{\varphi\in[0,\pi]} f(\varphi) \geq 0$$

The equation

$$\frac{df}{d\varphi} = 3\mu\eta^2\cos^2\varphi\sin\varphi - 4Q\eta(1 + \mu Q)\sin\varphi = 0$$

has the roots satisfying two equations

$$\sin\varphi = 0, \quad \cos^2\varphi = \frac{4Q\eta(1 + \mu Q)}{3\mu\eta^2}$$

Assume that $Q \sim \eta \sim 1$. Then for small μ the second equation does not have solutions and

$$\min_{\varphi\in[0,\pi]} f(\varphi) = f(\pi) = \mu\eta^2 - 4qQ\eta(1 + \mu Q) + 2Q^2(1 + 2\mu Q) \geq 0$$

The approximate expression $Q_0 + \mu Q_1$ for the dimensionless characteristic pressure Q_* is a root of the equation

$$4\mu Q^3 + (2 - 4\mu\eta)Q^2 - 4\eta Q + \mu\eta^2 = 0$$

Substituting $Q_0 + \mu Q_1$ into the last equation and equating the coefficients of the same power of the small parameter μ to zero, we find

$$Q_0 = 2\eta, \quad Q_1 = -\eta(4\eta + 1/4)$$

Table 3 lists the values of the dimensionless characteristic pressure Q_* for $\eta = 1$ and various values of the parameter μ. The second and the third columns contain the values Q_0 and $Q_0 + \mu Q_1$ found by the asymptotic approach. In the last column the values Q_* calculated by means of the numerical solution of boundary value problem

Table 3 Values of Q_* versus the parameter μ

μ	Q_0	$Q_0 + \mu Q_1$	Q_*
0.01	2.0	1.97	1.96
0.02	2.0	1.92	1.92
0.05	2.0	1.78	1.72
0.10	2.0	1.58	1.67
0.20	2.0	1.15	1.30

(9), (10) are given. The error of the asymptotic formula $Q_* \simeq Q_0 + \mu Q_1$ for $\mu = 0.1$ is 5 %.

7 Approximate Solution for Partly Stretched Membrane

Assume that for the pre-critical equilibrium state $t_2 > 0$ when $\varphi \in [0, \varphi_*]$ and $t_2 < 0$ when $\varphi \in [\varphi_*, \pi]$. Then in the domain $\varphi \in [0, \varphi_*]$ all fibers oriented along parallels are stretched and the part of the membrane surface $\varphi \in [\varphi_*, \pi]$ is covered by folds. At these assumptions equalities

$$t_2(\varphi_*) = 0, \quad \delta(\varphi_*) = 0 \tag{25}$$

take place.

At $\varphi \in [0, \varphi_*]$ the axisymmetric deformation of the toroidal membrane describe Eqs. (13), (14). The function α', β', γ', θ' and ζ' defined on the interval $\varphi \in [\varphi_*, \pi]$ satisfy the following equations:

$$\begin{aligned} &\frac{d\alpha'}{d\varphi} = 0, \quad \frac{d\theta'}{d\varphi} = \frac{Q}{\alpha'}[(1 + \mu\beta')(1 + \mu\gamma'), \\ &\frac{d\zeta'}{d\varphi} = (1 + \mu\beta') \cos\theta', \quad \frac{d\gamma'}{d\varphi} = -(1 + \mu\beta') \sin\theta' \end{aligned} \tag{26}$$

$$\mu\alpha' = (g_1 + g_3)(1 + \mu\beta') - \frac{g_1}{(1 + \mu\beta')^2} - \frac{g_3}{(1 + \mu\beta')^3(1 + \mu\gamma')^2} \tag{27}$$

We seek the solutions of Eqs. (13), (14) and (26), (27) satisfying the boundary conditions

$$\theta(0) = \zeta(0) = 0, \quad \theta'(\pi) = \pi, \quad \zeta'(\pi) = 0, \tag{28}$$

$$\alpha(\varphi_*) = \alpha'(\varphi_*), \quad \theta(\varphi_*) = \theta'(\varphi_*), \quad \gamma(\varphi_*) = \gamma'(\varphi_*), \quad \zeta(\varphi_*) = \zeta'(\varphi_*), \tag{29}$$

in the form (16). To find the approximate value of the unknown value φ_* we use the asymptotic expansion $\varphi_* = \varphi_0 + \mu\varphi_1$.

In the zeroth approximation we get the same results as in case when the toroidal membrane is completely stretched

$$\alpha_0 = \alpha_0' = Q, \quad \theta_0 = \theta_0' = \varphi, \quad \zeta_0 = \zeta_0' = \sin\varphi,$$
$$\gamma_0 = \gamma'_0 = \cos\varphi + a_0, \quad \beta_0 = \beta_0' = A_1 - A_2\gamma_0.$$

It follows from the second equality (25) that

$$a_0 = -\cos\varphi_0 - \frac{A_4}{A_3}$$

The value φ_0 will be found from the construction of the first approximation.

If $\varphi \leq \varphi_0$ then for the construction of the first approximation we can use Eq. (18). In case $\varphi \geq \varphi_0$ the equations of the first approximation have the form

$$\frac{d\alpha_1'}{d\varphi} = 0, \quad \frac{d\theta_1'}{d\varphi} = \beta_0 + \gamma_0 - \frac{\alpha_1'}{Q},$$
$$\frac{d\zeta_1'}{d\varphi} = \beta_0\cos\varphi - \theta_1'\sin\varphi, \quad \frac{d\gamma'_1}{d\varphi} = -\beta_0\sin\varphi - \theta_1'\cos\varphi. \tag{30}$$

The solutions of equations (18) and (30) satisfy the boundary conditions

$$\theta_1(0) = \zeta_1(0) = \theta_1'(\pi) = \zeta_1'(\pi) = 0, \tag{31}$$

$$\alpha_1(\varphi_0) = \alpha_1'(\varphi_0), \quad \theta_1(\varphi_0) = \theta_1'(\varphi_0), \quad \zeta_1(\varphi_0) = \zeta_1'(\varphi_0), \quad \gamma_1(\varphi_0) = \gamma'_1(\varphi_0) \tag{32}$$

In the considered case formulas (18) and (19) suit for definition of the functions $\alpha_1(\varphi)$ and $\theta_1(\varphi)$. Taking into account the first Eq. (30) and the first condition (32), we get

$$\alpha_1' = \alpha_1(\varphi_0) = -\frac{A_3}{2}\cos^2\varphi_0 + a_1 Q$$

It follows from the second Eq. (30), and from the third condition (31) that

$$\theta_1' = \left(A + 2c\cos^2\varphi_0\right)(\varphi - \pi) + (1 - A_2)\sin\varphi, \tag{33}$$

The substitution of expressions (19) and (33) into the second condition (32) gives the equality

$$(A/c + 2\cos^2\varphi_0)\pi = 3\varphi_0 - 5\sin\varphi_0\cos\varphi_0 + 2\varphi_0\cos^2\varphi_0 \tag{34}$$

Whereas for partly stretched toroidal membrane equality (20) is not valid, the expression for the function $\zeta_1(\varphi)$ differs from the expression obtained in the previous section

$$\zeta_1 = (A_1 - a_0 A_2)\sin\varphi - \frac{1}{2}(B + A_2)\varphi + \frac{B - A_2}{4}\sin 2\varphi + (A - 3c)(\varphi\cos\varphi - \sin\varphi) + c\sin^3\varphi \qquad (35)$$

The solution $\zeta_1'(\varphi)$ of the third Eq. (30) satisfying the last boundary condition (31) has the form

$$\zeta_1' = (A_1 - a_0 A_2)\sin\varphi + \frac{1}{2}(\pi - \varphi) + \frac{1 - 2A_2}{4}\sin 2\varphi + \left(A - 2c\cos^2\varphi_0\right)[(\varphi - \pi)\cos\varphi - \sin\varphi] \qquad (36)$$

Let's substitute relations (35) and (36) into the third condition (32). Then taking into account Eq. (34), after transformations we obtain the following equation for the evaluation of φ_0:

$$\sin\varphi_0 - \varphi_0\cos\varphi_0 = \pi p, \quad p = \frac{Q}{Q_0} \le 1 \qquad (37)$$

If we take $Q_0 = 2\eta$ in (37) we obtain the equation for the case $g_3 = 0$ and the linear-elastic material. Equation (37) has no a closed-form solution, however its root $\varphi_0 \in [0, \pi]$ depends only on one parameter $p \in [0, 1]$. Therefore for estimation of the value of this root it is possible to use Fig. 3.

If the parameter p is small the root φ_0 of equation (37) may be found by the approximate formula

$$\varphi_0 = (3p\pi)^{1/3} \qquad (38)$$

For $p < 0.1$ the relative error of Eq. (38) is less than 3 %.

The boundaries φ_* of the area covered by folds for the case $g_1 = g_2 = g_3 = 1/2$ and for different values of the dimensionless pressure Q are given in Table 4.

Fig. 3 Plot of the function $\varphi_0(p)$

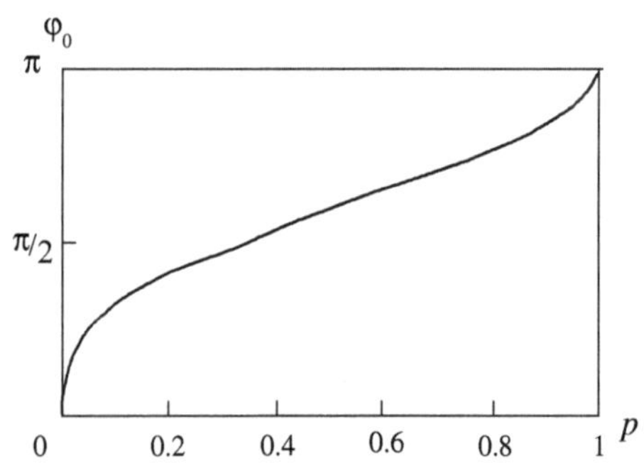

Table 4 Values of φ_* versus Q

	φ_*		
Q	$\mu = 0.1$	$\mu = 0.01$	Asymptotics
1	1.28	1.20	1.19
2	1.67	1.58	1.56
3	2.04	1.87	1.85
4	2.51	2.16	2.12

The second and third columns contain the values obtained by numerical solution of equations (6), (7). The last column contains the root φ_0 of Eq. (37). The error of the asymptotic results decreases with the parameters μ and Q.

8 Approximate Solution for Post-critical Equilibrium States

Numerical results show that for the post-critical equilibrium states the inequality $t_2 > 0$ is valid for all $\varphi \in [0, 2\pi]$, and for small values of μ the functions t_1, t_2, λ_1 and λ_2 are sufficiently large. Assume that $\mu \ll 1$ and substitute into Eqs. (6), (7) and boundary conditions (10) the asymptotic expansions

$$
\begin{aligned}
&t_1 = \alpha_0/\mu + \alpha_1, \quad \lambda_1 = \beta_0/\mu + \beta_1, \quad \lambda_2 = \gamma_0/\mu + \gamma_1, \\
&t_2 = \delta_0/\mu + \delta_1, \quad \theta = \theta_0 + \mu\theta_1, \quad \zeta = \zeta_0 + \mu\zeta_1
\end{aligned} \tag{39}
$$

In the zeroth approximation we get

$$
\alpha_0 = a_0, \quad \gamma_0 = B_1/Q, \quad \theta_0 = \varphi, \quad \zeta_0 = \frac{a_0}{B_1}\sin\varphi, \tag{40}
$$

where $B_1 = g_1 + g_3$. As earlier the arbitrary constant a_0 can be found from the second approximation construction.

The solutions of equations of the first approximation

$$
\begin{aligned}
&\frac{d\alpha_1}{d\varphi} = -B_2\gamma_0\sin\varphi, \quad \frac{d\gamma_1}{d\varphi} = -\frac{\alpha_0}{B_1}\sin\varphi, \\
&\frac{d\theta_1}{d\varphi} = \frac{\gamma_1 Q}{B_1} - \frac{B_2\gamma_0}{a_0}\cos\varphi, \quad \frac{dz_1}{d\varphi} = \frac{\alpha_1}{B_1}\cos\varphi - \frac{a_0}{B_1}\theta_1\sin\varphi,
\end{aligned} \tag{41}
$$

where $B_2 = g_2 + g_3$, satisfy the boundary conditions

$$
\theta_1(0) = 0, \quad \theta_1(\pi) = 0, \quad \zeta_1(0) = 0, \quad z_1(\pi) = 0 \tag{42}
$$

Two first Eq. (41) have the solutions

$$\alpha_1 = B_2\gamma_0 \cos\varphi + a_1, \quad \gamma_1 = \frac{a_0}{B_1}\cos\varphi + b_1, \tag{43}$$

where a_1 and b_1 are arbitrary constants. The substitution of expressions (43) into the third equation (41) in view of the first condition (42) leads to

$$\theta_1 = \frac{Qb_1}{B_1}\varphi + \left(\frac{Qa_0}{B_1^2} - \frac{B_1B_2}{Qa_0}\right)\sin\varphi \tag{44}$$

The second boundary condition (42) is carried out if $b_1 = 0$.

It follows from the fourth Eq. (41) and the third condition (42) that

$$\zeta_1 = \left(\frac{B_2}{Q} - \frac{Qa_0^2}{2B_1^3}\right)\varphi + \frac{a_1}{B_1}\sin\varphi + \frac{Qa_0^2}{4B_1^3}\sin 2\varphi$$

Taking into account fourth boundary condition (42), we obtain

$$a_0 = \frac{B_1}{Q}\sqrt{2B_1B_2}$$

To find a_1 we need to construct the third approximation. Using relations (39), (40) and (43), (44) we get, in particular, the following approximate formulas

$$\lambda_1 = \frac{a_0}{\mu B_1}, \quad \lambda_2 = \frac{B_1}{\mu Q} + \frac{a_0}{B_1}\cos\varphi, \quad \theta = \varphi + \mu\left(\frac{Qa_0}{B_1^2} - \frac{B_1B_2}{Qa_0}\right)\sin\varphi \tag{45}$$

Table 5 lists the values of the stretch ratios of parallels λ_2 for the case $g_1 = g_2 = g_3 = 1/2$, $Q = 1.5$. The second and third columns contain the values of λ_2 found by asymptotic formula (45). The last two columns contain the values calculated by numerical integration. The error of the asymptotic results decreases with the parameter μ and is less than 8 % even for $\mu = 0.3$.

Table 5 The values of $\lambda_2(\pi)$ and $\lambda_2(0)$ versus μ

μ	$\lambda_2(\pi)$	$\lambda_2(0)$	$\lambda_2(\pi)$	$\lambda_2(0)$
0.3	1.28	3.16	1.39	3.18
0.2	2.39	4.28	2.45	4.31
0.1	5.72	7.61	5.75	7.64

9 Conclusions

The application of asymptotic approach to the problem of toroidal membrane deformation under internal pressure permits to obtain the simple approximate solution. In particular, the approximate explicit expression Q_0 for the minimal dimensionless pressure Q_* at which folds on the shell are not formed is found.

In the case when the stiffness of the shell is small and material of threads is linear-elastic, Q_0 is proportional to the ratio η of the stiffnesses of the threads oriented along parallels and meridians. Unfortunately the error of the formula $Q_* \simeq Q_0 = 2\eta$ in comparison with the numerical solution is large even for parameter $\mu = 0.1$. The construction of the next approximation allows one to reduce the error and to expand the domain where the asymptotic formula holds. One can also construct the next approximation for the membrane composed of Neo-Hookean material taking into account the shell stiffness, but the asymptotic formulas become too complicated.

The equation for the boundary of the membrane area covered by folds φ_* is derived. This equation contains only one non-dimensional parameter Q/Q_0. For $\mu = 0.1$ and $Q < 3$ the error of the asymptotic estimate the boundary φ_* in comparison with numerical results is 10 % less.

To find post-critical equilibrium states in the case when the meridian radius is much smaller than the parallel radius the new asymptotic method is developed. The error of the asymptotic results is less than 8 % even for $\mu = 0.3$.

In the numerical solution of boundary value problem (6), (7), (10) the main difficulty consist in the choice of an initial approximation for λ_2^0 and t_1^0. The appropriate approximation may be found using the asymptotic formulas, presented in Sects. 6–8.

Acknowledgments This work was supported by RFBR (grant 13-01-00523) which is gratefully acknowledged.

References

1. Eriksson, A., Nordmark, A.: Instability of thin hyper-elastic space membranes under pressure loads. Comput. Methods Appl. Mech. Eng. **237–240**, 118–129 (2012)
2. Filippov, SB.: Finite post-critical deformation of an inflatable anisotropic toroidal membrane. In: 6th European Congress on Computational Methods in Applied Sciences and Engineering, Book of Abstracts, Vienna, p. 124 (2012)
3. Filippov, SB.: Finite axisymmetric deformation of an inflatable anisotropic toroidal membrane made of Neo-Hookean material. In: Shell and Membrane Theories in Mechanics and Biology: From Macro- to Nanoscale Structures, Proceedings of the International Scientific Conference, Minsk, pp. 22–24 (2013)
4. Jacques, N., Potier-Ferry, M.: On mode localisation in tensile plate buckling. C. R. Mecanique **333**, 804–809 (2005)
5. Kolesnikov, A.M., Zubov, L.M.: Large bending deformations of a cylindrical membrane with internal pressure. Z. Angew. Math. Mech. **89**, 288–305 (2009)
6. Kydoniefs, A.D., Spencer, A.J.M.: The finite inflating of an elastic toroidal membrane of circular cross section. Int. J. Eng. Sci. **5**, 367–391 (1967)

7. Li, X., Steigmann, D.J.: Finite deformation of a pressurized toroidal membrane. Int. J. Non-Linear Mech. **30**, 583–595 (1995)
8. Libai, A., Simmonds, J.G.: The Nonlinear Theory of Elastic Shells, 2nd edn. Cambridge University Press, Cambridge (1998)
9. Ng, B.S., Reid, W.H.: An initial-value method for eigenvalue problems using compound matrices. J. Comp. Phys. **30**, 125–136 (1979)
10. Oñate, E., Kröplin, B. (eds.): Textile Composites and Inflatable Structures II, Computational Methods in Applied Sciences, vol. 8. Springer, Dordrecht (2008)
11. Pietraszkiewicz, W.: Geometrically nonlinear theories of thin elastic shells. Adv. Mech. **12**, 51–130 (1989)
12. Tamadapu, G., DasGupta, A.: In-plane surface modes of an elastic toroidal membrane. Int. J. Eng. Sci. **60**, 25–36 (2012)
13. Tamadapu, G., DasGupta, A.: Finite inflation analysis of a hyperelastic toroidal membrane of initially circular cross-section. J. Nonlinear Mech. **49**, 31–39 (2013)

Simulation of Cardiac Cell-Seeded Membranes Using the Edge-Based Smoothed FEM

Ralf Frotscher, Matthias Goßmann, Hans-Jürgen Raatschen, Ayşegül Temiz-Artmann and Manfred Staat

Abstract We present an electromechanically coupled Finite Element model for cardiac tissue. It bases on the mechanical model for cardiac tissue of Hunter et al. that we couple to the McAllister-Noble-Tsien electrophysiological model of purkinje fibre cells. The corresponding system of ordinary differential equations is implemented on the level of the constitutive equations in a geometrically and physically nonlinear version of the so-called edge-based smoothed FEM for plates. Mechanical material parameters are determined from our own pressure-deflection experimental setup. The main purpose of the model is to further examine the experimental results not only on mechanical but also on electrophysiological level down to ion channel gates. Moreover, we present first drug treatment simulations and validate the model with respect to the experiments.

R. Frotscher · H.-J. Raatschen
Laboratory of Engineering Mechanics and FEM, Aachen University of Applied Sciences, Goethestraße 1, 52064 Aachen, Germany
e-mail: frotscher@fh-aachen.de

H.-J. Raatschen
e-mail: raatschen@fh-aachen.de

M. Goßmann · A. Temiz-Artmann
Laboratory of Medical and Molecular Biology, Institute of Bioengineering, Aachen University of Applied Sciences,
Heinrich-Mußmann-Straße 1, 52428 Jülich, Germany
e-mail: gossmann@fh-aachen.de

A. Temiz-Artmann
e-mail: a.artmann@fh-aachen.de

R. Frotscher · M. Staat (✉)
Biomechanics Lab, Institute of Bioengineering, Aachen University of Applied Sciences, Heinrich-Mußmann-Straße 1, 52428 Jülich, Germany
e-mail: m.staat@fh-aachen.de

R. Frotscher
e-mail: frotscher@fh-aachen.de

H. Altenbach and G.I. Mikhasev (eds.), *Shell and Membrane Theories in Mechanics and Biology*, Advanced Structured Materials 45, DOI 10.1007/978-3-319-02535-3_11

1 Introduction

Today the electrophysiological processes during contraction of myocardial cells are not sufficiently understood. Nevertheless huge efforts in research and in pharmaceutical industry have been made to characterize the effect of medication on those cells with respect to the incomplete knowledge of cellular processes. Therefore, quantitative assessments, selectivity and compatibility of medication needs to be viewed with caution.

In our labs one major challenge is to experimentally determine the effects of drugs on the contractility of myocardial tissue. The so-called CellDrum system, an inflatable thin silicone membrane with cultivated myocardial tissue, provides mechanical quantities that we can investigate and evaluate in order to determine the contractility of cells and the effect of drugs on them. This way we get macroscopic information about the cardiac tissue being investigated. Even more interesting than those information is the question: 'How do the drugs act on the gates that activate and inactivate cellular ion channels?' because this is the level where medication changes the cellular processes. The first simulations of the experiments suggest that we will be able to extract those information in the near future.

This article is about the establishment of a computational electromechanically coupled Finite Element model of cardiac tissue on the CellDrum. It uses the experimental results as input, simulates the contractile behaviour of the myocardial tissue on ion channel level and propagates the action potential to the mechanical (macroscopic) level to get a simulation result that is comparable with the experimental results. In the simulations we are able to investigate the influence of the drugs on ion channel level and therefore we are able to interpret the experimental results in much greater detail.

First, in Sect. 2 we present the plate formulation that we use to model the extremely thin geometry before we explain the so-called edge-based smoothed Finite Element Method that is well-suited especially for biomechanical simulations, in Sect. 3. The experimental setup will be explained in Sect. 4, followed by a description of the electromechanical constitutive model for the cardiac tissue in Sect. 5. We determine the mechanical material parameters ourselves and explain this procedure in Sect. 6. Closing the chapter, in Sect. 7 we present numerical in comparison with experimental results and discuss the model, its benefits and deficiencies as well as currently ongoing developments in Sect. 8.

2 Plate Formulation

In Fig. 1 the basic kinematic quantities of the formulation are shown. The displacement vector $\boldsymbol{u} = [u_1, u_2, u_3]^T$ linearly varies in thickness direction ξ_3 and at any point of the plate is given as

$$\boldsymbol{u} = \boldsymbol{v} + \xi_3 \boldsymbol{w}, \tag{1}$$

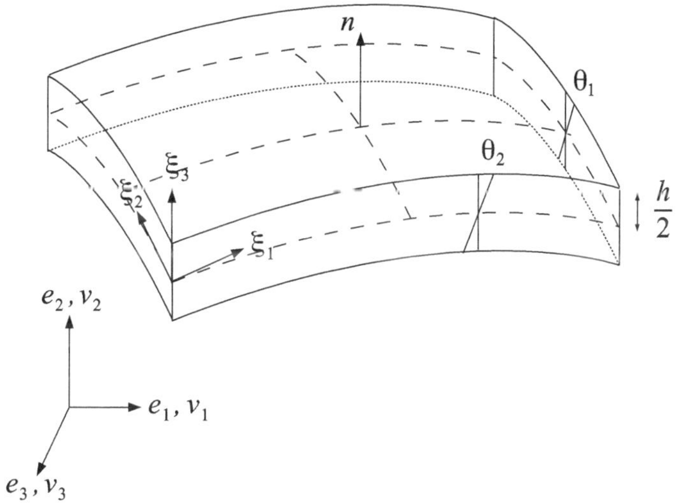

Fig. 1 Global and local coordinate systems $\boldsymbol{e}$ and $\boldsymbol{\xi}$, displacements v_i, rotations w_α and normal $\boldsymbol{n}$

with $\boldsymbol{v} = [v_1, v_2, v_3]^T$ representing the translational and $\boldsymbol{w}$ representing the rotational degrees of freedom. As depicted in Fig. 1, $\boldsymbol{w} = [w_1, w_2]^T = [\theta_2, -\theta_1]^T$, which leads to the detailed description of the displacements in a local coordinate system $\boldsymbol{\xi} = [\xi_1, \xi_2, \xi_3]$ tangential to the plate middle plane,

$$u_1(\xi_1, \xi_2, \xi_3) = v_1(\xi_1, \xi_2) + \xi_3\theta_2(\xi_1, \xi_2) \tag{2}$$

$$u_2(\xi_1, \xi_2, \xi_3) = v_2(\xi_1, \xi_2) - \xi_3\theta_1(\xi_1, \xi_2) \tag{3}$$

$$u_3(\xi_1, \xi_2, \xi_3) = v_3(\xi_1, \xi_2), \tag{4}$$

with the in-plane coordinates ξ_1 and ξ_2, and ξ_3 being the coordinate normal to the plane.

Following the Reissner–Mindlin approach we formulate the strains as follows

$$\boldsymbol{\varepsilon} = (\varepsilon_{11}, \varepsilon_{22}, \gamma_{12}, \gamma_{13}, \gamma_{23})^T = \begin{pmatrix} \boldsymbol{\varepsilon}^m \\ 0 \end{pmatrix} + \begin{pmatrix} \xi_3\boldsymbol{\varepsilon}^b \\ 0 \end{pmatrix} + \begin{pmatrix} 0 \\ \boldsymbol{\varepsilon}^s \end{pmatrix}, \tag{5}$$

consisting of a membrane strain part

$$\boldsymbol{\varepsilon}^m = \boldsymbol{\varepsilon}^l + \boldsymbol{\varepsilon}^{nl} = \begin{pmatrix} v_{1,1} \\ v_{2,2} \\ v_{1,2} + v_{2,1} \end{pmatrix} + \begin{pmatrix} \frac{1}{2}(v_{3,1})^2 \\ \frac{1}{2}(v_{3,2})^2 \\ v_{3,1}v_{3,2} \end{pmatrix}, \tag{6}$$

that in turn can be split into a linear and a nonlinear part. The nonlinear part accounts for the geometrical nonlinearity that arises from the assumption of large displacements. The linear bending and transverse shear parts read as

$$\boldsymbol{\varepsilon}^b = \begin{pmatrix} w_{1,1} \\ w_{2,2} \\ w_{1,2} + w_{2,1} \end{pmatrix} \quad \text{and} \quad \boldsymbol{\varepsilon}^s = \begin{pmatrix} v_{3,1} + w_1 \\ v_{3,2} + w_2 \end{pmatrix}. \tag{7}$$

The stress-strain relationship in general is nonlinear thus with $\sigma_{33} = 0$ it reads as

$$\boldsymbol{\sigma} = \begin{pmatrix} \sigma_{11} \\ \sigma_{22} \\ \sigma_{12} \\ \sigma_{13} \\ \sigma_{23} \end{pmatrix} = \boldsymbol{D}(\boldsymbol{\varepsilon}, \boldsymbol{\alpha}) \begin{pmatrix} \varepsilon_{11} \\ \varepsilon_{22} \\ \gamma_{12} \\ \gamma_{13} \\ \gamma_{23} \end{pmatrix} \tag{8}$$

in Voigt's notation and with $\boldsymbol{D}$ the constitutive tensor depending on the strain and internal variables $\boldsymbol{\alpha}$.

3 Basics of the Edge-Based Smoothed FEM

We use the so-called edge-based Smoothed Finite Element Method (ES-FEM) for the computations because this method shows a very good accuracy when applied to linear triangular elements, it is insensitive to element distortion and it overcomes shear locking naturally. We omit variational details and proofs here which can be found in Liu and Nguyen [1]. Besides ES-FEM there is a large number of different kinds of the smoothed FEM (S-FEM) with different benefits and drawbacks that are explained in detail in Liu and Nguyen [1] too. Moreover there is a number of publications covering different topics concerning S-FEM, like S-FEM for various 2D problems [2–6], its application on plates and shells [7–9], its application to the extended FEM [10–13], S-FEM for 3D problems [14] and many more.

All different kinds of S-FEM share the idea of smoothing the strain over so-called smoothing domains Ω_i^s

$$\bar{\boldsymbol{\varepsilon}}_i = \bar{\boldsymbol{\varepsilon}}(\boldsymbol{x}_i) = \int_{\Omega_i^s} W(\boldsymbol{x}_i - \boldsymbol{x}) \boldsymbol{\varepsilon}(\boldsymbol{x}) d\Omega, \tag{9}$$

with $\boldsymbol{\varepsilon}$ being the usual, possibly known (compatible), strain, W representing a scalar weighting function and $\bar{\boldsymbol{\varepsilon}}_i$ being the smoothed strain in Ω_i^s. Usually W is chosen in the way that the smoothed strain in Eq. (9) becomes an area-weighted average of the strain in the smoothing domain

$$W(\boldsymbol{x}_i - \boldsymbol{x}) = \begin{cases} \dfrac{1}{A_i^s}, & \boldsymbol{x} \in \Omega_i^s \\ 0, & \boldsymbol{x} \notin \Omega_i^s \end{cases}, \tag{10}$$

with A_i^s the area of the ith smoothing domain. Of course we can apply Eq. (9) to each strain part separately

$$\bar{\boldsymbol{\varepsilon}}_i^l = \frac{1}{A_i^s} \int\limits_{\Omega_i^s} \boldsymbol{\varepsilon}_i^l(\boldsymbol{x}) d\Omega, \tag{11}$$

$$\bar{\boldsymbol{\varepsilon}}_i^b = \frac{1}{A_i^s} \int\limits_{\Omega_i^s} \boldsymbol{\varepsilon}_i^b(\boldsymbol{x}) d\Omega, \tag{12}$$

$$\bar{\boldsymbol{\varepsilon}}_i^s = \frac{1}{A_i^s} \int\limits_{\Omega_i^s} \boldsymbol{\varepsilon}_i^s(\boldsymbol{x}) d\Omega, \tag{13}$$

$$\bar{\boldsymbol{\varepsilon}}_i^{nl} = \frac{1}{A_i^s} \int\limits_{\Omega_i^s} \boldsymbol{\varepsilon}_i^{nl}(\boldsymbol{x}) d\Omega, \tag{14}$$

with $\bar{\boldsymbol{\varepsilon}}_i^l, \bar{\boldsymbol{\varepsilon}}_i^b, \bar{\boldsymbol{\varepsilon}}_i^s, \bar{\boldsymbol{\varepsilon}}_i^{nl}$ the smoothed linear membrane, bending, shear and nonlinear strain parts, respectively.

In Eqs. (9) and (11)–(14) we smooth a known strain $\boldsymbol{\varepsilon}$. One way of getting a known strain field is to perform a standard FE computation and smooth the compatible strain afterwards. In this case the S-FEM is boiled down to a simple averaging procedure. As the reader will see below it is more advantageous to avoid the smoothing of the compatible strain and compute the smoothed strain directly instead. Here we derive the discretized equations for the smoothed linear membrane strain only. The derivation for the other smoothed strain parts is straightforward. To get to the discretized equations we first replace the compatible strain by an unknown strain that can be computed using the differential operator $\boldsymbol{L}_d$

$$\bar{\boldsymbol{\varepsilon}}^l(\boldsymbol{x}_i) = \int\limits_{\Omega_i^s} \boldsymbol{L}_d \bar{\boldsymbol{u}}(\boldsymbol{x}) W(\boldsymbol{x}_i - \boldsymbol{x}) d\Omega. \tag{15}$$

Using integration by parts we are able to transform the domain integral in Eq. (15) into an integral over the boundary Γ_i^s of the smoothing domain

$$\bar{\boldsymbol{\varepsilon}}^l(\boldsymbol{x}_i) = \int\limits_{\Gamma_i^s} \boldsymbol{L}_n(\boldsymbol{x}) \bar{\boldsymbol{u}}(\boldsymbol{x}) W(\boldsymbol{x}_i - \boldsymbol{x}) d\Gamma - \int\limits_{\Omega_i^s} \bar{\boldsymbol{u}}(\boldsymbol{x}) \underbrace{\dot{W}(\boldsymbol{x}_i - \boldsymbol{x})}_{=0} d\Omega \tag{16}$$

$$= \frac{1}{A_i^s} \int\limits_{\Gamma_i^s} \underbrace{\begin{bmatrix} n_1 & 0 \\ 0 & n_2 \\ n_2 & n_1 \end{bmatrix}}_{\boldsymbol{L}_n(\boldsymbol{x})} \bar{\boldsymbol{u}}(\boldsymbol{x}) d\Gamma \; \forall \boldsymbol{x} \in \Omega_i^s. \tag{17}$$

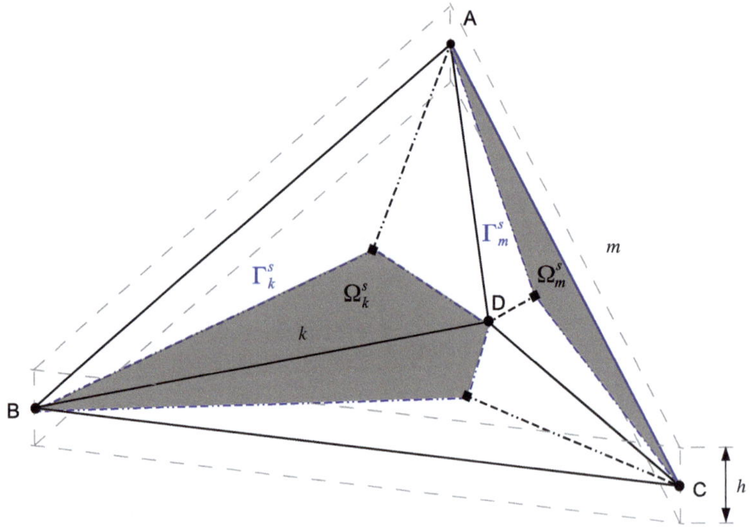

Fig. 2 Creation of smoothing domains

It is apparent that $\boldsymbol{L}_n$ contains the components of the outward normal vector on the boundary of the smoothing domain and the strains are no longer the symmetric part of the displacement gradient. The next step would be to derive the discretized equations but first we need to understand how the smoothing domains are created.

Figure 2 shows the middle plane of a plate that is meshed with three triangular elements based on four nodes A, B, C and D. Two smoothing domains Ω_k^s and Ω_m^s are highlighted here; the former one bases on edge k and the latter one on edge m of the standard finite element mesh. Both are formed using the two corner nodes of the respective edge and the barycenters of the element(s) connected to the respective edge. Please note that the barycenters are virtual nodes thus they do not carry additional degrees of freedom. This way we can fill the whole domain with non-gap and non-overlapping smoothing domains. Contrary to other S-FEM kinds, in the case of the ES-FEM the smoothing domains are created based on the edges of the FE mesh. All the following local computations are performed on each smoothing domain instead of each element.

Besides the fact that the local quantities now lie on smoothing domains the discretization of Eq. (17) looks quite familiar:

$$\bar{\boldsymbol{\varepsilon}}^l(\boldsymbol{x}_i) = \sum_{I=1}^{N_n} \bar{\boldsymbol{B}}_I^l(\boldsymbol{x}_i)\bar{\boldsymbol{d}}_I, \tag{18}$$

with N_n being the number of nodes related to the smoothing domain, $\bar{\boldsymbol{B}}_I^l$ being the respective smoothed strain-displacement matrix and $\bar{\boldsymbol{d}}_I$ being the nodal displacement at node I. Please note that for the computation of the global smoothed displacement field, we employ the standard nodal shape functions $\boldsymbol{\Phi}_I$, thus

$$\bar{\boldsymbol{u}}(\boldsymbol{x}) = \sum_{I=1}^{N} \boldsymbol{\Phi}_I(\boldsymbol{x})\bar{\boldsymbol{d}}_I. \tag{19}$$

N is the total number of nodes in the FE mesh. From Eqs. (17), (18) and (19) it is now possible to construct

$$\bar{\boldsymbol{B}}_I^l = \frac{1}{A_i^s}\int_{\Gamma_i^s} \boldsymbol{L}_n(\boldsymbol{x})\boldsymbol{\Phi}_I(\boldsymbol{x})d\Gamma = \begin{bmatrix} \bar{b}_{I1} & 0 & 0\,0\,0 \\ 0 & \bar{b}_{I2} & 0\,0\,0 \\ \bar{b}_{I2} & \bar{b}_{I1} & 0\,0\,0 \end{bmatrix}, \tag{20}$$

with

$$\bar{\boldsymbol{b}}_I(\boldsymbol{x}_i) = \frac{1}{A_i^s}\int_{\Gamma_i^s} \boldsymbol{\Phi}_I(\boldsymbol{x})\boldsymbol{n}_i(\boldsymbol{x})d\Gamma. \tag{21}$$

Placing Gauss Points in the center of each boundary segment of the smoothing domain (e.g. Γ_k^s and Γ_m^s in Fig. 2) we can discretize the remaining boundary integral to get

$$\bar{\boldsymbol{b}}_I(\boldsymbol{x}_i) = \frac{1}{A_i^s}\sum_{k=1}^{N_s} \boldsymbol{\Phi}_I(\boldsymbol{x}_k^{GP})\boldsymbol{n}_i(\boldsymbol{x}_k^{GP})l_k, \tag{22}$$

with N_s being the number of boundary segments, $\boldsymbol{x}_k^{GP}$ the Gauss Point at the center of segment Γ_k, $\boldsymbol{n}_i$ the outward normal vector and l_k the length of segment Γ_k.

It has been mentioned earlier that computing the smoothed strain from Eq. (18) rather than using a known compatible strain field is beneficial. The first reason is that the computation of $\bar{\boldsymbol{B}}_I$ does not require derivatives of shape functions. The shape functions themselves are sufficient due to the boundary integral formulation. Secondly, with the choice of linear triangular elements we compute all quantities in physical coordinates thus avoid isoparametric mappings.

The given procedure applies to the other parts of $\bar{\boldsymbol{B}}$ too thus the smoothed strain-displacement matrix for smoothing domain Ω_i^s reads as

$$\bar{\boldsymbol{B}}_i = \begin{bmatrix} \bar{\boldsymbol{B}}_i^l + \bar{\boldsymbol{B}}_i^{nl} \\ \bar{\boldsymbol{B}}_i^b \\ \bar{\boldsymbol{B}}_i^s \end{bmatrix}, \tag{23}$$

with $\bar{\boldsymbol{B}}_i = [\bar{\boldsymbol{B}}_1, \ldots, \bar{\boldsymbol{B}}_{N_n}]$. $\bar{\boldsymbol{B}}_I^l$ has already been given in Eq. (20). The other parts of the smoothed strain are

$$\bar{\boldsymbol{B}}_I^b = \begin{bmatrix} 0\,0\,0 & 0 & \bar{b}_{I1} \\ 0\,0\,0 & -\bar{b}_{I2} & 0 \\ 0\,0\,0 & -\bar{b}_{I1} & \bar{b}_{I2} \end{bmatrix}, \tag{24}$$

$$\bar{\boldsymbol{B}}_I^s = \begin{bmatrix} 0\,0\,\bar{b}_{I1} & \frac{1}{N_n}\sum\limits_{j=1}^{N_n} \Phi_I(\boldsymbol{x}_j) & 0 \\ 0\,0\,\bar{b}_{I2} & 0 & \frac{1}{N_n}\sum\limits_{j=1}^{N_n} \Phi_I(\boldsymbol{x}_j) \end{bmatrix}, \tag{25}$$

$$\bar{\boldsymbol{B}}_I^{nl} = \sum_{J=1}^{N_n} \begin{bmatrix} \bar{b}_{J1}\bar{\omega}_J & 0 \\ 0 & \bar{b}_{J2}\bar{\omega}_J \\ \bar{b}_{J2}\bar{\omega}_J & \bar{b}_{J1}\bar{\omega}_J \end{bmatrix} \begin{bmatrix} 0\,0\,\bar{b}_{I1}\,0\,0 \\ 0\,0\,\bar{b}_{I2}\,0\,0 \end{bmatrix}, \tag{26}$$

with $\bar{\omega}_J$ the nodal deflection. The sum in Eq. (26) represents the smoothed displacement gradient. The smoothing of which is similar to that of the strain and can be found in Dai and Liu [3]. Figure 3 shows the different coordinate systems that are needed for the computation besides the global coordinate system. We introduce the element coordinate systems $\hat{\boldsymbol{\xi}}^1$ and $\hat{\boldsymbol{\xi}}^2$ as well as a smoothing domain coordinate system $\bar{\boldsymbol{\xi}}^1$ as shown.

With respect to Fig. 3 we can sum up the computation of $\bar{\boldsymbol{B}}_i$ as follows. We compute both smoothed strain-displacement matrices $\hat{\boldsymbol{B}}_1$ and $\hat{\boldsymbol{B}}_2$ on element level in the respective element coordinate system and transform them into the smoothing domain coordinate system to add up the contributions to $\bar{\boldsymbol{B}}_i$ from both elements. Realizing that the case is even simpler if the smoothing domain is built around a boundary edge of the computation domain (cf. edge m in Fig. 2) the smoothed strain-displacement matrix on smoothing domain Ω_i^s reads as

$$\bar{\boldsymbol{B}}_i^* = \frac{1}{A_i^s} \sum_{k=1}^{N_e} A_k \bar{\boldsymbol{R}}_{*k}^1 \hat{\boldsymbol{R}}_{*k}^2 \hat{\boldsymbol{B}}_k^* \hat{\boldsymbol{T}}^k, \tag{27}$$

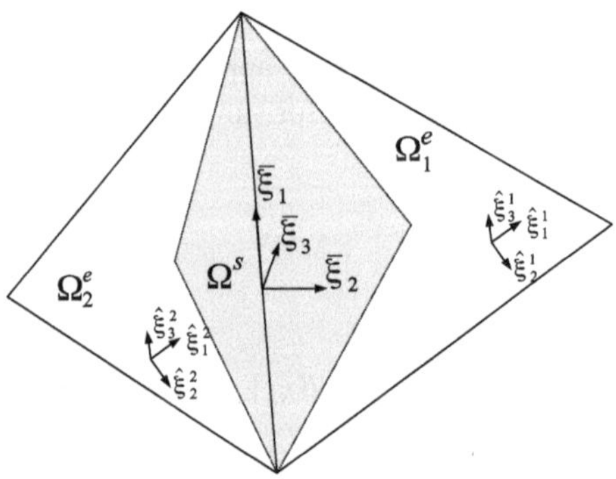

Fig. 3 Element and smoothing domain coordinate systems

with $\hat{\boldsymbol{T}}^k$ being the transformation matrix from global coordinates to element local coordinates, $\hat{\boldsymbol{R}}^2_{*k}$ a rotation matrix from element to global coordinates, $\bar{\boldsymbol{R}}^1_{*k}$ a rotation matrix from global to smoothing domain coordinates, N_e the number of elements around edge i of the FE mesh and the asterisk indicating the applicability to each strain part. For details concerning the transformation and rotation matrices the reader is referred to Cui et al. [7].

We remark that, instead of Eq. (25), we compute the transverse shear part using the so-called *Discrete Shear Gap* method [15] to avoid transverse shear locking. The respective formulae are given in Cui et al. [7].

The global smoothed tangent stiffness matrix in the smoothing domain coordinate system then is given by

$$\bar{\boldsymbol{K}}^t_{IJ} = \sum_{i=1}^{N_s} \left(\bar{\boldsymbol{B}}_i\right)_I^T \boldsymbol{D} \left(\bar{\boldsymbol{B}}_i\right)_J A_i + \sum_{i=1}^{N_s} \left(\bar{\boldsymbol{G}}_i\right)_I^T \breve{\boldsymbol{N}} \left(\bar{\boldsymbol{G}}_i\right)_J A_i, \tag{28}$$

with N_s the number of smoothing domains, i.e. edges in the mesh. The second term in Eq. (28) describes the prestress and consists of

$$\left(\bar{\boldsymbol{G}}_i\right)_I = \begin{bmatrix} 0 & 0 & \bar{b}_{I1} & 0 & 0 \\ 0 & 0 & \bar{b}_{I2} & 0 & 0 \end{bmatrix} \tag{29}$$

and the matrix of the membrane stress resultants

$$\breve{\boldsymbol{N}} = \begin{bmatrix} N_{11} & N_{12} \\ N_{12} & N_{22} \end{bmatrix}. \tag{30}$$

With the right hand side of the equation system also given in coordinate system $\bar{\boldsymbol{\xi}}$ we arrive at

$$\bar{\boldsymbol{K}}^t \boldsymbol{u} = \bar{\boldsymbol{f}}. \tag{31}$$

The assembly process for the stiffness matrix and the internal force vector is similar to the one of standard FEM.

4 Experimental Setup and Methods

Since its invention in 2001 there have been a number of publications concerning the experimental device called CellDrum^{TM}, e.g. Linder et al. [16], Trzewik et al. [17], Trzewik et al. [18]. It consists of a circular, 4 μm thin silicone membrane with a diameter of 1.6 cm = 16,000 μm. On top of the membrane we cultivate and chemically fix myocardiac cells in a collagen monolayer. A culture medium keeps the cells alive and as in the real heart the tissue is autonomously beating. This composite material is clamped in a fixed ring like shown in Fig. 4. The CellDrum^{TM} then can be placed

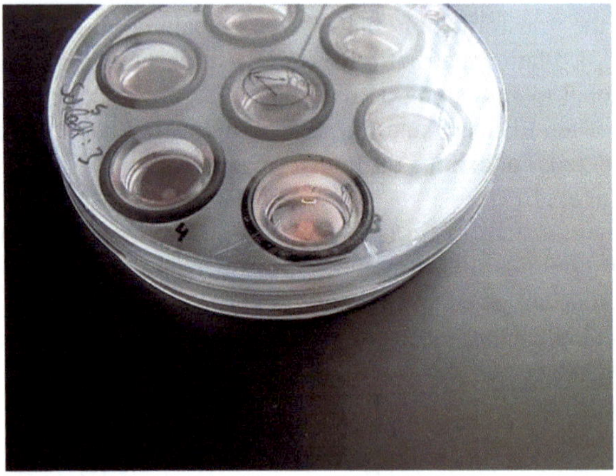

Fig. 4 Seven CellDrums

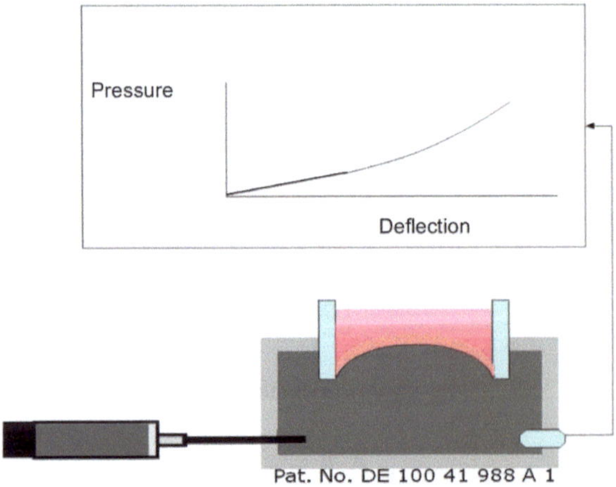

Fig. 5 Schematic drawing of the inflation experiment

into an inflation setup as depicted in Fig. 5. The inflation setup generates a pressure and therefore inflates the composite tissue. The corresponding deflection is measured using a laser sensor resulting in a pressure-deflection curve as shown in the same figure.

Currently the investigations are limited to so-called 2D or monolayered tissue with an approximate thickness of around 4 μm although in general the experimental setup allows for 3D tissue with a thickness of hundreds of microns. Due to the circular shape of the tissue the myocytes do not align in a common preferred direction as can be seen in Fig. 6. That gives rise to the assumption that the tissue is isotropic.

Figure 7 shows the deflection of the composite in time at constant pressure. In this case the cells are beating with a frequency of approximately 1 Hz and during

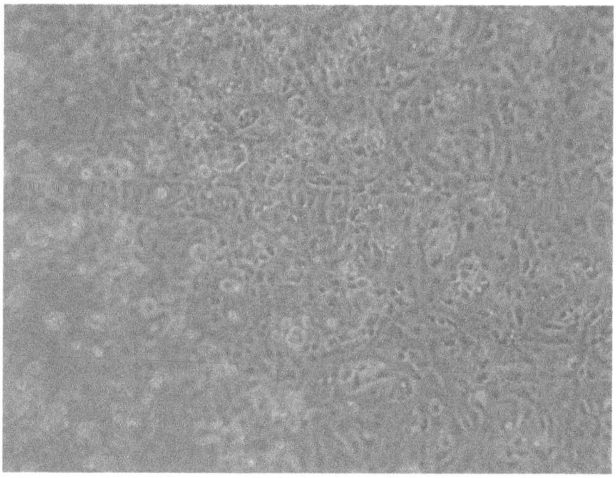

Fig. 6 1 mm × 1 mm microscopic cutout of the tissue showing isotropy

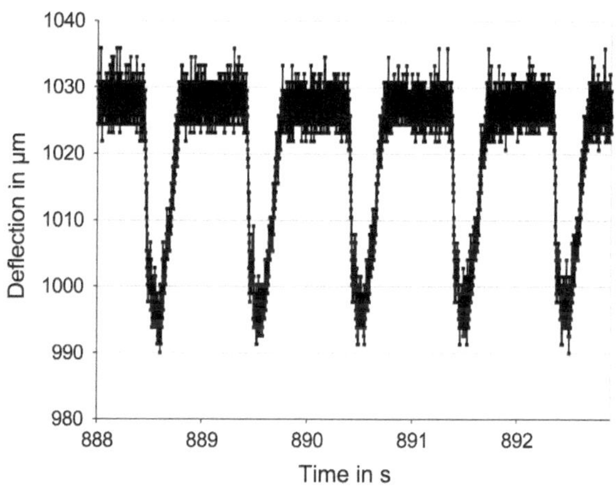

Fig. 7 Experimental data (incl. noise) showing the change in deflection due to contraction

contraction they change the deflection of the composite by approximately 37 μm. Depending on the medication that we apply to the tissue, the change in deflection due to the contraction will be different and the beating frequency might change because the action potential of the cells differs from the normal situation.

The total thickness of a monolayer composite is 8 μm. We are producing the silicone membranes with a thickness of 4 μm. Additionally we determine the average monolayer thickness using a laser scanning microscope (LSM) and find it to be 4 μm.

5 Constitutive Model

In order to appropriately model the passive and active mechanical behaviour of cardiac tissue it is necessary to employ electromechanically coupled models. In this work we combine existing electrophysiological and mechanical models on the constitutive level.

5.1 Mechanical Model

A very widely used mechanical model has been proposed by Hunter et al. [19]. The main idea is to additively split the Cauchy stress $\boldsymbol{\sigma}$ into a passive part $\boldsymbol{\sigma}_p$ and an active part $\boldsymbol{\sigma}_a$

$$\boldsymbol{\sigma} = \boldsymbol{\sigma}_p + \boldsymbol{\sigma}_a = 2J^{-1}\boldsymbol{B}\frac{\partial \Psi}{\partial \boldsymbol{B}} - p\boldsymbol{I} + T(t, \boldsymbol{B})\boldsymbol{a} \otimes \boldsymbol{a}, \tag{32}$$

with Ψ the strain energy function, J the determinant of the deformation gradient, $\boldsymbol{B}$ the left Cauchy-Green tensor, p the hydrostatic pressure, $\boldsymbol{I}$ the identity tensor, T a scalar related to the active stress, t the time and $\boldsymbol{a}$ a vector oriented in the direction of the cardiac muscle fibres. In view of the isotropy shown in Fig. 6 we assume that $\boldsymbol{a} \otimes \boldsymbol{a} = \boldsymbol{I}$. Using the transformation

$$\boldsymbol{S} = J\boldsymbol{F}^{-1}\boldsymbol{\sigma}(\boldsymbol{F}^{-1})^T \tag{33}$$

with the the deformation gradient $\boldsymbol{F}$ we obtain the 2nd Piola-Kirchhoff stress $\boldsymbol{S}$ as

$$\boldsymbol{S} = \boldsymbol{S}_p + \boldsymbol{S}_a = 2\frac{\partial \Psi}{\partial \boldsymbol{C}} - p\boldsymbol{C}^{-1} + T(t, \boldsymbol{C})\boldsymbol{C}^{-1}, \tag{34}$$

in terms of the right Cauchy-Green tensor $\boldsymbol{C}$. The passive component of the stress is not in the focus of this work. Depending on the complexity of the pressure-deflection curves one observes, one might choose a proper quasi-incompressible hyperelastic constitutive law like Neo–Hookean, Mooney-Rivlin or Varga. For a comprehensive overview about this topic the reader is referred to Holzapfel [20].

We follow the approach of Hunter et al. [19] to appropriately model the active contribution to the stress, $T(t, \boldsymbol{C})$. Firstly it can be simplified to $T(t, \lambda_1) = T(t, \lambda_2) = T(t, \lambda)$ because we only work with isotropic circular (2D) monolayers. Using this approximation Hunter et al. establish a tension-length relation

$$T(\lambda) = T_{ref}(1 + \beta_0(\lambda - 1))z, \tag{35}$$

with T_{ref} being the reference tension at $\lambda = 1$, T being the tension,[1] β_0 introducing myofilament cooperativity and $z \in [0, 1]$ being an activation parameter ranging from no activation ($z = 0$) to full activation ($z = 1$). Under steady state conditions, i.e. at a given sarcomere length, $z = z_{SS}$ which is given as

$$z_{SS} = \frac{(Ca_b)^n}{(Ca_b)^n + (Ca_{50})^n}, \tag{36}$$

where Ca_b is the concentration of Calcium bound to so-called troponin C resulting in increased contractility. Ca_{50} is the Calcium needed to have 50 % availability of the muscle ($z_{SS} = 0.5$) and n is the so-called Hill parameter. Both parameters are length dependent and can be computed via

$$n = n_{ref}(1 + \beta_1(\lambda - 1)), \tag{37}$$

$$pCa_{50} = pCa_{50ref}(1 + \beta_2(\lambda - 1)), \tag{38}$$

with $Ca_{50} = 10^{6-pCa_{50}}$. In Eq. (36) Ca_b is still unknown. In the steady state it can be computed using

$$\frac{Ca_b}{Ca_{bmax}} = \frac{Ca_i}{Ca_i + \frac{\rho_1}{\rho_0}\left(1 - \frac{T}{\gamma T_0}\right)}. \tag{39}$$

Herein Ca_i is the free calcium concentration inside the cell. The attachment of inner calcium to the troponin C bindings is governed by a constant rate $\rho_0 = 100\,\mathrm{s}^{-1}\,\mu\mathrm{M}^{-1}$ and ρ_1 governs the detachment of Ca_i from troponin C at zero tension and under several assumptions (cf. [19]) it can be computed to be $\rho_1 = 163\,\mathrm{s}^{-1}$.

Ca_i is the main driver for active tension in the mechanical model. Thus it is important to describe its time course as realistic as possible. Before we come to this point we sum up the material parameters needed for the mechanical part of the model in Table 1.

5.2 Electrophysiological Model

In Hunter et al. [19] the inner calcium concentration is represented via a time-dependent exponential function. This approach does not fit our needs because we want to model drug treatment appropriately including the heart rate (frequency). Drugs act on ion channel level thereby changing the action potential of the specific cell type, namely auto-contractile cardiac cells. Thus we require a system of differential equations representing the ionic currents that diffuse through the cell membrane and that determine Ca_i, rather than computing it explicitly. There are plenty of models for different kinds of cardiac cells available, ranging from simple

[1] More exactly T is the tension in case that the muscle is at rest. Including a history of cross-bridge bindings one could establish an integral equation for the computation of T.

Table 1 Parameter values used in the model

Parameter	Value
Ca_0	0.01 μM
Ca_{max}	1 μM
τ_{Ca}	0.06 s
Ca_{bmax}	2.26 μM
ρ_0	$100\,s^{-1}\,\mu M^{-1}$
ρ_1	$163\,s^{-1}$
γ	2.6
pCa_{50ref}	6.2 μM
n_{ref}	6.9
β_0	1.45
β_1	1.95
β_2	0.31
T_{ref}	0.5808 kPa

models (2 parameters) to very complex models (60 parameters) [21, 22]. All of them are based on the assumption that a limited number of different ions diffuses through respective ion channels in the cell membrane thus creating a cell membrane potential V_m that can be described by an ordinary differential equation

$$\frac{\partial V_m}{\partial t} = \frac{1}{C_m}\left(I_{stim} - \sum_{i=1}^{N} I_i(g_1, g_2, \ldots, g_{M_i})\right). \tag{40}$$

Therein C_m is the electrical capacitance of the cell membrane, I_{stim} is a (potentially zero) electrical stimulus, I_i are ionic currents related to ion i and N is the total number of ionic currents in the model. Each current depends on M_i (usually one or two) gate variables g_j that control the opening and closure of the ionic channel where ion i diffuses through.

The gates themselves are described using decoupled ordinary differential equations

$$\frac{\partial g_j(V_m)}{\partial t} = \alpha_j^+(V_m)(1 - g_j) + \alpha_j^-(V_m)g_j, \tag{41}$$

with α_j^+ and α_j^- being experimentally obtained parameters defining the opening and closure rates of the gate respectively.

We employ the McAllister-Noble-Tsien model [23] which introduces nine different ionic currents and therefore is not too complex to investigate basic properties of the model. It models self-contractile purkinje fibre cells. We omit the details of the model here because its basic structure is given by Eqs. (40) and (41) and an explanation of the model in detail as well as a model summary can be found in

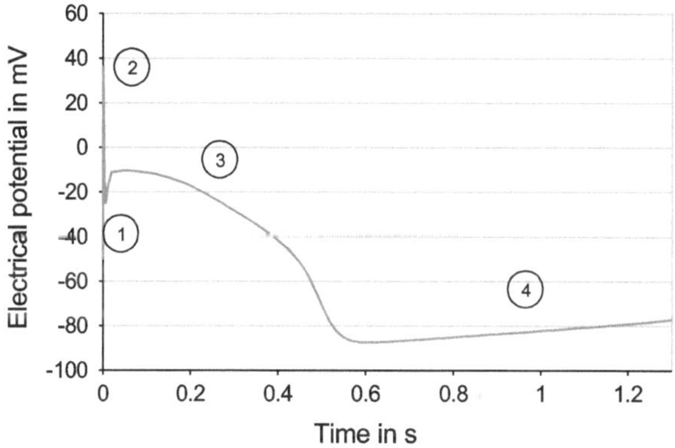

Fig. 8 Action potential V_m produced by the McAllister-Noble-Tsien model with phases 1–4 as discussed in the text

McAllister et al. [23]. Nevertheless, in order to understand the action of drugs, it is necessary to explain the nine ion currents represented in the model.

Figure 8 shows an action potential like it is computed by the McAllister-Noble-Tsien model. The depolarization (phase 1) is mainly driven by a fast sodium inflow I_{Na} that culminates in hyperpolarization (phase 2). A chloride current I_{qr} leads to a quick repolarization that is followed by the first plateau (phase 3) which is assumed to be governed by a much slower secondary inward sodium current I_{si} and a potassium current I_{x1} that together balance the plateau. The second plateau (phase 4) is mainly driven by another potassium current I_{x2}. Purkinje cells are self-contractile which is accounted for via a pacemaker potassium current I_{K2}. Further there are three background currents, a sodium current I_{Nab}, a potassium current I_{K1} and a chloride current I_{Clb}.

Each of the currents is controlled by either one or two gates described by Eq. (41) or by one analytical function. To apply drugs to the model one needs to determine a so-called IC_{50} value that represents the concentration of the respective drug needed for half-blocking the respective gate. Knowing the IC_{50} value and the drug concentration D we simply introduce a scaling factor [24]

$$\frac{1}{1+\dfrac{D}{IC_{50}}} \tag{42}$$

into the respective gate differential equation, Eq. (41).

Up to now we still did not introduce the computation of the inner calcium concentration Ca_i. It is mainly driven by the slow inward current I_{si} that results from slow diffusion of sodium and calcium ions through the cell membrane. The idea goes back to Beeler and Reuter [25] who propose another ordinary differential equation to determine the calcium concentration from I_{si}

$$\frac{\partial Ca_i}{\partial t} = -1.848 \cdot 10^{-4} I_{si} + 0.07(10^{-4} - Ca_i). \tag{43}$$

In the current model I_{si} comprises multiple cellular processes like the sodium-calcium exchanger and the calcium induced calcium release of the sarcoplasmic reticulum, the latter of which is known to deliver the largest portion of calcium.

On both, the mechanical and the electrophysiological level, the model bases on a vast amount of material parameters. The mechanical material parameters are listed in Table 1 and the parameters of the electrophysiological part of the model are listed in McAllister et al. [23]. From our experimental setup we are able to determine some mechanical material parameters as described in the next section.

6 Parameter Fitting

Currently we model the passive material response with the Neo–Hookean material. Its strain energy function Ψ in terms of the principal stretch $\boldsymbol{\lambda} = (\lambda_1, \lambda_2, \lambda_3)^T$ reads as

$$\Psi(\lambda_1, \lambda_2, \lambda_3) = C_{10}(\lambda_1^2 + \lambda_2^2 + \lambda_3^2 - 3), \tag{44}$$

with its material constant C_{10} that defines the slope of the nonlinear stress-strain curve. Taking into account the active component of the constitutive law we can specialize (34) using (35) to

$$S^{11} = 2C_{10} - p\frac{1}{\lambda_1^2} + T_{ref}(1 + \beta_0(\lambda_1 - 1))\frac{1}{\lambda_1^2}, \tag{45}$$

at maximum activation $z = 1$. The main mechanical parameters in this model are the Neo–Hookean material constant C_{10} and the tissue tension at rest T_{ref}, both determined by a parameter fitting. To this end we minimize

$$2C_{10} + \frac{1}{\lambda_1^2}(1 + \beta_0(\lambda_1 - 1))T_{ref} - S_e \rightarrow Min, \tag{46}$$

with respect to C_{10} and T_{ref} with the stress S_e obtained from the experiments. The experimental stress values can be determined by simple geometric considerations that are shown in Fig. 9. It is obvious that for the circular isotropic membrane

$$\lambda_1 = \lambda_2 = \lambda = \frac{d_2}{d_1} \tag{47}$$

holds with d_1 being the initial diameter of the membrane and d_2 the arc length in the deflected state. $\lambda_3 = \lambda^{-2}$ can be obtained from the incompressibility condition $J = 1$ that has been introduced in Eqs. (45) and (46). The arc length can be computed

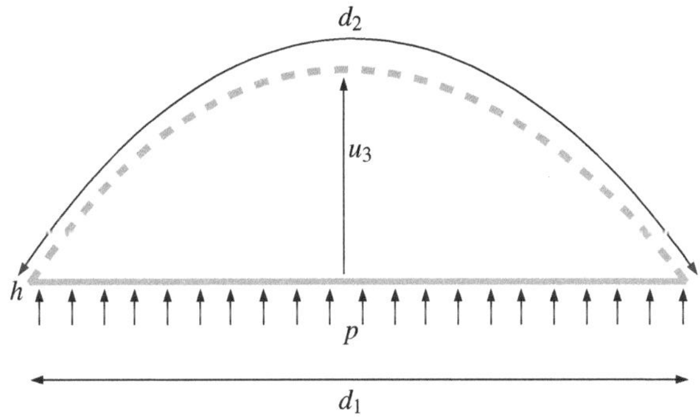

Fig. 9 Illustration of the deflected CellDrum

from the center deflection u_3 via

$$d_2 = \frac{\arctan\left(\frac{2u_3}{d_1}\right)(4u_3^2 + d_1^2)}{2u_3}. \tag{48}$$

Using a modified Laplace formula for thin-walled structures [17] the experimentally determined 2nd Piola-Kirchhoff stress is given as

$$S_e = \frac{p\left(\frac{d_2}{2}\right)^2}{4hu_3}, \tag{49}$$

with pressure p and membrane thickness h. As usual in soft tissue biomechanics the Poisson ratio ν is assumed to be close to 0.5.

With respect to the described parameter fitting the data for the membrane-tissue composite yields $C_{10} = 0.0838284$ and $T_{ref} = 0.5808$ kPa. Computing the active stress directly using the simple analytical approach as given in Eq. (49) we find a value of around 0.37 kPa. In the extensive literature T_{ref} values in between 1–125 kPa are reported for single cells, e.g. in Niederer et al. [26] (56.2 kPa), van der Velden [27] (51 ± 8 kPa) and van der Velden [28] (29.6 ± 4.5 kPa). Here we apply the model to a tissue and therefore T_{ref} is considered as the reference active tension of the whole tissue.

7 Numerical Methods and Results

The parameter fitting has been implemented in SCILAB whereas the analysis of the raw experimental data has been done using the software package R. For the FE model we use the open source industrial software *Code_Aster* from the French utility Electricité de France (EDF). The plate ES-FEM as well as the constitutive law including the electrophysiological system of ODEs have been implemented in *Code_Aster*. The nonlinear equation system in Eq. (31) is solved using a Newton-Raphson algorithm with a relative error of $e_{rel} = 10^{-6}$ at discrete times t_i. At each Gaussian point we solve the stiff system of ODEs with an adaptive time stepping scheme in the way that the Newton-Raphson time interval $[t_i, t_{i+1}]$ is further divided into m time steps because the electrical processes are much faster than the mechanical response. A fourth order singly-diagonal implicit Runge-Kutta solver with adaptive time stepping [29] which has been shown to be unconditionally A-stable is used for the solution of the system of ODEs.

7.1 Numerical Model

Due to isotropy and symmetry we model only a quarter of the CellDrum and apply appropriate boundary conditions as depicted in Fig. 10. We discretize this simple geometry with around 6,600 3-noded triangular elements. The fineness of the discretization is mainly due to the extremely large deflection of the CellDrum. It has a diameter of 16 mm and in the case of a cell monolayer it has a thickness of only 0.008 mm. Therefore we have to deal with radius-thickness ratios of around 1,000 and a maximum deflection of the membrane in the range of 2.5–3 mm. This computationally demanding situation requires a large number of time steps. Especially at low pressures we need a very fine time discretization to achieve convergence.

7.2 Simulation of Pressure-Deflection Curves

We first validate the model by simulating the experimental pressure-deflection curves. The contraction of the cells produces an active stiffness contribution that reduces the deflection of the membrane as observed in Fig. 7. This change in deflection Δu_3^c due

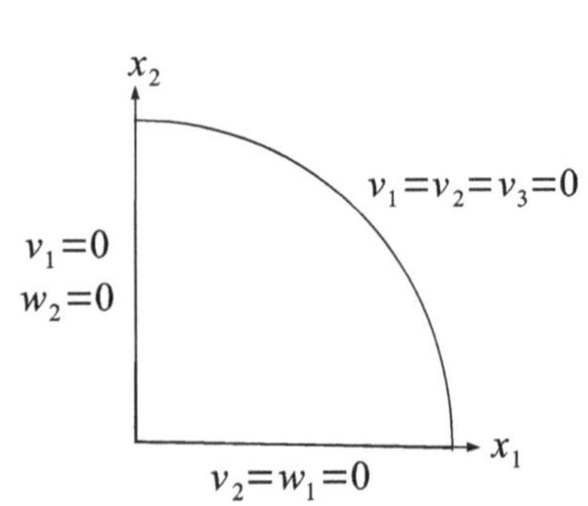

Fig. 10 Symmetry boundary conditions for the quarter CellDrum model

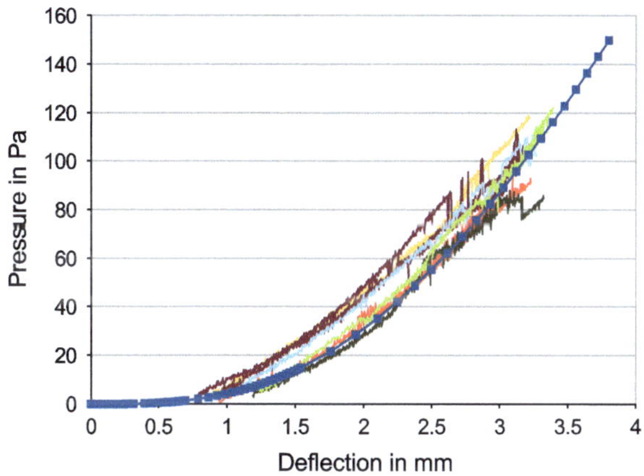

Fig. 11 Comparison of seven experimental (*continuous lines*) and simulation (*continuous line with markers*) results

to contraction is in the range of 10–40 μm and therefore is much smaller than the usual deflections of 1–2 mm at which we investigate the membrane. Consequently in those simulations we do not compute the electrophysiological cell processes but simply assume maximum activation of the cells (i.e. $z = 1$). Figure 11 shows seven experimental pressure-deflection curves of 4 μm thick silicone membranes with a 4 μm thick monolayer of cultivated cardiac tissue. We use those curves for the parameter fitting and with the resulting parameters we simulate the shown pressure-deflection curve that perfectly fits the experimental results.

7.3 Simulation of Cell Contraction

To simulate single or multiple contractions we start at a certain point of the simulation result of Sect. 7.2. This starting point can be chosen depending on the experimental results as follows.

From the experiments we are able to extract the deflection of the membrane when the cells are inactivated. For clarity Fig. 12 shows only one experimental result that has a deflection at rest of 1.23 mm and compares it to a simulation result. Of course different experimental results will show different deflections at rest in the order of approximately 0.05 mm. Therefore we can select a time step of the pressure-deflection simulation in Sect. 7.2 at which the membrane has a deflection that is close to the one shown in the experiments. If that doesn't provide sufficient accuracy we can simulate some more pressure to get to the proper deflection at rest.

Starting from this resting state when the cells are inactive, we now simulate one or multiple contractions using all parts of the model, i.e. computing the electrophysiological processes on cell level and use the resulting free inner calcium concentration

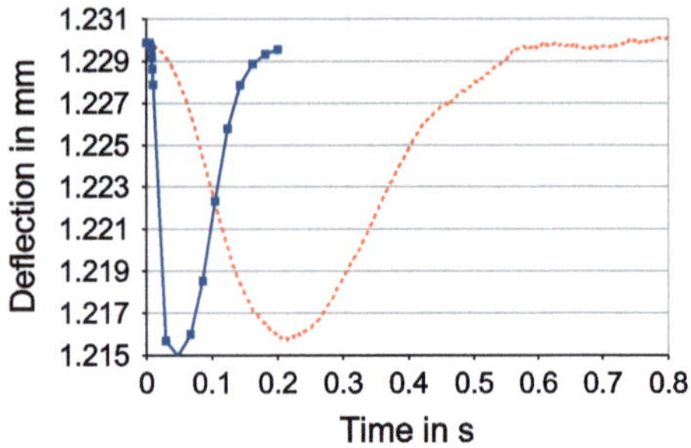

Fig. 12 Comparison of Simulation (*continuous line*) and Experiment (*dashed line*) with respect to Δu_3^c

as an input to the mechanical part of the constitutive model at time t_i. The time steps $\Delta t_i = t_{i+1} - t_i$ need to be chosen in the way so that the mechanical response of the model is sufficiently accurate. The electrophysiological processes are much faster and therefore are computed in adaptively subdivided time steps.

Figure 12 compares one contraction of the CellDrum with a simulation using the McAllister-Noble-Tsien model [23] for purkinje fibres. It is obvious that the change in deflection is nearly the same in both cases. Thus the contractility of the cells can be captured very well by the model. We also observe that the duration of the deflection, i.e. the contraction time is much shorter in the simulation (0.2 s vs. 0.6 s). This effect as well as other deficiencies of the current model will be discussed in Sect. 8.

7.4 Simulation of Drug Treatment

We usually measure 3–5 CellDrums at 5 different concentrations of a drug each. As in the previous Sect. 7.3 we compare the simulation to only one experimental curve for clarity.

Figure 13 shows the same two curves as Fig. 12 together with one experimental and one simulation result of Lidocaine treatment. We applied 100 μM Lidocaine in the experiment as well as in the simulation and recognized a decay in contractility

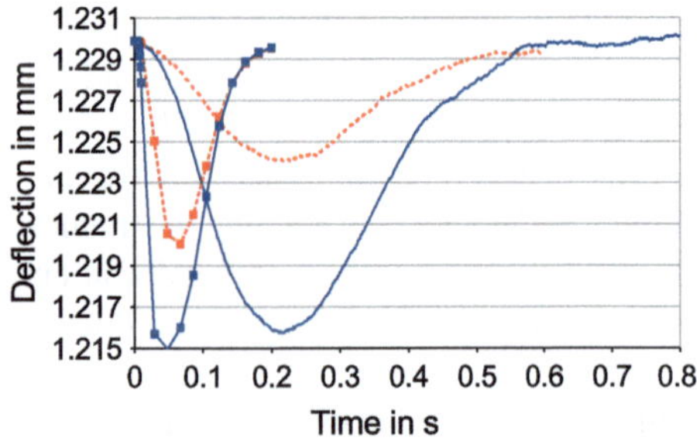

Fig. 13 Comparison of control group (*continuous lines*) and treatment with 100 μM of Lidocaine (*dashed lines*) in experiment (*without markers*) and simulation (*with markers*), respectively

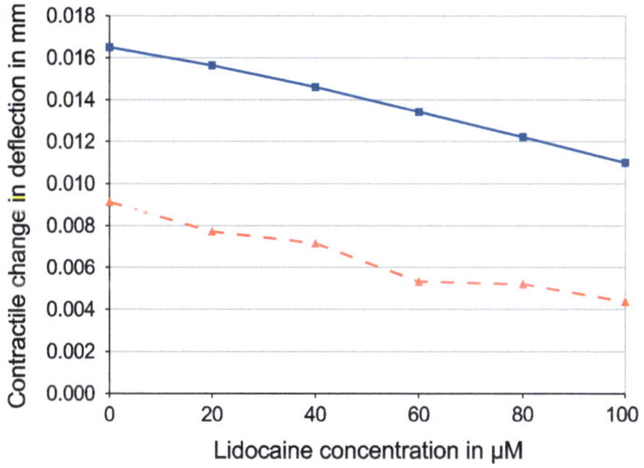

Fig. 14 Lidocaine affecting contractility in experiment (*triangular markers*) and simulation (*rectangular markers*); sample size $m = 3$

as expected. Although the model can represent the action of Lidocaine on the tissue qualitatively there is a quantitative difference that can be explained by Fig. 14.

It displays Δu_3^c at different Lidocaine concentrations and shows that the model is able to capture the qualitative effect of Lidocaine on the tissue appropriately. The quantitative difference originates from the experiment. There is a significant difference in the experimentally determined Δu_3^c without drug treatment in Figs. 13 ($\Delta u_3^c \approx 15\,\mu\text{M}$) and 14 ($\Delta u_3^c \approx 9\,\mu\text{M}$) i.e. the control group in Fig. 14 does not show the behaviour of the control group in Fig. 13. This can be explained by the fact that the experimental results show significant differences depending on whether the cells come out of different batches, depending on their age and on their degree of specialization. On the CellDrum the cells can survive up to four months and the experimental results differ if one experiment is done in week 2 of cultivation and the other one is done in week 12. Therefore a proper quantitative analysis of drug effects is not yet possible. Nevertheless Fig. 14 clearly shows that the slopes of the simulation and the experiments are nearly identical. Currently the experimental basis is too small to draw significant conclusions but there is enough evidence that the model and its future developments can help to interpret the experimental results coming from the CellDrum and possibly other devices in much more detail.

Figure 15 compares the change in beating frequency with respect to the concentration of the applied Lidocaine between experiment and simulation. Again, the control groups show different beating frequencies thus a quantitative comparison of the results is not yet possible. Possible reasons for these differences already have been explained above. Nevertheless it is obvious that from the qualitative point of view the model is able to predict the change in beating frequency due to the application of Lidocaine.

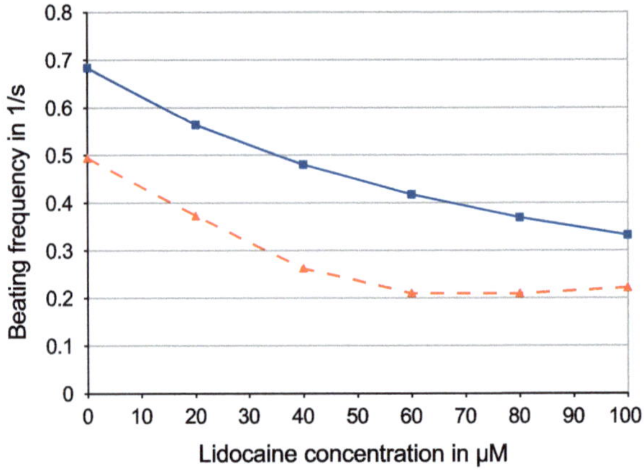

Fig. 15 Lidocaine affecting beating frequency in experiment (*triangular markers*) and simulation (*rectangular markers*); sample size $m = 3$

8 Discussion and Outlook

The results in the previous section clearly show that the computational model captures the contractile behaviour of cardiac tissue. Also it qualitatively predicts the effects of medication on contractility and beating frequency. First simulation results have been shown in comparison to experimental results indicating that the described model is very promising for tasks like the enhanced interpretation of experimental results and the computational investigation of drug effects on cardiac tissue (cf. [30]).

It is obvious that the model still needs some development in order to reach the objective of quantitative comparability of experiment and simulation. In the future an improved model can be applied to identify a drug that is acting on a tissue, to identify unknown effects of drugs on cardiac tissue and to replace animal testing or other expensive and time-consuming experiments by simulations. Finally we discuss the different parts of the model and its upcoming developments that partially are already work in progress.

8.1 Edge-Based Smoothed Finite Element Method

The edge-based plate FE model has been proven to be well suited for biomechanical applications. Not only for simple geometries like the one discussed in this article but especially for more complex geometries the ES-FEM is a good choice. In soft tissue simulations elements often become distorted due to large deformations and this is even worse if the elements are distorted from the beginning because of a poor mesh for a complex geometry. Moreover, we could demonstrate that ES-FEM produces highly accurate results even for linear triangular elements.

8.2 Mechanical Part of Constitutive Model

The Hunter et al. model for active stress in cardiac tissue is well suited for our investigations. We have not even implemented the full model by ignoring the time-dependent relaxation of the muscle cells due to cross-bridge detachment. Niederer et al. further developed this model and performed an extensive literature research in order to find proper material parameters for their model [26]. As those parameters significantly differ from the original parameter set in Hunter et al. [19] one of the next steps will be the modification of our model in those terms. This will include very recent parameter adjustments as done by Weise et al. to model human cardiac cells more appropriately [31].

8.3 Electrophysiological Part of Constitutive Model

The current electrophysiological model simulates the action potential of purkinje fibre cells which is necessary in terms of self-contractility. In reality the tissue construct consists of predominantly ventricular-like cells and a few nodal cells grown in a monolayer from induced pluripotent stem (iPS) cells. That would give reason to a bidomain model but assuming that the drugs act on the cells only and not on the extracellular matrix we will stick with a monodomain model. Anyway we need to replace the current purkinje fibre cell model by a more appropriate ventricular model of human cells, like the one proposed in ten Tusscher et al. [32]. Still, the current McAllister model or a similar, more detailed model will provide the pacemaker for the simulations.

Some simulation results already show that the current model is not detailed enough to simulate certain drugs like for instance Verapamil because they are acting on ion channels that are not represented in the model. Especially the time course of the intracellular calcium concentration can be modelled more appropriately by splitting up the slow inward current I_{si} into parts that have an electrophysiological counterpart in the cell, like the sodium-calcium exchanger and the calcium release of the sarcoplasmic reticulum, as done in ten Tusscher et al. [32]. We will choose the new cell model in the way that at least all the drugs can be simulated, which we currently are investigating or that we are planning to investigate.

8.4 Action Potential Propagation

The model presented in this chapter considers the contraction of cardiac cells cultured on circular flexible membranes to be in accurate synchrony. The underlying experimental results indicate that this is an appropriate assumption in respect to the duration of the contraction-relaxation-cycle. Nevertheless, cardiac tissue contraction is determined by spatial propagation of the action potential modelled in Sect. 5.2. This applies for both the situation in vivo as well as the in vitro model represented by the CellDrum Technology. The propagation of the action potential from cell to cell is realized by so-called gap junctions. These intercellular protein complexes provide a direct and open-door-like connection between adjacent cells. Once one cell undergoes an action potential, involved ions will diffuse through the gap junctions to proximate cells eliciting a new action potential in the latter. The resulting delay, which is not yet covered by the model presented above, is rather short as compared to the total duration of the action potential. This applies especially to our in vitro model with its maximal propagation distance of 16 mm. However, for a better coverage of the in vivo situation, in a next step the spatial action potential propagation will be implemented into the model by a generalized diffusion equation. Considering the propagation the model stays prepared for the simulation of larger and differently shaped geometries as well as for simulations at different spatial scales, e.g. at organ level. The parabolic differential equation describing the membrane potential V_m then reads as

$$C_m \frac{\partial V_m}{\partial t} = \nabla \left(\boldsymbol{G}(\boldsymbol{C}) \nabla V_m \right) - I_m, \tag{50}$$

with $\boldsymbol{G}$ the strain-dependent diffusion tensor and I_m the sum of all membrane currents.

Acknowledgments The first two authors have partially been financed by the project Cardiakytos and gratefully thank.

EUROPÄISCHE UNION

Europe—Investment in our future
The project has been selected from the operational program for NRW in 'Ziel 2 Regionale Wettbewerbsfähigkeit und Beschäftigung' 2007–2013 which is co-financed by EFRE.

References

1. Liu, G., Nguyen, T.: Smoothed Finite Element Methods. CRC Press, Boca Raton (2010)
2. Cui, X., Liu, G., Li, G., Zhang, G., Sun, G.: Analysis of elastic-plastic problems using edge-based smoothed finite element method. Int. J. Press. Vessel. Pip. **86**(10), 711–718 (2009)
3. Dai, K., Liu, G.: Free and forced vibration analysis using the smoothed finite element method (SFEM). J. Sound Vib. **301**(3–5), 803–820 (2007)
4. Dai, K., Liu, G., Nguyen, T.: An n-sided polygonal smoothed finite element method (nSFEM) for solid mechanics. Finite Elem. Anal. Des. **43**(11–12), 847–860 (2007)
5. Liu, G., Nguyen, T., Dai, K., Lam, K.: Theoretical aspects of the smoothed finite element method (SFEM). Int. J. Numer. Meth. Eng. **71**(8), 902–930 (2007)
6. Liu, G., Nguyen, T., Nguyen-Xuan, H., Lam, K.: A node-based smoothed finite element method (NS-FEM) for upper bound solutions to solid mechanics problems. Comput. Struct. **87**(1–2), 14–26 (2009)
7. Cui, X., Liu, G., Li, G., Zhang, G., Zhang, G.: Analysis of plates and shells using an edge-based smoothed finite element method. Comput. Mech. **45**(2–3), 141–156 (2009b)

8. Frotscher, R., Staat, M.: Effectiveness of the edge-based smoothed finite element method applied to soft biological tissues. In: Holzapfel, G., Ogden, R. (eds.) 8th European Solid Mechanics Conference, Verlag d. Technischen Universität Graz, Graz, Austria (2012)
9. Nguyen-Xuan, H., Liu, G., Thai-Hoang, C., Nguyen-Thoi, T.: An edge-based smoothed finite element method (ES-FEM) with stabilized discrete shear gap technique for analysis of Reissner-Mindlin plates. Comput. Method Appl. Mech. **199**(9–12), 471–489 (2010)
10. Bordas, S., Rabczuk, T., Hung, N.X., Nguyen, V., Natarajan, S., Bog, T., Hiep, N.: Strain smoothing in FEM and XFEM. Comput. Struct. **88**(23–24), 1419–1443 (2010)
11. Chen, L., Liu, G., Jiang, Y., Zeng, K., Zhang, J.: A singular edge-based smoothed finite element method (ES-FEM) for crack analyses in anisotropic media. Eng. Fract. Mech. **78**(1), 85–109 (2011)
12. Liu, G., Nourbakhshnia, N., Zhang, Y.: A novel singular ES-FEM method for simulating singular stress fields near the crack tips for linear fracture problems. Eng. Fract. Mech. **78**(6), 863–876 (2011)
13. Nix, Y., Frotscher, R., Staat, M.: Implementation of the edge-based smoothed extended finite element method. In: Eberhardsteiner, F., Böhm, J., Rammerstorfer, H. (eds.) CD-ROM Proceedings of the 6th European Congress on Computational Methods in Applied Sciences and Engineering. Austria, Vienna (2012)
14. Nguyen-Thoi, T., Liu, G., Lam, K., Zhang, G.: A face-based smoothed finite element method (FS-FEM) for 3D linear and geometrically non-linear solid mechanics problems using 4-node tetrahedral elements. Int. J. Numer. Meth. Eng. **78**(3), 324–353 (2009)
15. Bletzinger, K., Bischoff, M., Ramm, E.: A unified approach for shear-locking-free triangular and rectangular shell finite elements. Comput. Struct. **75**(3), 321–334 (2000)
16. Linder, P., Trzewik, J., Rüffer, M., Artmann, G., Digel, I., Kurz, R., Temiz-Artmann, A.: Contractile tension and beating rates of self-exciting monolayers and 3D-tissue constructs of neonatal rat cardiomyocytes. Méd. Biol. Eng. Comput. **48**(1), 59–65 (2010)
17. Trzewik, J.: Experimental analysis of biaxial mechanical tension in cell monolayers and cultured three-dimensional tissues. Ph.D. thesis, Fakultät Informatik und Automatisierung, Technische Universität Ilmenau (2008)
18. Trzewik, J., Temiz-Artmann, A., Linder, P., Demirci, T., Digel, I., Artmann, G.: Evaluation of lateral mechanical tension in thin-film tissue constructs. Ann. Biomed. Eng. **32**(9), 1243–1251 (2004)
19. Hunter, P., McCulloch, A., ter Keurs, H.: Modelling the mechanical properties of cardiac muscle. Prog. Biophys. Mol. Biol. **69**(2–3), 289–331 (1998)
20. Holzapfel, G.A.: Nonlinear Solid Mechanics: A Continuum Approach for Engineering. Wiley, Chichester (2000)
21. Fenton, F., Cherry, E.: Models of cardiac cell. Scholarpedia. http://www.scholarpedia.org/article/Models_of_cardiac_cell (2008)
22. Nickerson, D.: Cardiac electro-mechanics: From cellml to the whole heart. Ph.D. thesis, Bioengineering, University of Auckland (2004)
23. McAllister, R., Noble, D., Tsien, R.: Reconstruction of the electrical activity of cardiac purkinje fibres. J. Physiol. **251**, 1–59 (1975)
24. Obiol-Pardo, C., Gomis-Tena, J., Sanz, F., Saiz, J., Pastor, M.: A multiscale simulation system for the prediction of drug-induced cardiotoxicity. J. Chem. Inf. Model. **51**(2), 483–492 (2011)
25. Beeler, B., Reuter, H.: Reconstruction of the action potential of ventricular myocardial fibres. J. Physiol. **268**, 177–210 (1977)
26. Niederer, S., Hunter, P., Smith, N.: A quantitative analysis of cardiac myocyte relaxation: a simulation study. Biophys. J. **90**(5), 1697–1722 (2006)
27. van der Velden, J., Klein, L., van der Bijl, M., Huybregts, M., Stooker, W., Witkop, J., Stienen, G.: Force production in mechanically isolated cardiac myocytes from human ventricular muscle tissue. Cardiovasc. Res. **38**(2), 414–423 (1998)
28. van der Velden, J., Papp, Z., Zaremba, R., Boontje, N., de Jong, J., Owen, V., Stienen, G.: Increased Ca2+-sensitivity of the contractile apparatus in end-stage human heart failure results from altered phosphorylation of contractile proteins. Cardiovasc. Res. **57**(1), 37–47 (2003)

29. Hairer, E., Wanner, G.: Solving ordinary differential equations. Stiff and Differential-Algebraic Problems, vol. II. Springer, Berlin (1991)
30. Frotscher, R., Koch, J.P., Raatschen, H.J., Staat, M.: Evaluation of a computational model for drug action on cardiac tissue. In: Oñate, E., Oliver, J., Huerta, A. (eds.) Proceedings 11th World Congress on Computational Mechanics (WCCM XI), 5th European Conference on Computational Mechanics (ECCM V), 6th European Conference on Computational Fluid Dynamics (ECFD VI). Barcelona, Spain (2014)
31. Weise, L., Panfilov, A.: A discrete electromechanical model for human cardiac tissue: Effects of stretch-activated currents and stretch conditions on restitution properties and spiral wave dynamics. PloS One **8**(3), e59317 (2013)
32. ten Tusscher, K., Noble, D., Noble, P., Panfilov, A.: A model for human ventricular tissue. Am. J. Physiol. Heart Circ. **286**(4), H1573–H1589 (2004)

Determining the Modulus of Elasticity for Polymer Materials by Numerical Testing Thin-Walled Double-Layer Circular Shells

Sergej Gluhih, Andrejs Kovalovs and Andris Chate

Abstract The paper describes the method of determining the elasticity modulus of the external layer of the double-layer circular shell consisting of multiple layers with various elasticity modules. The internal layer is produced of the material, which elasticity module is known to experimenters. The outer layer is made of a softer material with unknown elasticity modulus. The method is based on the numerical solution of the nonlinear contact task of compressing the single-layer circular shell, which has been modified to provide solution for the double-layer circular shell. As a result of the research the ratios were obtained, which allow to determine the elasticity module of the external layer given the known value of elasticity module of the internal layer. The method is useful for testing the time and external factors dependant properties of the materials.

1 Introduction

Due to extensive use of polymer film materials in various economic sectors their overall production volumes demonstrate steady growth. Polymer films have widespread application in many industries and life spheres. With the development of new technologies extra thin films and nanofilms appear and gain universal currency alongside with the compositions, which include "nano" elements [5].

Creation of new films with targeted properties and high performance characteristics is one of the prospective lines of development. Therefore, the main requirements specified to films are good physical, mechanical (especially high resistance to

S. Gluhih (✉) · A. Kovalovs · A. Chate
Institute of Materials and Structures,
Riga Technical University,
1 Kalku Str., Riga 1658, Latvia
e-mail: s_gluhih@inbox.lv

A. Kovalovs
e-mail: kovalev@mail.lv

A. Chate
e-mail: and_cate@latnet.lv

H. Altenbach and G.I. Mikhasev (eds.), *Shell and Membrane Theories in Mechanics and Biology*, Advanced Structured Materials 45, DOI 10.1007/978-3-319-02535-3_12

deformation—tensile strength) and chemical properties. Polymers are characterised by a wide range of mechanical characteristics that have high dependency on their structure [2].

In the course of continuous operation the structure of the material may change. Biological, chemical and thermal effect may cause material destruction, which results in the changes of its mechanical properties. Therefore, it becomes essential to define the material properties both at the stage of its preparation as well as during operation [1].

In many application tasks where the properties of a material are being tested, the determination of the modulus of elasticity is of primary importance. Different types of plastic materials are often compared on the basis of tensile strength, elongation, and tensile modulus data. Therefore, the data obtained by this method cannot be considered valid for applications involving load-time scales or environments widely different from this method [7].

There is a need to develop new tools and methods for determining and examining the properties of polymer films through synthesis of experimental and theoretical research.

As an example, the authors would like to mentioned the work by Yakupov, in which the author suggests a two-dimensional approach to the research of the properties of thin coatings by blowing the round membrane. The elasticity modulus as well as the deformation pattern are determined here with the help of experimental data, by using the correlation of shell theories in elastic and plastic areas [3, 13]. Interest causes another research involving use of ultrasonic equipment for determining the elastic and plastic properties of polymer films [9]. Atsumi Ohtsuki suggested his own method for determining the Young's modulus. He provided a new mechanical testing method for measuring the Young's modulus in a flexible multi-layered material. The method is based on the nonlinear deformation theory that takes into account large deformation behaviour (i.e., postbuckling behaviour) of multi-layered materials. Exact analytical solutions are obtained in terms of elliptic integrals. By just measuring the horizontal or the vertical displacement, every Young's modulus in a thin flexible multi-layered material can be easily obtained [10].

The goal of the present work is the non-destructive method of determining the modulus of elasticity of the external layer of a double-layer cylindrical shell according to finite element solution. The inner layer of the tested specimen is made of the material with the known elasticity modulus. The modulus of elasticity of the material of the external layer in unknown. The suggested method allows to research polymer material with the possibility of measuring the modulus of elasticity as a function of various processes affecting the structure and properties of a polymeric material. The numerical solution is found by applying the finite elements method with the help of the programme ANSYS. It is based on the solution of a geometrically non-linear problem of transverse deformation of thin-walled circular cylindrical shells due to large displacements and rotations. Contact zone of deformation occurs when the ring is also responsible for nonlinear problem to be solved. Adding non-linear effects allows using a considerably broader range of the load curve in the elastic area of deformation, thus making sure that determination of the elasticity modulus is performed with a higher accuracy.

The task makes use of the solution of a geometrical non-linear problem by deforming the thin-walled round cylindrical shells between the two planes, which was earlier discussed by the authors [4, 6]. A similar problem was considered earlier by [8] in a geometrically nonlinear statement for a linear material obeying Hooke's law. The solution was obtained by using the variational method of minimization of potential energy on the assumption of the predominant role of bending energy over the energy of compression and shear. At the start of the paper the authors briefly review the algorithm of determining the elasticity modulus of a single-layer circular shell; then, the above mentioned solution technique is applied for determining the elasticity modulus of a double-layer circular shell.

2 Determination of the Elasticity Modulus for One Layer of Ring Specimens

Let us consider a thin homogeneous isotropic cylindrical shell under the action of a force that is transmitted through the undeformed upper plane as shown in Fig. 1. The dependence of the force P on the displacement Δ is now called the loading diagram. The dimensionless parameters of displacement α and load β are presented as:

$$\alpha = \frac{\Delta}{2R}, \qquad \beta = \frac{PR^2}{EI}, \tag{1}$$

where E—elasticity modulus of the material, $J = lt^3/12$ moment of inertia, R—radius of the cylinder, l—width of the specimen, t—thickness of the specimen, P—force, Δ—displacement.

The unified loading diagram in the $\alpha - \beta$ coordinates allows us to solve the inverse problem on identification of the elastic modulus. At the first stage, the finite element model of thin circular shells has been used to model the unified loading diagram with the initial assumption of parameter values. The initial assumption of parameter values has been determined by using the method of planning of experiments.

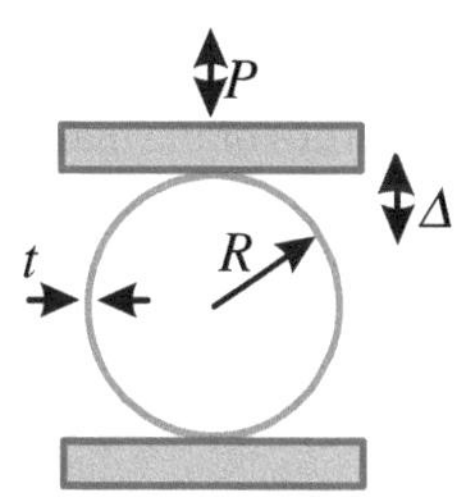

Fig. 1 Contact compression of a circular shell

The criterion for elaboration of the plans of the experiment is to be independent from the mathematical model of the designed object Rikards [11], Rikards et al. [12]. The initial information for development of the plan is the number of factors n and the number of experiments k. The points of experiments in the domain of factors are distributed as regularly as possible. For this reason, the following criterion is used:

$$\Phi = \sum_{i=1}^{k} t_i \times \sum_{j=i+1}^{k} t_i \times \left(\frac{1}{l_{ij}^2}\right) \Rightarrow \min, \tag{2}$$

where l_{ij} is the distance between the points having numbers i and j (i≠1). Physically it is equal to the minimum of potential energy of repulsive forces for the points with unity mass if the magnitude of these repulsive forces is in inverse proportion to the distance between the points.

The plan of the experiment is characterized by the matrix of plan B_{ij}. The domain of the experiments is determined as $x_j \in g\lfloor x_j^{\min}; x_j^{\max} \rfloor$ and the points of the experiments are calculated by the following expression:

$$x_j^{(i)} = x_j^{\min} + \left(\frac{1}{k-1}\right)(x_j^{\max} - x_j^{\min})(B_{ij} - 1), \tag{3}$$

Here $i = 1, 2, \ldots, k$ and $j = 1, 2, \ldots, n$.

The plan of the experiment was formulated for four design parameters and 30 experiments. The upper and lower limits of the identification parameters were taken as follows:

$$\begin{gathered} 1000 \text{ MPa } \leq E \leq 5000 \text{ MPa} \\ 0.05 \text{ mm } \leq t \leq 0.25 \text{ mm} \\ 20 \text{ mm } \leq R \leq 60 \text{ mm} \\ 0.1D\ \% \leq \Delta \leq 0.9D\ \% \end{gathered}$$

where D is diameter of the cylinder.

Then, finite element analysis was performed at the points of the experiments' plan. The finite-element model constructed in the ANSYS program has the form of a quarter of a circular ring due to symmetry. The boundary conditions correspond to the symmetry conditions. Since we look at thin-walled shells, which can be subjected to great displacements and rotations, for different forms of elastic potentials, as a finite element we took a SHELL181 shell element, which corresponds to the requirements adopted. In the contact zone, a CONTA174 contact element is used. The elastic properties of the material were modelled by Hooke's relations. The problem is solved with account of the friction in the contact zone.

After a series of finite element calculations performed for these shells, the forces obtained are converted into a relative loading β according to Eq. (1). The unified loading diagram in dimensionless coordinates (Fig. 2) in the given range of parameters

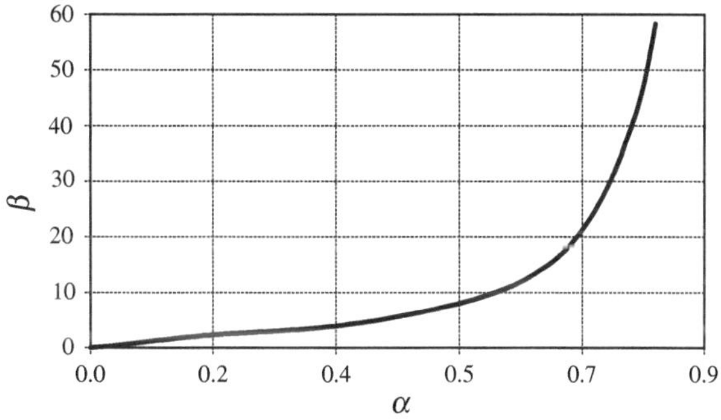

Fig. 2 Unified loading diagram

is approximated by the sixth-degree polynomial:

$$\beta = 3650.8\alpha^6 - 7254.8\alpha^5 + 5496.6\alpha^4 - 1879.5\alpha^3 + 272.65\alpha^2 \\ -0.7882\alpha - 0.0001 \tag{4}$$

Approximation is a universal solution for thin-walled ring specimens within the above mentioned range of parameters, which allows, by comparing with the result of the experiment on compression of the ring specimen, to solve the inverse problem of determining the elasticity modulus at the 5–8 selected points on the loading diagram. The algorithm of deriving of the elasticity modulus from unified loading diagram can be described as follows. Five or eight points of the loading diagram are measured in the range of relative displacements $\alpha = 0.2 - 0.8$ on the test bench. The high value of α allows using a considerably wider range of the loading curve in the elastic area of deformation and thus increasing the accuracy of results. The relation between forces and displacements is recalculated in the dimensionless $\alpha - \beta$ coordinates. By comparing the resulting data with the unified loading diagram, we find the modulus for each point of the particular loading diagram. Over all measurement points, the average value of is determined; then the deviation of the modulus from its average value is calculated at each point, and the average value of error $\mid \delta \mid$ is found.

Using 0.2 mm-thick polyvinyl chloride (PVC) film, three identical series of ring specimens with the radius R = 15, 30, and 45 mm and width l = 10, 20, 30, and 40 mm were prepared for each value of R. An example of such procedure is given in Table 1. An error $\mid \delta \mid$ not exceeding 5 % points to a satisfactory accuracy of the obtained result.

The aim was to determine the modulus of the material according to the suggested method and to compare the resulting data with those derived during the tension test. Let us consider the problem of compression of circular specimens. Results of the experiment are provided in Table 2. The tensile experiments were carried out on a Zwick-100 testing machine according to ASTM D638M. The average value of the elasticity modulus in tension was 2,670 MPa, which was 5 % higher or less than the value found for the elastic modulus in bending.

Table 1 Data processing for a ring specimen with $R = 29.5\,\text{mm}$, $l = 40\,\text{mm}$, and $t = 0.2\,\text{mm}$

Δ	P (N)	α	β	E (MPa)	$\mid \delta \mid$ (%)
7.5	0.14	0.13	1.65	2,840	6.8
11.0	0.19	0.19	2.31	2,636	0.9
14.0	0.23	0.24	2.67	2,798	5.2
23.5	0.34	0.40	4.37	2,570	3.3
33.0	0.64	0.56	8.58	2,450	7.9
Average				2,659	4.8

Table 2 Experimental values of E [MPa] during compression of ring specimens

R (mm)	$l = 10$ (mm)	$l = 20$ (mm)	$l = 30$ (mm)	$l = 40$ (mm)
15	2,585	2,600	2,623	2,655
30	2,595	2,590	2,638	2,605
45	2,678	2,696	2,663	2,653

3 Determination of the Elasticity Modulus for External Layer of a Double-Layer Circular Shell

Let us consider a double-layer circular shell consisting of the layers with different elasticity moduli. The internal layer of the specimen is made of the stiff polymer with a relatively high modulus $E_1 = 10^3\,\text{MPa}$, layer thickness t_1. The external layer is made of a softer polymer with the modulus of elasticity $E_2 = 10$, 100, 1,000 MPa, layer thickness t_2. Poisson's ratios for both layers are the same and equal to 0.35.

Let us assume that we know the values of the average radius of the circular shell, R, width l, thickness of both layers t_1 and t_2 and the modulus of elasticity of the inner layer E_1. Our target value will be the value of the modulus of elasticity of external layer E_2. We will apply the above described method to determine the modulus of elasticity of the second layer made of soft polymer. For this purpose we shall introduce the notion of the so-called reduced modulus of elasticity E_{reduced}. This modulus may be treated as the value adjusted to a single-layer shell with the thickness $t = t_1 + t_2$. This modulus is calculated with the help of a universal loading diagram in the joint coordinate system (see Eq. 2). The adjusted modulus ensures the equality of the loading diagrams of the double-layer shell and the corresponding single-layer shell.

Let us develop the task model for compressing the double-layer shell as it was done for the single-layer shell. For finite element modelling in the programme ANSYS the element SHELL181 is used, which permits presence of multiple layers. The layers are assumed to be homogeneous and isotropic. Let us perform a range of calculations for the shell radius $R = 50\,\text{mm}$, $t = 1\,\text{mm}$ and $t_1 = t_2 = 0.5\,\text{mm}$. The shell width l is assumed to be a unit quantity. The modulus of elasticity of the first

Table 3 Loading diagrams

No	Δ (mm)	P (N)				
		$E_2 = 1000$ (MPa)	$E_2 = 100$ (MPa)	$E_2 = 10$ (MPa)	K_1	K_2
1	4	0.02	0.0056	0.0029	3.63	7.05
2	12	0.0547	0.0151	0.0078	3.62	7.00
3	20	0.0843	0.0232	0.0120	3.63	7.03
4	28	0.1104	0.0305	0.0157	3.62	7.02
5	36	0.1404	0.0388	0.0200	3.62	7.01
6	44	0.1847	0.0510	0.0265	3.62	6.97
7	52	0.2542	0.0700	0.0366	3.63	6.94
8	60	0.3719	0.1022	0.0534	3.64	6.96
9	68	0.5916	0.1633	0.0844	3.62	7.01
10	76	1.0925	0.2979	0.1551	3.67	7.05

Here $K_1 = P(E_2 = 1,000\,\text{MPa})/P(E_2 = 100\,\text{MPa})$, $K_2 = P(E_2 = 1,000\,\text{MPa})/P(E_2 = 10\,\text{MPa})$

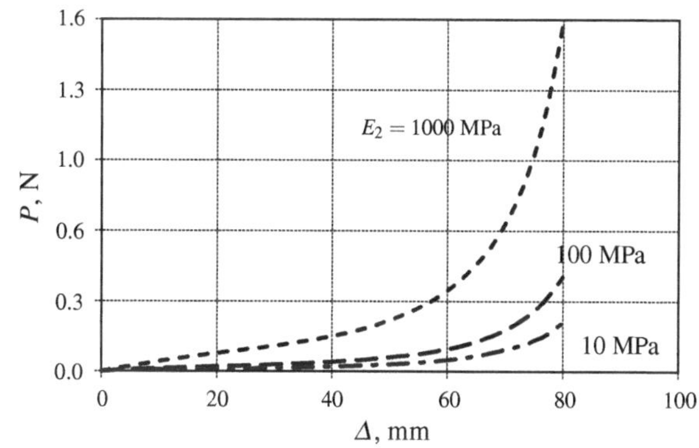

Fig. 3 Loading diagrams for three different values of the modulus of elasticity of external layer E_2

layer is $E_1 = 1,000\,\text{MPa}$; for the second layer we are going to assume three levels $E_2 = 10$, 100 and 1,000 MPa. The results of calculation are presented in Table 3 and graphically shown in (Fig. 3).

The obtained loading results are used to construct the loading diagrams of the modulus of elasticity of the second layer E_2. Moduli K_1 and K_2 have been obtained by dividing the ordinates of the loading diagram for the mentioned single-layer shell by the corresponding values of the double-layer shell at $E_2 = 100$ and 10 MPa. The average values of $K_1 = 3.63$ and $K_2 = 7.01$ have the tolerance of less than 2 %. By knowing K_1 and K_2 we can now determine the adjusted modulus of elasticity E_{reduced} for the double-layer shells. The results of table values are presented in the diagram (Fig. 3). The loading diagram actually corresponds to the single-layer shell with the elasticity modulus $E_2 = E_1 = 1000\,\text{MPa}$.

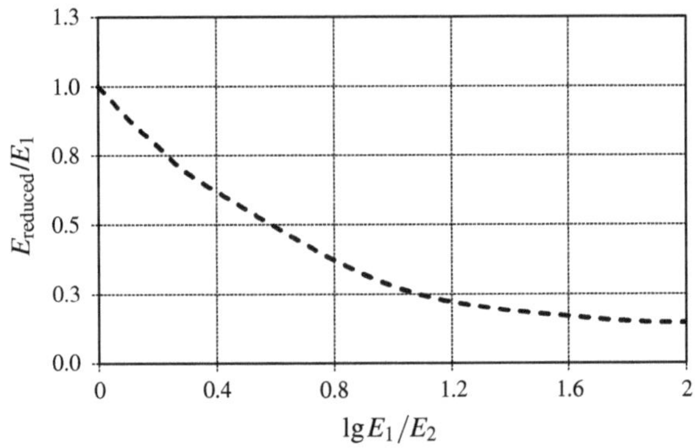

Fig. 4 Results for $t_2/t = 0.5$ and $R/t = 50$

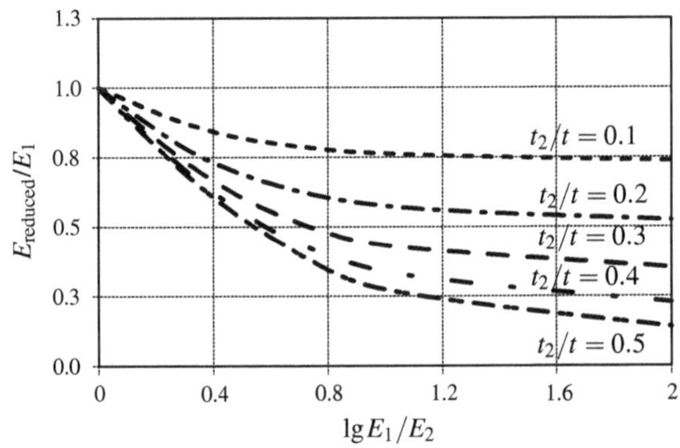

Fig. 5 Results for five levels $t_2/t = 0.1 - 0.5$ and $R/t = 50$

The diagram presented below (Fig. 4), allows us to determine the target modulus of elasticity of the external layer. Procedure of determination the moduli is the following: the adjusted elasticity modulus is determined according to the above described method. The obtained value of the modulus is fixed on the Y-axis. The value marked on the X-axis corresponds to the target modulus E_2. Results thus obtained for $t_2/t = 0.5$ and $R/t = 50$ are shown in Fig. 4.

Now, let us evaluate the effect of these parameters. Let us consider the relation between the layer gages. The calculation results are given for five ratios $t_2/t = 0.1, 0.2, 0.3, 0.4$ and 0.5. Calculation results are shown in Fig. 5.

Let us examine the effect of the geometrical parameter R/t. To achieve this, we need to look at the shells within the range of relative radius $R/t = 25 - 200$. As shown in Table 4, the effect of this parameter proved to be infinitesimal within the 0.3 % range, on average. Thus, the above presented diagrams of the dependence of elasticity moduli on the layer gage ratio (see Fig. 5) can be used across the whole of the studied range of dependence $R/t = 25 - 200$.

Table 4 Results for different values of R/t

R/t	$E_1/E_{reduced}$		$\vert\delta\vert$ (%)	
	$E_2 = 100$ (MPa)	$E_2 = 10$ (MPa)	$E_2 = 100$ (MPa)	$E_2 = 10$ (MPa)
25	3.61	6.99	0.49	0.54
35	3.62	7.00	0.35	0.39
50	3.63	7.02	0.10	0.20
60	3.63	7.03	0.06	0.06
75	3.64	7.04	0.17	0.18
100	3.64	7.06	0.26	0.38
200	3.65	7.07	0.45	0.62
Average	3.629	7.029	0.27	0.34

4 Conclusion

The research suggests the method of determination of the elasticity modulus of the external layer of a double-layer circular shell given the known value of this modulus of the internal layer of the shell. The method is based on the solution of a nonlinear contact problem for a circular compressed specimen within the finite element package ANSYS. First, the solution is demonstrated and the procedure of determining the modulus of elasticity of a single-layer shell is described, which is supported by experimental verification of the results.

After modification of the given method numerical solution of the double-layer shell is shown. Based on the results of parametric study of double-layer ring specimens the solution for the modulus of elasticity of the external layer is suggested.

The suggested non-destructive method allows to take advantage of a “one specimen” approach, that is the possibility to measure the elasticity modulus depending on the processes occurring inside a specimen and changing the structure of the polymer material. For example, deterioration of the material, sorption and desorption by the material of liquid-saturated media and steam, temperature fluctuations.

Acknowledgments This study was financially supported by the Latvian Ministry of Education and Science according to Contract No. V7909.1.

References

1. Abdel-Bary, E. (ed.): Handbook of Plastic Films. Smithers Rapra Technology, Shawbury (2003)
2. Briston, J.: Plastics films, 2nd edn. Longman Scientific, Harlow (Essex) (1986)
3. Galimov, N., Yakupov, N., Yakupov, S.: Experimental-theoretical method for determining mechanical characteristics of spherical films and membranes of complex structure. Mech. Solids **46**(3), 380–386 (2011)

4. Gluhih, S., Kovalov, A., Tishkunov, A., Akishin, A., Chate, A., Auzins, E., Kalnins, M.: Identification of the elastic modulus of polymeric materials by using thin-walled cylindrical specimens. Mech. Compos. Mater. **48**(1), 57–64 (2012)
5. Knoll, W., Advincula, R.: Functional Polymer Films, vol. 1 and 2. Wiley, Weinheim (2011)
6. Kovalovs, A., Gluhih, S., Chate, A.: Young's modulus identification by using cylindrical specimens. Key Eng. Mater. **559**, 75–79 (2013)
7. Kutz, M. (ed.): Handbook of Materials Selection. Wiley, New York (2002)
8. Lavendel, E., Dirba, D.: Problems of dynamics and strength (in Russ.) **20**, 195–200 (1970)
9. Lellinger, D., Tadjbach, S., Alig, I.: Determination of the elastic moduli of polymer films by a new ultrasonic reflection method. Macromol. Symp. **184**, 203–213 (2002)
10. Ohtsuki, A.: An innovative measuring method of young's modulus using postbuckling behavior. In: Tandon, G., Tekalur, S., Carter, R., Sottos, N., Blaiszik, B. (eds.) Experimental Mechanics of Composite, Hybrid, and Multifunctional Materials. Conference proceedings of the society for experimental mechanics series, vol. 6, pp. 247–254. Springer International Publishing, New York (2014)
11. Rikards, R.: Elaboration of optimal design models for objects from data of experiments. In: Pedersen, p (ed.) Optimal Design with Advanced Materials, pp. 149–162. Elsevier, Oxford (1993)
12. Rikards, R., Bledzki, A.K., Eglajs, V., Chate, A., Kurek, K.: Elaboration of optimal design models for composite materials from data of the experiments. Mekhanika Kompozitnykh Materialov **4**, 435–445 (1992)
13. Yakupov, N., Nurullin, R., Yakupov, S.: Mechanical properties of thin films and nanofilms. Russ. Eng. Res. **26**(6), 571–574 (2009)

Three-Dimensional Exact Analysis of Functionally Graded Laminated Composite Plates

Gennady M. Kulikov and Svetlana V. Plotnikova

Abstract A paper focuses on the implementation of the method of sampling surfaces (SaS) to three-dimensional (3D) exact solutions for functionally graded (FG) laminated composite plates. According to the SaS method, we introduce inside the nth layer I_n not equally spaced SaS parallel to the middle surface of the plate in order to choose the displacements of these surfaces as basic plate variables. This permits the presentation of the proposed FG laminated plate formulation in a very compact form. It is important that the SaS are located inside each layer at Chebyshev polynomial nodes. This fact allows one to uniformly minimize the error due to Lagrange interpolation, i.e. the use of Lagrange polynomials of high degree becomes possible. As a result, the SaS method can be applied efficiently to 3D exact solutions of elasticity for FG laminated plates with a specified accuracy utilizing the sufficient number of SaS.

1 Introduction

Nowadays, the functionally graded (FG) materials are widely used in mechanical engineering due to their advantages compared to traditional laminated materials [5, 11]. However, the study of FG plates is not a simple task because the material properties depend on the thickness coordinate and some specific assumptions regarding their variations in the thickness direction are required [9]. This fact restricts the implementation of the Pagano approach [24, 29] for the 3D exact analysis of FG simply supported rectangular plates [31]. As concerned other two popular approaches to 3D exact solutions, namely, the state space approach and the asymptotic approach they can be applied efficiently to FG plates, see Alibeigloo [1], Alibeigloo and Liew [2], Cheng and Batra [8], Reddy and Cheng [26], Zhong and Shang [33]. A new approach

G.M. Kulikov (✉) · S.V. Plotnikova
Tambov State Technical University, Sovetskaya Street, 106, Tambov 392000, Russia
e-mail: gmkulikov@mail.ru

S.V. Plotnikova
e-mail: plotnikovasvetlana62@gmail.com

H. Altenbach and G.I. Mikhasev (eds.), *Shell and Membrane Theories in Mechanics and Biology*, Advanced Structured Materials 45, DOI 10.1007/978-3-319-02535-3_13

to closed-form elasticity solutions for FG isotropic and transversely isotropic plates is presented in [10, 30]. These solutions are based on the general solution of the equilibrium equations of inhomogeneous elastic media [25]. The efficient approach to the 3D exact analysis of thermoelasticity has been proposed by Vel and Batra [27, 28]. They studied the static and transient thermoelastic problems for FG simply supported plates whose material properties and basic variables are presented by Taylor series expansions through the thickness. Ootao [22], Ootao and Ishihara [23], Ootao and Tanigawa [21] obtained the 3D exact solutions for the transient thermoelastic response of FG strips and rectangular plates with simply supported edges under nonuniform heating on outer surfaces. The original approach to analytical solutions for the FG beams and plates was developed in Altenbach and Eremeyev [4], Birsan et al. [6]. This approach is based on the so-called theory of directed curves and surfaces [3, 32], which follows the original idea of Cosserat brothers. However, the 3D exact solutions for FG laminated plates through the use of the recently developed concept of sampling surfaces (SaS) [15, 16] can not be found in the current literature. The present paper serves to fill the gap of knowledge in this research area.

The SaS approach has been already applied efficiently to 3D exact solutions of elasticity for laminated composite plates and shells [17–20]. In accordance with this method, we choose inside the nth layer I_n not equally spaced SaS $\Omega^{(n)1}$, $\Omega^{(n)2}, \ldots$, $\Omega^{(n)I_n}$ parallel to the middle surface of the plate and introduce the displacement vectors $\mathbf{u}^{(n)1}$, $\mathbf{u}^{(n)2}$, ..., $\mathbf{u}^{(n)I_n}$ of these surfaces as basic plate variables, where $I_n \geq 3$. Such choice of unknowns in conjunction with the use of Lagrange polynomials of degree $I_n - 1$ in the thickness direction permits the presentation of governing equations of the proposed FG plate formulation in a very compact form.

An idea of using the SaS can be traced back to Kulikov [12], Kulikov and Carera [13] in which three, four and five equally spaced SaS are employed. These contributions describe the SaS approach applied to the approximate solution of 3D shell problems. It is important to note that the more general SaS approach with the arbitrary number of equally spaced SaS [14, 15] developed later by the authors does not work properly with Lagrange polynomials of high degree because of the Runge's phenomenon [7], which yields the wild oscillation at the edges of the interval when the user deals with any specific functions. If the number of equally spaced nodes is increased then the oscillations become even larger. However, the use of Chebyshev polynomial nodes inside each layer can help to improve significantly the behavior of Lagrange polynomials of high degree because such a choice allows one to minimize uniformly the error due to Lagrange interpolation.

2 Kinematic Description of Undeformed Plate

Consider a FG laminated plate of the thickness h. Let the middle surface Ω be described by Cartesian coordinates x_1 and x_2 . The coordinate x_3 is oriented in the thickness direction. The transverse coordinates of SaS inside the nth layer are defined as

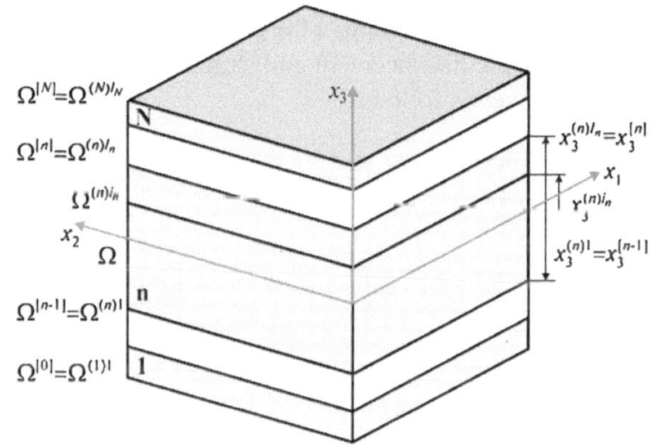

Fig. 1 Geometry of the laminated plate

$$x_3^{(n)1} = x_3^{[n-1]}, \quad x_3^{(n)I_n} = x_3^{[n]}, \tag{1}$$

$$x_3^{(n)m_n} = \frac{1}{2}\left(x_3^{[n-1]} + x_3^{[n]}\right) - \frac{1}{2} h_n \cos\left(\pi \frac{2m_n - 3}{2\,(I_n - 2)}\right), \tag{2}$$

where $x_3^{[n-1]}$ and $x_3^{[n]}$ are the transverse coordinates of layer interfaces $\Omega^{[n-1]}$ and $\Omega^{[n]}$ depicted in Fig. 1; $h_n = x_3^{[n]} - x_3^{[n-1]}$ is the thickness of the nth layer; I_n is the number of SaS corresponding to the nth layer; the index n identifies the belonging of any quantity to the nth layer and runs from 1 to N; N is the total number of layers; the index m_n identifies the belonging of any quantity to the inner SaS of the nth layer and runs from 2 to $I_n - 1$, whereas the indices i_n, j_n, k_n to be introduced later for describing all SaS of the nth layer run from 1 to I_n. Besides, the tensorial indices i, j, k, l range from 1 to 3 and Greek indices α, β range from 1 to 2. It is worth noting that the transverse coordinates of inner SaS Eq. (2) coincide with the coordinates of Chebyshev polynomial nodes [7]. This fact has a great meaning for the convergence of the SaS method [16–18].

3 Kinematic Description of Deformed Plate

The components of the strain tensor ε_{ij} are written as

$$2\varepsilon_{ij} = u_{i,j} + u_{j,i}, \tag{3}$$

where u_i are the displacements of the plate.

We start now with the first assumption of the proposed FG laminated plate formulation. Let us assume that the displacement and strain fields are distributed through the thickness of the nth layer as follows:

$$u_i^{(n)} = \sum_{i_n} L^{(n)i_n} u_i^{(n)i_n}, \quad x_3^{[n-1]} \leq x_3 \leq x_3^{[n]}, \tag{4}$$

$$\varepsilon_{ij}^{(n)} = \sum_{i_n} L^{(n)i_n} \varepsilon_{ij}^{(n)i_n}, \quad x_3^{[n-1]} \leq x_3 \leq x_3^{[n]}, \tag{5}$$

where $u_i^{(n)i_n}(x_1, x_2)$ are the displacements of SaS of the nth layer $\Omega^{(n)i_n}$, $\varepsilon_{ij}^{(n)i_n}$ (x_1, x_2) are the strains of the same SaS; $L^{(n)i_n}(x_3)$ are the Lagrange polynomials of degree $I_n - 1$ defined as

$$u_i^{(n)i_n} = u_i\left(x_3^{(n)i_n}\right), \tag{6}$$

$$\varepsilon_{ij}^{(n)i_n} = \varepsilon_{ij}\left(x_3^{(n)i_n}\right), \tag{7}$$

$$L^{(n)i_n} = \prod_{j_n \neq i_n} \frac{x_3 - x_3^{(n)j_n}}{x_3^{(n)i_n} - x_3^{(n)j_n}}. \tag{8}$$

The use of relations (3), (6) and (7) yields

$$2\varepsilon_{\alpha\beta}^{(n)i_n} = u_{\alpha,\beta}^{(n)i_n} + u_{\beta,\alpha}^{(n)i_n}, \tag{9}$$

$$2\varepsilon_{\alpha 3}^{(n)i_n} = \beta_\alpha^{(n)i_n} + u_{3,\alpha}^{(n)i_n}, \tag{10}$$

$$\varepsilon_{33}^{(n)i_n} = \beta_3^{(n)i_n}, \tag{11}$$

$$\beta_i^{(n)i_n} = u_{i,3}\left(x_3^{(n)i_n}\right), \tag{12}$$

where $\beta_i^{(n)i_n}(x_1, x_2)$ are the values of derivatives of displacements with respect to transverse coordinate x_3 at SaS defined according to Eq. (4) as

$$\beta_i^{(n)i_n} = \sum_{j_n} M^{(n)j_n}\left(x_3^{(n)i_n}\right) u_i^{(n)j_n}, \tag{13}$$

where $M^{(n)j_n} = L_{,3}^{(n)j_n}$ are the derivatives of Lagrange polynomials, which are calculated at SaS as follows:

$$M^{(n)j_n}\left(x_3^{(n)i_n}\right) = \frac{1}{x_3^{(n)j_n} - x_3^{(n)i_n}} \prod_{k_n \neq i_n, j_n} \frac{x_3^{(n)i_n} - x_3^{(n)k_n}}{x_3^{(n)j_n} - x_3^{(n)k_n}} \quad \text{for} \quad j_n \neq i_n,$$

$$M^{(n)i_n}\left(x_3^{(n)i_n}\right) = -\sum_{j_n \neq i_n} M^{(n)j_n}\left(x_3^{(n)i_n}\right). \tag{14}$$

This means that the key functions $\beta_i^{(n)i_n}$ of the proposed FG laminated plate formulation are represented as a linear combination of displacements of SaS of the nth layer $u_i^{(n)j_n}$.

4 Variational Formulation

The variational equation for the FG laminated plate in the case of conservative loading can be written as

$$\delta \Pi = 0, \tag{15}$$

where Π is the total potential energy given by

$$\Pi = \frac{1}{2} \iint_{\Omega} \sum_n \int_{x_3^{[n-1]}}^{x_3^{[n]}} \boldsymbol{\sigma}_{ij}^{(n)} \varepsilon_{ij}^{(n)} dx_1 dx_2 dx_3 - W, \tag{16}$$

$$W = \iint_{\Omega} \left(p_i^+ u_i^{[N]} - p_i^- u_i^{[0]}\right) dx_1 dx_2 + W_\Sigma, \tag{17}$$

where $\boldsymbol{\sigma}_{ij}^{(n)}$ are the components of the stress tensor of the nth layer; $u_i^{[0]} = u_i^{(1)1}$ and $u_i^{[N]} = u_i^{(N)I_N}$ are the displacements of bottom and top surfaces $\Omega^{[0]}$ and $\Omega^{[N]}$; p_i^- and p_i^+ are the loads acting on bottom and top surfaces; W_Σ is the work done by external loads applied to the edge surface Σ.

Substituting the strain distribution (5) in (16) and introducing stress resultants

$$H_{ij}^{(n)i_n} = \int_{x_3^{[n-1]}}^{x_3^{[n]}} \boldsymbol{\sigma}_{ij}^{(n)} L^{(n)i_n} dx_3, \tag{18}$$

one obtains

$$\Pi = \frac{1}{2} \iint_{\Omega} \sum_{n} \sum_{i_n} H_{ij}^{(n)i_n} \varepsilon_{ij}^{(n)i_n} dx_1 dx_2 - W. \tag{19}$$

The second assumption of the FG laminated plate formulation is quite standard and consists in the following:

$$\boldsymbol{\sigma}_{ij}^{(n)} = C_{ijkl}^{(n)} \varepsilon_{kl}^{(n)}, \quad x_3^{[n-1]} \leq x_3 \leq x_3^{[n]}, \tag{20}$$

where $C_{ijkl}^{(n)}$ are the elastic constants of the nth layer.

Next, we introduce the third and last assumption. Let us assume that elastic constants are distributed through the thickness of a plate as follows:

$$C_{ijkl}^{(n)} = \sum_{i_n} L^{(n)i_n} C_{ijkl}^{(n)i_n}, \tag{21}$$

where $C_{ijkl}^{(n)i_n}$ are the values of elastic constants on SaS of the nth layer.

Substituting constitutive Eq. (20) in (18) and taking into account the through-thickness distributions (5) and (21), we have

$$H_{ij}^{(n)i_n} = \sum_{j_n, k_n} \Lambda^{(n)i_n j_n k_n} C_{ijkl}^{(n)j_n} \varepsilon_{kl}^{(n)k_n}, \tag{22}$$

where

$$\Lambda^{(n)i_n j_n k_n} = \int_{x_3^{[n-1]}}^{x_3^{[n]}} L^{(n)i_n} L^{(n)j_n} L^{(n)k_n} dx_3. \tag{23}$$

5 3D Exact Solution for Functionally Graded Laminated Plates

In this section, we study a simply supported FG laminated orthotropic rectangular plate. The edge boundary conditions of the plate are written as

$$\begin{aligned} \boldsymbol{\sigma}_{11}^{(n)} &= u_2^{(n)} = u_3^{(n)} = 0 \quad \text{at} \quad x_1 = 0 \quad \text{and} \quad x_1 = a, \\ \boldsymbol{\sigma}_{22}^{(n)} &= u_1^{(n)} = u_3^{(n)} = 0 \quad \text{at} \quad x_2 = 0 \quad \text{and} \quad x_2 = b, \end{aligned} \tag{24}$$

where a and b are the plate dimensions. To satisfy boundary conditions, we search the analytical solution of the problem by a method of the double Fourier series expansion

$$
\begin{aligned}
u_1^{(n)i_n} &= \sum_{r,s} u_{1rs}^{(n)i_n} \cos\frac{r\pi x_1}{a} \sin\frac{s\pi x_2}{b},\\
u_2^{(n)i_n} &= \sum_{r,s} u_{2rs}^{(n)i_n} \sin\frac{r\pi x_1}{a} \cos\frac{s\pi x_2}{b},\\
u_3^{(n)i_n} &= \sum_{r,s} u_{3rs}^{(n)i_n} \sin\frac{r\pi x_1}{a} \sin\frac{s\pi x_2}{b},
\end{aligned} \tag{25}
$$

where r, s are the wave numbers in plane directions. The external loads are also expanded in double Fourier series.

Substituting (25) and Fourier series corresponding to mechanical loading into the total potential energy (17) and (19) with $W_\Sigma = 0$ and allowing for relations (9), (10), (11), (13) and (22), one obtains

$$
\Pi = \sum_{r,s} \Pi_{rs}\left(u_{irs}^{(n)i_n}\right). \tag{26}
$$

Invoking the variational Eqs. (15), (26), we arrive at the system of linear algebraic equations

$$
\frac{\partial \Pi_{rs}}{\partial u_{irs}^{(n)i_n}} = 0 \tag{27}
$$

of order

$$
3\left(\sum_n I_n - N + 1\right).
$$

The linear system (27) can be easily solved by using a method of Gaussian elimination.

The described algorithm was performed with the Symbolic Math Toolbox, which incorporates symbolic computations into the numeric environment of MATLAB. Such a technique gives the possibility to derive the exact solutions of 3D elasticity for FG laminated orthotropic plates with a specified accuracy.

5.1 Single-Layer Square Plate

Consider a FG single-layer square plate subjected to transverse sinusoidal loading acting on its top surface

$$
p_3^+ = p_0 \sin\frac{\pi x_1}{a} \sin\frac{\pi x_2}{b}, \tag{28}
$$

where $p_0 = 1$ Pa and $a = b = 1$ m.

It is assumed that the elastic modulus is distributed in the thickness direction according to the exponential law:

$$E = E^+ e^{\alpha(z-0.5)}, \quad z = x_3/h, \tag{29}$$

where E^+ is the Young modulus on the top surface; α is the material gradient index defined by

$$\alpha = \ln\left(\frac{E^+}{E^-}\right), \tag{30}$$

where E^- is the Young modulus on the bottom surface, whereas the Poisson ratio ν is constant through the thickness [10]. The material parameters are taken to be $E^+ = 10^7$ Pa and $\nu = 0.3$. To compare the results derived with closed-form solutions [10, 29], the following dimensionless variables are introduced:

$$\bar{u}_1 = G^+ u_1(0, a/2, z)/hp_0, \quad \bar{u}_3 = G^+ u_3(a/2, a/2, z)/hp_0,$$
$$\bar{\sigma}_{11} = \sigma_{11}(a/2, a/2, z)/p_0, \quad \bar{\sigma}_{12} = \sigma_{12}(0, 0, z)/p_0,$$
$$\bar{\sigma}_{13} = \sigma_{13}(0, a/2, z)/p_0, \quad \bar{\sigma}_{33} = \sigma_{33}(a/2, a/2, z)/p_0,$$

where $G^+ = E^+/(2(1+\nu))$ is the shear modulus on the top surface.

Tables 1, 2, 3, 4 show the results of a convergence study due to increasing the number of SaS. As it turned out, the SaS method provides 15 right digits for all basic variables (in fact, the better accuracy is possible) utilizing 13 inner SaS inside the plate body. It should be noted that herein the bottom and top surfaces are not included into a set of SaS because the use of Chebyshev polynomial nodes allows one to minimize uniformly the error due to Lagrange interpolation. Figure 2 displays the distributions of transverse stresses in the thickness direction for the slenderness ratio $a/h = 1$ employing 11 SaS. These results demonstrate convincingly the high potential of the proposed FG plate formulation because boundary conditions on the bottom and top surfaces of the plate for transverse stresses are satisfied exactly.

5.2 Three-Layer Square Plate

Here, we study a three-layer square plate subjected to sinusoidally distributed transverse loading (28). The outer layers with equal thicknesses $h_1 = h_3 = h/4$ are composed of the graphite-epoxy composite with Young moduli $E_1 = 172.72$ GPa, $E_2 = E_3 = 6.909$ GPa, shear moduli $G_{12} = G_{13} = 3.45$ GPa, $G_{23} = 1.38$ GPa and Poisson ratios $\nu_{12} = \nu_{13} = \nu_{23} = 0.25$. It is assumed that fibers of the bottom and top layers are oriented respectively in x_1- and x_2-directions. The central layer is made of the FG transversely isotropic material whose elastic constants are distributed in the thickness direction according to a power law:

Table 1 Results for a single-layer square plate with $a/h = 3$ and $\alpha = 0$

I_1	$-\bar{u}_1(0.5)$	$\bar{u}_3(0.5)$	$\bar{\sigma}_{11}(0.5)$	$-\bar{\sigma}_{12}(0.5)$	$\bar{\sigma}_{13}(0)$	$\bar{\sigma}_{33}(0)$
3	0.3986372494504680	1.278878243389591	1.999302146854837	0.8349039028805547	0.4977436151168003	0.4752837277628003
7	0.4358933937131948	1.342554953466513	2.124032428288413	0.9129329889584641	0.7022762666060094	0.4943950643281928
11	0.4358933942603120	1.342554689543095	2.124018410314048	0.9129329901043438	0.7023022083223538	0.4944039935419638
15	0.4358933942603121	1.342554689542491	2.124018410193782	0.9129329901043439	0.7023022084767580	0.4944039936052152
19	0.4358933942603120	1.342554689542491	2.124018410193780	0.9129329901043437	0.7023022084767578	0.4944039936052150
Exact[a]	0.4358933942603120	1.342554689542491	2.124018410193781	0.9129329901043437	0.7023022084767578	0.4944039936052149

[a] The exact results have been obtained by the authors using the Vlasov's closed-form solution

Table 2 Results for a single-layer square plate with $a/h = 3$ and $\alpha = 0.1$

I_1	$-\bar{u}_1(0.5)$	$\bar{u}_3(0.5)$	$\bar{\sigma}_{11}(0.5)$	$-\bar{\sigma}_{12}(0.5)$	$\bar{\sigma}_{13}(0)$	$\bar{\sigma}_{33}(0)$
3	0.4158336502652652	1.347977241257294	2.073239906807475	0.8700097684529745	0.4983226799021384	0.4596565434470269
7	0.4536977979133792	1.414636043682728	2.193265858989844	0.9502222750248108	0.7020700339720654	0.4877173446168515
11	0.4536977984142576	1.414635771310962	2.193270258459039	0.9502224469642998	0.7020957676355946	0.4877129579452970
15	0.4536977984142576	1.414635771310368	2.193270258650021	0.9502224469653965	0.7020957677904856	0.4877129578319663
19	0.4536977984142575	1.414635771310368	2.193270258650021	0.9502224469653963	0.7020957677904854	0.4877129578319657
Exact		1.41464				

Table 3 Results for a single-layer square plate with $a/h = 10$ and $\alpha = 0$

I_1	$-\bar{u}_1(0.5)$	$\bar{u}_3(0.5)$	$\bar{\sigma}_{11}(0.5)$	$-\bar{\sigma}_{12}(0.5)$	$\bar{\sigma}_{13}(0)$	$\bar{\sigma}_{33}(0)$
3	16.69442044348862	112.4420178073405	19.91690124596434	10.48941372424062	1.706202131658078	0.4978329738513218
7	16.81014395019634	113.1720941024937	20.04394635519068	10.56212494794475	2.383373139214406	0.4999497585886086
11	16.81014395019644	113.1720941004914	20.04394633189751	10.56212494794481	2.383373791176881	0.4999497657893389
15	16.81014395019644	113.1720941004913	20.04394633189750	10.56212494794481	2.383373791176909	0.4999497657893389
19	16.81014395019645	113.1720941004914	20.04394633189751	10.56212494794481	2.383373791176910	0.4999497657893400
Exact[a]	16.81014395019644	113.1720941004914	20.04394633189751	10.56212494794481	2.383373791176911	0.4999497657893391

[a] The exact results have been obtained by the authors using the Vlasov's closed-form solution

Table 4 Results for a single-layer square plate with $a/h = 10$ and $\alpha = 0.1$

I_1	$-\bar{u}_1(0.5)$	$\bar{u}_3(0.5)$	$\bar{\sigma}_{11}(0.5)$	$-\bar{\sigma}_{12}(0.5)$	$\bar{\sigma}_{13}(0)$	$\bar{\sigma}_{33}(0)$
3	17.27349261247991	118.2552720004223	20.59552826078969	10.84973560354474	1.706517604171541	0.4818430337132246
7	17.39081000813009	119.0218924804116	20.72151070605207	10.92696818763486	2.383067071421124	0.4936497163566589
11	17.39081000813025	119.0218924788537	20.72151235713497	10.92696819230357	2.383067798660550	0.4936485236803348
15	17.39081000813025	119.0218924788537	20.72151235713520	10.92696819230357	2.383067798660592	0.4936485236802035
19	17.39081000813026	119.0218924788537	20.72151235713521	10.92696819230358	2.383067798660592	0.4936485236802040

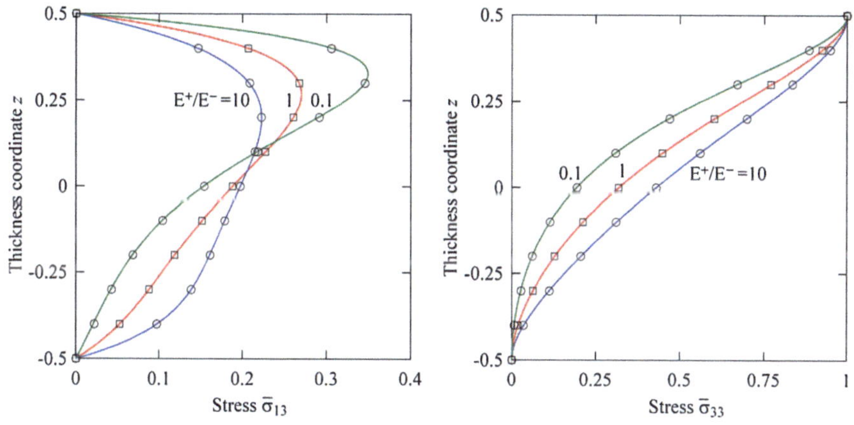

Fig. 2 Distributions of transverse stresses through the thickness of the plate with $a/h = 1$: present analysis (-) for $I_1 = 11$ and analytical solutions [10] (○) and [29] (□)

$$C_{ijkl} = C^-_{ijkl} V^- + C^+_{ijkl} V^+,$$
$$V^+ = 1 - V^-, \quad V^- = (0.5 - 2z)^\gamma, \quad -0.25 \leq z \leq 0.25, \tag{31}$$

where C^-_{ijkl} and C^+_{ijkl} are the values of elastic constants on the bottom and top surfaces; γ is the material gradient index; $z = x_3/h$ is the dimensionless thickness coordinate. The elastic constants on the bottom surface are considered to be the same as those in Woodward and Kashtalyan [30]:

$$C^-_{1111} = C^-_{2222} = 41.3\,\text{GPa}, \quad C^-_{1122} = 14.7\,\text{GPa}, \quad C^-_{1133} = C^-_{2233} = 10.1\,\text{GPa},$$

$$C^-_{3333} = 36.2\,\text{GPa}, \quad C^-_{1313} = C^-_{2323} = 10.0\,\text{GPa}, \quad C^-_{1212} = 13.3\,\text{GPa},$$

whereas the elastic constants on the top surface are taken as $C^+_{ijkl} = 2C^-_{ijkl}$. The distribution of elastic constants through the thickness of the FG layer for three values of the material gradient index $\gamma = 1$, 2 and 5 is shown in Fig. 3.

The geometric parameters of the plate are chosen to be $a = b = 1\text{m}$, $h_1 = h_3 = h/4$ and $h_2 = h/2$. To analyze the derived results efficiently, we introduce the following dimensionless variables:

$$\bar{u}_1 = E_1 u_1(0, a/2, z)/hp_0, \quad \bar{u}_2 = E_1 u_2(a/2, 0, z)/hp_0,$$
$$\bar{u}_3 = E_1 u_3(a/2, a/2, z)/hp_0,$$
$$\bar{\sigma}_{11} = \sigma_{11}(a/2, a/2, z)/p_0, \quad \bar{\sigma}_{22} = \sigma_{22}(a/2, a/2, z)/p_0,$$

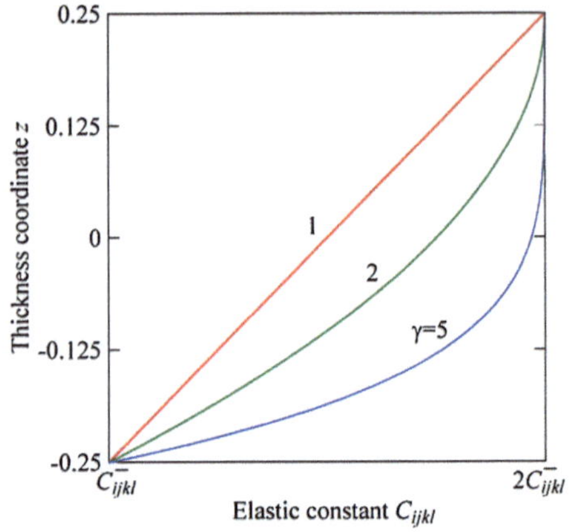

Fig. 3 Distribution of elastic constants through the thickness of the central FG layer

Table 5 Results for a FG three-layer square plate for $a/h = 2$ and $\gamma = 2$

I_n	$-\bar{u}_1(0.5)$	$\bar{u}_3(0.5)$	$\bar{\sigma}_{11}(0.5)$	$\bar{\sigma}_{22}(0.5)$	$-\bar{\sigma}_{12}(0.5)$	$\bar{\sigma}_{13}(0)$	$\bar{\sigma}_{23}(0)$	$\bar{\sigma}_{33}(0)$
3	3.1521	12.771	0.48242	2.4846	0.14094	0.55566	0.49464	0.43914
5	3.2011	12.835	0.47688	2.5322	0.14363	0.61857	0.57944	0.45240
7	3.2012	12.835	0.47647	2.5318	0.14364	0.61804	0.57621	0.45157
9	3.2012	12.835	0.47646	2.5318	0.14364	0.61796	0.57638	0.45156
11	3.2012	12.835	0.47646	2.5318	0.14364	0.61797	0.57637	0.45156
13	3.2012	12.835	0.47646	2.5318	0.14364	0.61797	0.57637	0.45156

Table 6 Results for a FG three-layer square plate for $a/h = 10$ and $\gamma = 2$

I_n	$-\bar{u}_1(0.5)$	$\bar{u}_3(0.5)$	$\bar{\sigma}_{11}(0.5)$	$\bar{\sigma}_{22}(0.5)$	$-\bar{\sigma}_{12}(0.5)$	$\bar{\sigma}_{13}(0)$	$\bar{\sigma}_{23}(0)$	$\bar{\sigma}_{33}(0)$
3	418.69	1969.8	5.9874	46.783	3.5268	2.8096	2.2369	0.48787
5	418.93	1970.7	5.9825	46.799	3.5288	2.9551	2.5383	0.50494
7	418.93	1970.7	5.9825	46.799	3.5288	2.9655	2.5325	0.50435
9	418.93	1970.7	5.9825	46.799	3.5288	2.9655	2.5329	0.50436
11	418.93	1970.7	5.9825	46.799	3.5288	2.9655	2.5329	0.50436

$$\bar{\sigma}_{33} = \sigma_{33}(a/2, a/2, z)/p_0, \quad \bar{\sigma}_{12} = \sigma_{12}(0, 0, z)/p_0,$$
$$\bar{\sigma}_{13} = \sigma_{13}(0, a/2, z)/p_0, \quad \bar{\sigma}_{23} = \sigma_{23}(a/2, 0, z)/p_0.$$

Tables 5 and 6 show that the SaS method allows the derivation of the 3D exact solution of elasticity for thick FG laminated composite plates with a prescribed

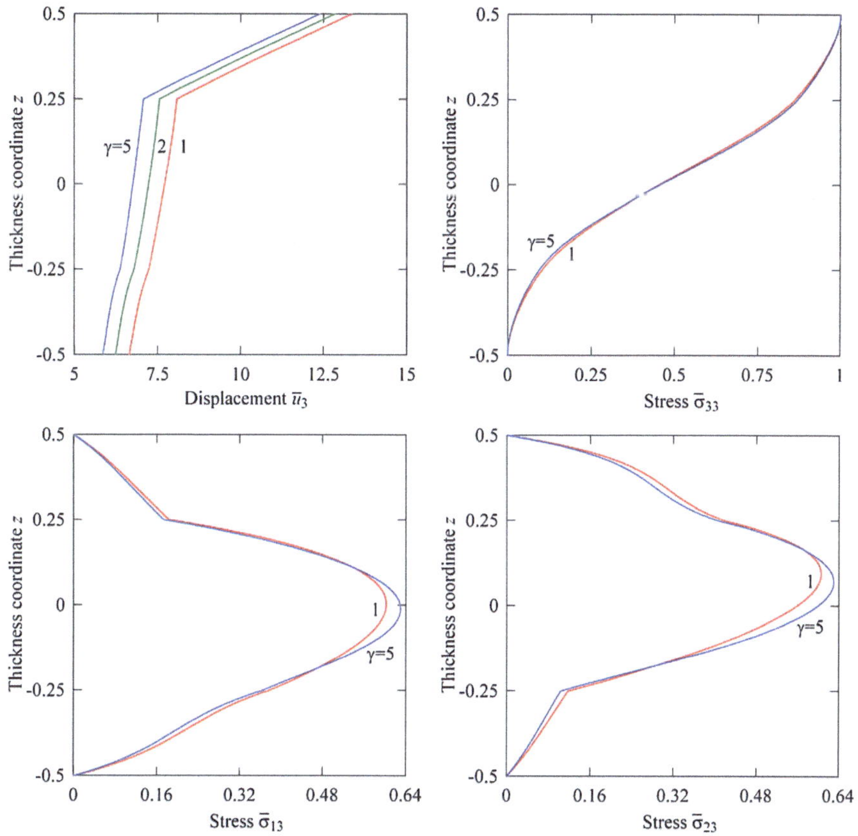

Fig. 4 Distributions of the transverse displacement and transverse stresses through the thickness of the FG three-layer plate with $a/h = 2$ and $\gamma = 1, 2, 5$

accuracy by using the large number of SaS inside each layer. Figures 4, 5 and 6 present the distributions of the transverse displacement and transverse stresses through the thickness of the plate for different values of the material gradient index γ employing 13 SaS for each layer. As can be seen, the boundary conditions on outer surfaces are satisfied again properly. The similar conclusion can be made concerning the continuity conditions at layer interfaces.

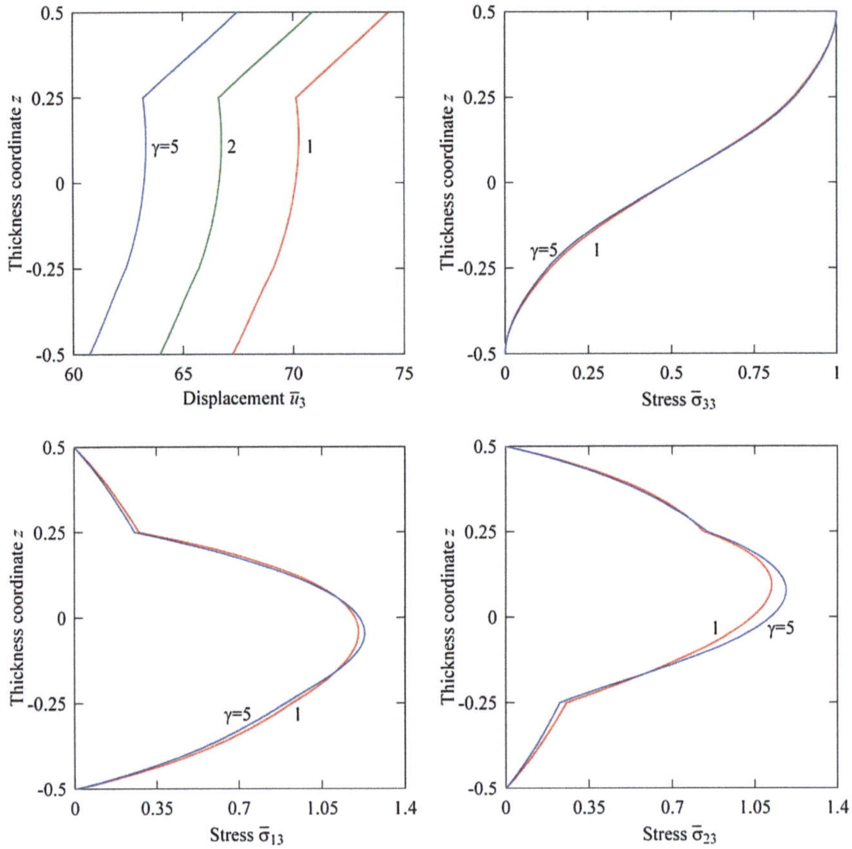

Fig. 5 Distributions of the transverse displacement and transverse stresses through the thickness of the FG three-layer plate with $a/h = 4$ and $\gamma = 1, 2, 5$

6 Conclusions

An efficient approach to 3D elasticity solutions for FG laminated plates has been proposed. It is based on the recently developed method of SaS located at Chebyshev polynomial nodes inside each layer of the plate and layer interfaces as well. The stress analysis is based on the 3D constitutive equations of elasticity and gives an opportunity to obtain the 3D exact solutions for thick and thin FG laminated composite plates with a specified accuracy.

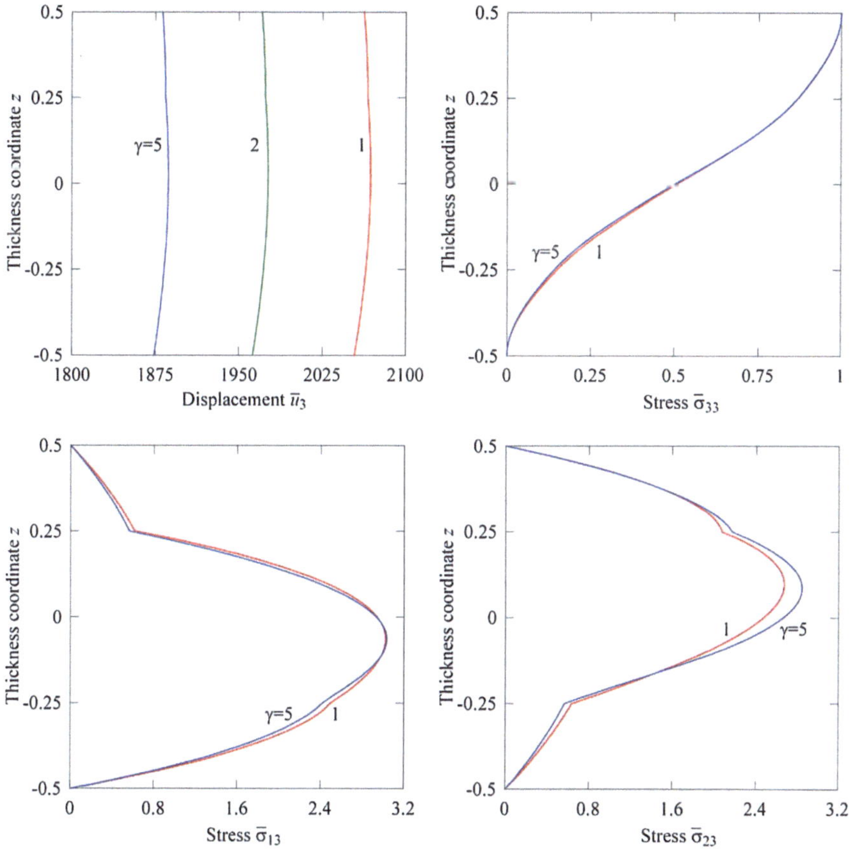

Fig. 6 Distributions of the transverse displacement and transverse stresses through the thickness of the FG three-layer plate with $a/h = 10$ and $\gamma = 1, 2, 5$

Acknowledgments This work was supported by Russian Ministry of Education and Science under Grant No. 9.137.2014K and by Russian Foundation for Basic Research under Grant No 13-01-00155.

References

1. Alibeigloo, A.: Exact solution for thermo-elastic response of functionally graded rectangular plates. Compos. Struct. **92**, 113–121 (2010)
2. Alibeigloo, A., Liew, K.M.: Thermoelastic analysis of functionally graded carbon nanotube-reinforced composite plate using theory of elasticity. Compos. Struct. **106**(3), 873–881 (2013)
3. Altenbach, H., Naumenko, K., Zhilin, P.A.: A direct approach to the formulation of constitutive equations for rods and shells. In: Pietraszkiewicz, W., Szymczak, C. (eds.) Shell Structures:

Theory and Applications, pp. 87–90. Taylor and Francis, London (2006)
4. Altenbach, H., Eremeyev, V.A.: Direct approach-based analysis of plates composed of functionally graded materials. Arch. Appl. Mech. **78**, 775–794 (2008)
5. Birman, V., Byrd, L.W.: Modeling and analysis of functionally graded materials and structures. Appl. Mech. Rev. **60**, 195–216 (2007)
6. Birsan, M., Altenbach, H., Sadowski, T., Eremeyev, V.A., Pietras, D.: Deformation analysis of functionally graded beams by the direct approach. Compos. B **43**, 1315–1328 (2012)
7. Burden, R.L., Faires, J.D.: Numerical Analysis, 9th edn. Brooks/Cole, Boston (2010)
8. Cheng, Z.Q., Batra, R.C.: Three-dimensional thermoelastic deformations of a functionally graded elliptic plate. Compos. B **31**, 97–106 (2000)
9. Jha, D.K., Kant, T., Singh, R.K.: A critical review of recent research on functionally graded plates. Compos. Struct. **96**, 833–849 (2013)
10. Kashtalyan, M.: Three-dimensional elasticity solution for bending of functionally graded rectangular plates. Eur. J. Mech. A. Solids **23**, 853–864 (2004)
11. Koizumi, M.: FGM activities in Japan. Compos. B **28**, 1–4 (1997)
12. Kulikov, G.M.: Refined global approximation theory of multilayered plates and shells. J. Eng. Mech. **127**, 119–125 (2001)
13. Kulikov, G.M., Carrera, E.: Finite deformation higher-order shell models and rigid-body motions. Int. J. Solids Struct. **45**, 3153–3172 (2008)
14. Kulikov, G.M., Plotnikova, S.V.: On the use of a new concept of sampling surfaces in a shell theory. Adv. Struct. Mater. **15**, 715–726 (2011a)
15. Kulikov, G.M., Plotnikova, S.V.: Solution of static problems for a three-dimensional elastic shell. Dokl. Phys. **56**, 448–451 (2011b)
16. Kulikov, G.M., Plotnikova, S.V.: On the use of sampling surfaces method for solution of 3D elasticity problems for thick shells. ZAMM **92**, 910–920 (2012b)
17. Kulikov, G.M., Plotnikova, S.V.: Exact 3D stress analysis of laminated composite plates by sampling surfaces method. Compos. Struct. **94**, 3654–3663 (2012a)
18. Kulikov, G.M., Plotnikova, S.V.: Advanced formulation for laminated composite shells: 3D stress analysis and rigid-body motions. Compos. Struct. **95**, 236–246 (2013a)
19. Kulikov, G.M., Plotnikova, S.V.: Three-dimensional exact analysis of piezoelectric laminated plates via a sampling surfaces method. Int. J. Solids Struct. **50**, 1916–1929 (2013c)
20. Kulikov, G.M., Plotnikova, S.V.: A sampling surfaces method and its application to three-dimensional exact solutions for piezoelectric laminated shells. Int. J. Solids Struct. **50**, 1930–1943 (2013b)
21. Ootao, Y., Tanigawa, Y.: Three-dimensional solution for transient thermal stresses of an orthotropic functionally graded rectangular plate. Compos. Struct. **80**, 10–20 (2007)
22. Ootao, Y.: Transient thermoelastic analysis for a multilayered thick strip with piecewise exponential nonhomogeneity. Compos. B **42**, 973–981 (2011)
23. Ootao, Y., Ishihara, M.: Three-dimensional solution for transient thermoelastic problem of a functionally graded rectangular plate with piecewise exponential law. Compos. Struct. **106**, 672–680 (2013)
24. Pagano, N.J.: Exact solutions for rectangular bidirectional composites and sandwich plates. J. Compos. Mater. **4**, 20–34 (1970)
25. Plevako, V.P.: On the theory of elasticity of inhomogeneous media. J. Appl. Math. Mech. **35**, 806–813 (1971)
26. Reddy, J.N., Cheng, Z.Q.: Three-dimensional thermomechanical deformations of functionally graded rectangular plates. Eur. J. Mech. A. Solids **20**, 841–855 (2001)
27. Vel, S.S., Batra, R.C.: Exact solution for thermoelastic deformations of functionally graded thick rectangular plates. AIAA J. **40**, 1421–1433 (2002)
28. Vel, S.S., Batra, R.C.: Three-dimensional analysis of transient thermal stresses in functionally graded plates. Int. J. Solids Struct. **40**, 7181–7196 (2003)
29. Vlasov, B.F.: On the bending of a rectangular thick plate (in Russ.). Vestnik. Moskov. Univ. Ser. Mat. Mekh. **2**, 25–31 (1957)

30. Woodward, B., Kashtalyan, M.: Three-dimensional elasticity solution for bending of transversely isotropic functionally graded plates. Eur. J. Mech. A. Solids **30**, 705–718 (2011)
31. Wu, C., Chiu, K., Wang, Y.: A review on the three-dimensional analytical approaches of multilayered and functionally graded piezoelectric plates and shells. Comput. Mater. Continua. **8**, 93–132 (2008)
32. Zhilin, P.: Mechanics of deformable directed surfaces. Int. J. Solids Struct. **12**, 635–648 (1976)
33. Zhong, Z., Shang, E.: Three-dimensional exact analysis of a simply supported functionally gradient piezoelectric plate. Int. J. Solids Struct. **40**, 5335–5352 (2003)

Prediction of Eigenfrequencies of the Middle Ear Oscillating System After Tympanoplasty and Stapedotomy

Gennadi I. Mikhasev, Irina Slavashevich and Kirill Yurkevich

Abstract The mathematical model of the reconstructed middle ear subjected to tympanoplasty and stapedotomy is proposed. The biomechanical system consists of a restored tympanic membrane made from cartilage and the total prosthesis replacing the malleus-incus-stapes chain. The reconstructed eardrum coupled to the circular prosthesis plate is modeled as an annular circular elastic plate. The prosthesis stem is considered as a rigid cylindrical rod that projects through a small perforation in the fixed footplate stapes into the vestibule of the inner ear. Resting in the perilymph, the piston-like stem is under static and dynamic forces acting from the cochlear liquid or/and additional spring element. The static forces assert stable position of the prosthesis but result in initial membrane stresses in the reconstructed eardrum. The eardrum motion is governed by the system of three differential equations written in terms of normal and tangential displacements taking into account the initial stresses. Free low-frequency vibrations of the prestressed biomechanical system are studied using two methods: when the initial stresses are small, a perturbation method is applied; if not, the finite element approach is utilized to predict the natural frequencies.

1 Introduction

The normal middle ear consists of the tympanic membrane (TM) and three hinged ossicles (malleus, incus and stapes), as illustrated in Fig. 1a. The vibrations of the TM due to external sound waves are transferred by the ossicles to the stapes footplate and the stapes footplate vibration induces sound pressure in the perilymphatic fluid of the inner ear. Due to the area difference and structural difference between the TM and

G.I. Mikhasev (✉) · I. Slavashevich · K. Yurkevich
Department of Bio- and Nanomechanics, Belarusian State University, Minsk, Belarus
e-mail: mikhasev@bsu.by

I. Slavashevich
e-mail: slavashevichi@yandex.ru

K. Yurkevich
e-mail: kirill.bsu@gmail.com

H. Altenbach and G.I. Mikhasev (eds.), *Shell and Membrane Theories in Mechanics and Biology*, Advanced Structured Materials 45, DOI 10.1007/978-3-319-02535-3_14

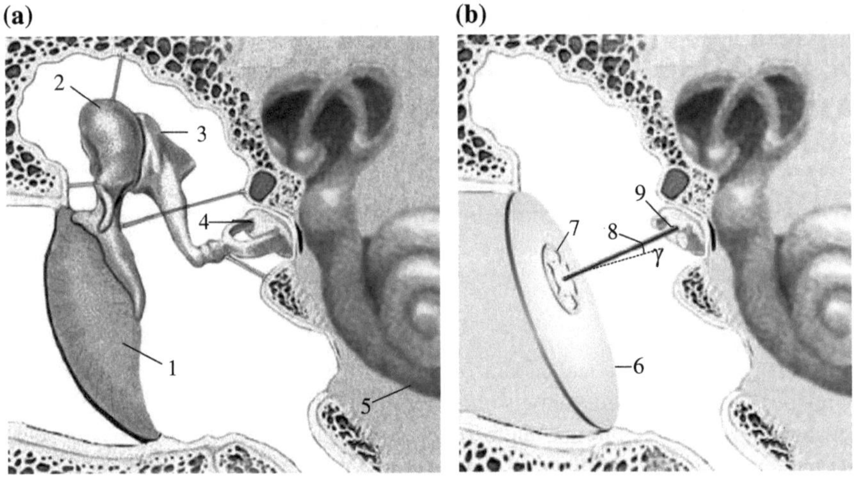

Fig. 1 Profiles of normal (**a**) and reconstructed middle ear (**b**): *1* TM, *2* malleus, *3* incus, *4* stapes, *5* cochlear, *6* reconstructed TM, *7* prosthesis plate, *8* prosthesis shaft, *9* oval window fistula (opening of the inner ear where the stapes footplate rests), γ is deflection angle of the steam from the initial straight position

the stapes footplate, the middle ear provides force intensification and thus impedance matching between air and fluid (for more details, refer to Kobrak [1]). Mechanical injuries of the middle ear elements, different diseases and pathologic changes of the middle-ear structures can result in a hearing loss of up to 60 dB sound pressure level (SPL). In particular, mechanical injuries of the TM can decrease hearing at low frequencies (less than 2 kHz) by 10–30 dB SPL [2]. Another example: the salt-deposition process in the middle-ear cavity may be the cause of such pathology as the otosclerotic ankylosis of the ossicular joints (i.e. fixation of the incudomalleal or/and the incudostapedial joint [3, 4]) that may result in the loss of an acoustic transfer up to 42 dB SPL [5]. Common treatment for these and other pathological diseases is the surgical reconstruction of the middle ear to achieve normal sound conduction. The rare and worst case of this procedure results in replacement of all elements of the middle ear. The surgery is performed in two steps: at the first stage a surgeon performs tympanoplasty using cartilage transplants and at the second step he inserts the total ossicular replacement prosthesis (TORP) bridging a gap in the ossicular chain [6]. If the stapes footplate is immobilized by bone growth around the oval window, a small hole is drilled manually or by a laser beam in the footplate (stapedotomy). Then the piston-like total cylindrical prosthesis is glued by its plate on the restored TM, and inserted into the vestibule through the drilled hole (see Fig. 1b).

The negative consequence of the stapedotomy surgery is a reduction in the total stiffness of the middle-ear structure [4] and, as a result, a drop in the natural frequencies of the biomechanical oscillating system [7]. The quality of sound signal transmission through the reconstructed middle ear (RME) into the cochlea depends strongly on the spectrum of natural frequencies and eigenforms of the RME, which should match the original one. For similar reconstructions, including tympanoplasty and stapedotomy, the outcome is influenced by the cartilage transplant thickness, the prosthesis mass and geometrical design (length, diameter, the stem alignment) [7] as well as its position with respect to the TM [6].

This study relates to the case when the middle ear is subjected to both tympanoplasty and stapedotomy. The main objective is to propose the mathematical model of the RME based on the theory of thin isotropic circular plate and make a detailed theoretical analysis of the natural frequencies and modes versus the geometrical and physical parameters of the cartilage transplants and prosthesis. Another goal is to study the dependence of the natural frequencies from a position of the prosthesis at the eardrum and alignment of its stem.

2 Mathematical Model of the Reconstructed Middle Ear

2.1 Model Development

There are five different TM reconstruction techniques [8]. The most difficult case of tympanoplasty relates to use of the cartilage plate technique implying the total replacement of the membrane with the cartilage transplants. According to this technique, the reconstructed TM may be represented by almost circular thin plate made of conchal or tragal cartilage. This type of cartilage may be considered as elastic and isotropic material. Zahnert et al. [9] showed that the generalized stiffness of the cartilage is primarily influenced by its thickness, so that the thin cartilage plate under reasonable load may be characterized by Young's modulus. Here, Young's modulus E is taken from 2.8 to 3.4 N/mm^2 for a plate made of conchal and tragal cartilage with a thickness varying from 270 to 1,000 mm [9].

The tympanic ring of the eardrum in normal is a curve close to an ellipse with the semiaxes $a_1, a_2 (a_1 < a_2)$, where the horizontal semiaxis a_1 varies from 0.4 to 0.45 cm, and the vertical one, a_2, is about 0.45–0.5 cm [10]. These data are typical of an adult and may vary slightly. When developing the simplified mathematical model, the reconstructed TM will be considered as the circular plate of the radius a. In contrast, applying the FEM approach, we will consider the real sizes of the eardrum closed to an ellipse.

We introduce a global Cartesian coordinate system $Oxyz$ related to the TM (see Fig. 2), where the point of origin O coincides with the TM center, and the y-axis is orientated along the manubrium mallei of the natural eardrum (see Fig. 1a).

The stapes footplate spatial orientation may be defined by introducing a local Cartesian coordinate system $S\xi\eta\zeta$ as illustrated in Fig. 2a, where the point S is the center of the footplate. The orientation of the $S\xi\eta\zeta$ coordinate system in regard to the global coordinate system is assumed to be known [11]. Let H be a point at

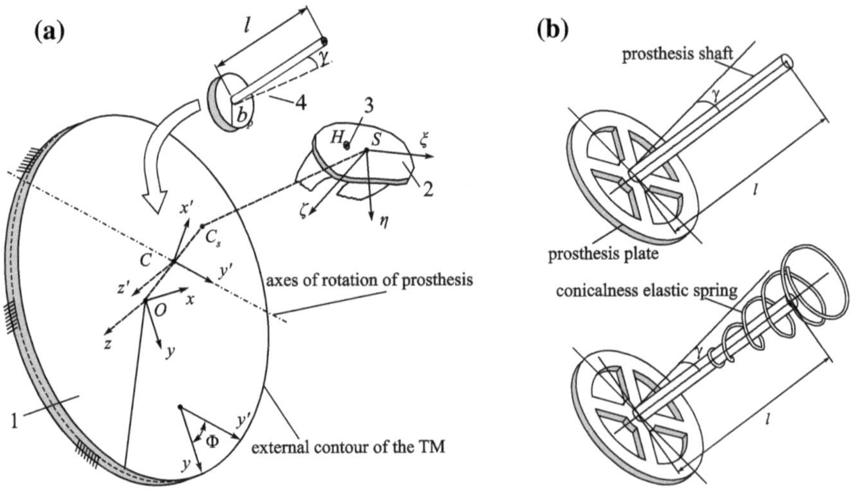

Fig. 2 Components of the RME: **a** *1* reconstructed TM, *2* stapes footplate, *3* hole in the stapes footplate, *4*. geometrical model of the prosthesis with the inclined shaft. Cartesian coordinate systems ($Oxyz$ is a global coordinate system related to the TM; $Cx'y'z'$ and $S\xi\eta\zeta$ are local coordinate systems related to the prosthesis plate and stapes footplate, respectively), ϕ is an angle between the y- and the y'-axes. Point C is the contact point of the prosthesis plate and the TM, point H is the perforation center. **b** The prosthesis like TORP with the pliable shaft (*upper*), and the modified prosthesis with the conicalness elastic spring attached to the shaft (*lower*)

the footplate where a small hole is drilled. Its position with respect to the global coordinate system is also defined.

The prostheses like TORP are usually made of titanium or gold (see also at http://www.kurzmed.de). Due to the thin and pliable shaft (see Fig. 2b), the plate at the upper end of the prosthesis (called prosthesis plate) can be positioned according to the plane of the reconstructed TM. There are several contact points at the eardrum (the center, posterior and inferior points [11]) where the prosthesis plate is usually located. According to [12], better results are obtained when the prosthesis is attached near to the center of the membrane than to the posterior or inferior points. Similar conclusion has been made [11] when studying strain-stress state in the middle-ear system after tympanoplasty and insertion of a malleus-incus prosthesis. Let x_C, y_C, z_C be the coordinates of the contact point C in the global coordinate system (here $z_C = 0$). In our mathematical model the point C coincides with the center O, though its position can be any one in the FEM simulation.

Although the prosthesis plate structure is complex (see Fig. 2b), it is treated as a perfectly rigid thin circular plate of a radius b_p. We consider also rigid coupling between the prosthesis plate and the reconstructed TM, so that the TM may be modelled as an annular plate having radii a and $b_p (a > b_p)$ with rigid attachment conditions to the prosthesis plate.

After attachment of the prosthesis plate to the TM, the prosthesis is adapted in its shape to individual middle-ear anatomy by adjusting the angle between its shaft and plate. Then the shaft is inserted into the vestibule through a small perforation. We assume that the perforation has the form of a cylinder of the radius r_p coinciding exactly with the radius of the prosthesis cylindrical shaft. Thus, the prosthesis can perform only the longitudinal oscillations along the perforation axis. It should be noted that the shape of the curved shaft influences the moments of inertia of the prosthesis. However, in the case of its longitudinal oscillations, the choice of the prosthesis profile is not crucial. Therefore, the prosthesis shaft will be considered straight rigid rod inclined to the plate with the inclination angle $\pi/2-\gamma$ (see Fig. 2). Resting in the perilymph, the stem end experiences static forces from the cochlear liquid. The magnitude of these forces depends on the mechanical properties of the cartilage transplant and the membrane of the round window. If the round window membrane stiffness is too small and the prosthesis stem is too short, the prosthesis may dislodge from the surgically created oval window fistula after the sudden changes of the external sound pressure. To avoid this scenario, a modified version of the TORP (see Fig. 2b) is proposed for this kind of reconstruction. The new prosthesis contains the conicalness elastic spring, with its narrow end clamping to the prosthesis shaft and its wide edge resting against the fixed footplate of the stapes. The attached spring provides a stable position of the prosthesis in the middle-ear cavity and stiffens the RME as well.

Let l be the prosthesis length, and l_{min} the distance between the contact point C at the TM and the footplate perforation point H (see Fig. 2a). Then the following inequality $\Delta l = l - l_{min} > 0$ is required for prosthesis fixation. Indeed, after inserting the prosthesis, it will be compressed by both forces acting from the cochlear liquid (or the spring element attached to the prosthesis shaft) and shear forces from the TM. But these forces, providing a stable position of the prosthesis, result in the initial stresses in the reconstructed TM. The magnitude of these stresses can influence the natural frequencies of the oscillating system and should be taken into account in our simulations.

2.2 Initial Equilibrium Position

Let us consider the initial equilibrium position of the system after inserting the prosthesis. When ignoring the friction force between the stem and internal surface of the perforation, it is characterized by the force F° acting from the cochlear liquid and the stresses $T_i^\circ, S_i^\circ, Q_i^\circ, M_i^\circ, M_{12}^\circ (i = 1,2)$ in the TM. Here, T_i°, S_i° and Q_i° are the membrane and shear stresses respectively, and M_i°, M_{12}° are the bending and twisting moments. In particular, in the polar coordinate system r, φ with the center in the contact point O (see Fig. 3a), the membrane stresses are

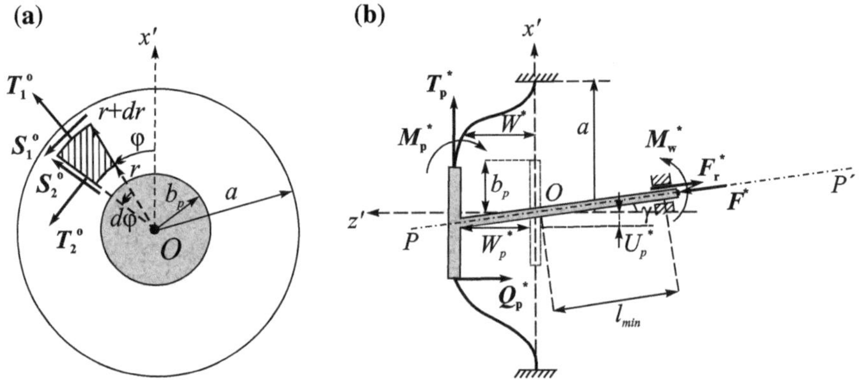

Fig. 3 Two sections of the geometrical model of the RME: **a** the annular plate, the polar coordinate system r, φ with the center in the contact point $O(C)$, and the initial membrane stresses T_i°, S_i°, **b** the plate deformation, the prosthesis displacements in the $x' - z'$ *plane* and forces appearing in the oscillating system

written as follows [11]

$$\begin{aligned} T_1^{\circ} &= K(1-\nu)\frac{-2c_2 + r^2[2c_1(1+\nu)r^2 + c_3(9-\nu^2)]}{r^3}\cos\varphi, \\ T_2^{\circ} &= K(1-\nu)\frac{2c_2 + r^2[6c_1(1+\nu)r^2 - c_3(3-4\nu+\nu^2)]}{r^3}\cos\varphi, \\ S^{\circ} &= S_i^{\circ} = K(1-\nu)\frac{-2c_2 + r^2[2c_1(1+\nu)r^2 - c_3(3-4\nu+\nu^2))]}{r^3}\sin\varphi, \end{aligned} \tag{1}$$

where $K = Eh/(1-\nu^2)$ is the plate stiffness, E, ν are Young's modulus and Poisson ratio of the cartilage, h is the TM thickness, and c_j are constants, which are found from the boundary conditions. The initial force F° depends on a penetration length δ_p of the prosthesis stem into the vestibule and is found as

$$F^{\circ} = k^*\delta_p. \tag{2}$$

Here

$$k^* = k_{rw}^* = 8\pi T_{rw}(S_{ps}/S_{rw})^2, \tag{3}$$

where T_{rw} is the tensile force of the round window membrane, and S_{ps}, S_{rw} are the areas of the stem cross-section and the round window membrane. If the additional spring with the stiffness k_{sp}^* is attached to the prosthesis shaft, then $k^* = k_{rw}^* + k_{sp}^*$.

Values of the parameter δ_p and the initial stresses T_i°, S_i° depend on the prosthesis length increment $\Delta l = l - l_{min} > 0$. We introduce the following magnitudes

$$T_c^{\circ} = \frac{Eh\delta_p \sin\gamma}{12a(1-\nu^2)}, \quad \varepsilon = \frac{a\delta_p \sin\gamma}{h^2}, \tag{4}$$

where T_c° is the characteristic value of the initial membrane stresses in the TM, and ε is a dimensionless parameter. As previously shown [7], for this kind of the reconstruction technique and real sizes of the cartilage transplant (here $h \approx 0.5$ mm) and prosthesis, the parameter δ_p varies in the interval from 0.01 to 0.05 mm. [13] indicates that in stapedotomy a mean penetration length of the prosthesis stem into the cochlear liquid is about 0.5 mm. Thus, in any case a parameter $\varepsilon \ll 1$ may be considered here as a small one.

2.3 Governing Equations

Small flexural vibrations of the prestresses TM are governed by the following equation [14]

$$\Delta^2 W - \varepsilon \Delta_T W - \lambda W = 0 \tag{5}$$

written in the dimensionless form. Here Δ is the Laplace operator in the polar coordinate system $r = a\rho, \varphi\ (b = b_p/a < \rho < 1,\ 0 \leq \varphi < 2\pi)$ with the center in the contact point O, Δ_T is the differential operator introduced as follows

$$\Delta_T W = \frac{1}{\rho}\left\{\frac{\partial}{\partial \rho}\left(\rho t_1^\circ \frac{\partial W}{\partial \rho} + s^\circ \frac{\partial W}{\partial \varphi}\right) + \frac{\partial}{\partial \varphi}\left(\frac{t_2^\circ}{\rho}\frac{\partial W}{\partial \varphi} + s^\circ \frac{\partial W}{\partial \rho}\right)\right\}, \tag{6}$$

$W^* = aW$ is the normal displacement of the plate, $t_i^\circ = T_i^\circ/T_c, s^\circ = S^\circ/T_c$ are dimensionless membrane stresses, λ is a required parameter associated with the natural frequency ω by the equation $\omega = \lambda^{1/2}\omega_c$, where $\omega_c = hE^{1/2}a^{-2}[12\sigma(1-\nu^2)]^{-1/2}$ is the characteristic frequency and σ is the density of the cartilage transplant respectively.

Let $U^* = aU,\ V^* = aV$ be the plate displacements in the radial and circumferential directions respectively. We introduce another dimensionless parameter $\varepsilon_t = h^2/(12a^2)$. For ordinary used cartilage transplants (at $h \approx 0.5$ mm, $a \approx 5$ mm), ε_t is a small parameter so that $\varepsilon_t \sim \varepsilon^2$. We assume here $\varepsilon_t = \tau\varepsilon^2$, where $\tau \sim 1$. Then in-plane motion of the plate will be governed by the following equations

$$(\mathbf{L}_p + \varepsilon^2 \tau \lambda)\mathbf{X}^T = 0, \tag{7}$$

written in the dimensionless form, where $\mathbf{X} = (U,\ V)$ is the 2-vector, the superscript T denotes a transposition, and $\mathbf{L}_p$ is the 2×2 matrix which elements are the differential operators

$$l_{11} = \rho\frac{\partial^2}{\partial \rho^2} + \frac{\partial}{\partial \rho} - \frac{1}{\rho} + \frac{1-\nu}{2\rho}\frac{\partial^2}{\partial \varphi^2}, \quad l_{12} = \frac{1+\nu}{2}\frac{\partial^2}{\partial \rho \partial \varphi} - \frac{3-\nu}{2\rho}\frac{\partial}{\partial \varphi},$$
$$l_{21} = \frac{1+\nu}{1-\nu}\frac{\partial^2}{\partial \rho \partial \varphi} + \frac{3-\nu}{(1-\nu)\rho}\frac{\partial}{\partial \varphi}, \quad l_{22} = \rho\frac{\partial^2}{\partial \rho^2} + \frac{\partial}{\partial \rho} - \frac{1}{\rho} + \frac{2}{(1-\nu)\rho}\frac{\partial^2}{\partial \varphi^2}.$$

Although the flexural and tangential vibrations governed by Eqs. (5) and (7) look like uncoupled ones here, the displacements W and U, V are linked by the boundary conditions at the inner mobile circular counter $\rho = b$:

$$W = w_p, \quad \frac{\partial W}{\partial \rho} = 0, \quad U = u_p \cos\varphi, \quad V = u_p \sin\varphi \quad \text{at} \quad \rho = b, \tag{8}$$

where w_p and $u_p = \tan\gamma w_p$ are the dimensionless displacements of the plate prosthesis in the Oz'-axis and Ox'-axis directions respectively (see Fig. 3b). It is assumed that the external counter of the TM is clamped so that

$$W = \frac{\partial W}{\partial \rho} = U = V = 0 \quad \text{at} \quad \rho = 1. \tag{9}$$

Taking into account the boundary conditions Eqs. (5), (7), and (8) describe the coupled flexural and tangential vibrations of the annular prestresses plate.

We aim to study low-frequency vibrations here. Therefore, the hydrodynamic forces acting on the prosthesis shaft from the perilymph [13] are ignored. It is also assumed that the friction forces between the shaft and perforation wall, and the viscosity force form the perilymph are too small. Then the equation governing oscillations of the prosthesis along the PP'-axis (see Fig. 3b) will be as follows

$$\cos^2\gamma\, Q_p(\lambda, w_p) + \sin\gamma\cos\gamma T_p - k\, w_p + \lambda m_p w_p = 0. \tag{10}$$

Here

$$Q_p = \frac{1-\nu^2}{Eha} Q_p^*, \qquad T_p = \frac{1-\nu^2}{Eha} T_p^*, \qquad m_p = \frac{mh}{12\sigma a^4}, \qquad k = \frac{1-\nu^2}{Eh} k^*, \tag{11}$$

where $k^* = k_{rw}^* + k_{sp}^*$, Q_p^* and T_p^* are the resultant shear and membrane stresses appearing along the internal edge of the annular plate, m is the prosthesis mass, and

$$F^* = k^* a w_p / \cos\gamma \tag{12}$$

is the resultant force acting from the cochlear liquid and the spring element attached to the prosthesis shaft.

Besides the forces appearing in Eq. (10), there are others acting on the prosthesis: the resultant reaction force F_r^* and moment M_w^* acting from the perforation wall, and the resultant bending moment M_p^* from the prosthesis plate. These forces are easily found, but they do not have influence on the prosthesis oscillation along the PP'-axis.

Equations (5), (7), and (10), with the boundary conditions (8), (9), describe small low-frequency free vibrations of the prestressed bio-mechanical system of the reconstructed middle ear.

Since the magnitudes Q_p, T_p are directly proportional to w_p, Eq. (10) holds at $w_p = 0$. Let Ω be the natural frequency spectrum of the boundary-value problem

(5), (7)–(9). Then $\Omega = \Omega_1 \bigcup \Omega_2$, where Ω_1 contains the eigen-frequencies for which $w_p \neq 0$, and the subset Ω_2 consists of frequencies in the case when the prosthesis is dead so that is $w_p = 0$. The second part of the spectrum Ω_2 is of no interest for the current study. Let $w_p \sim 1$ at $\varepsilon \to 0$. Then

$$(W, U, V) = (w,\ u \tan\gamma,\ v \tan\gamma) w_p, \tag{13}$$

where $w(\rho, \varphi)$ is the solution of Eq. (5) with the boundary conditions

$$w = \frac{\partial w}{\partial \rho} = 0 \quad \text{at} \quad \rho = 1, \quad \text{and} \quad w = 1,\ \frac{\partial w}{\partial \rho} = 0 \quad \text{at} \quad \rho = b, \tag{14}$$

and $u(\rho, \varphi)$, $v(\rho, \varphi)$ satisfy Eq. (7) with the boundary conditions

$$u = v = 0 \quad \text{at} \quad \rho = 1, \quad \text{and} \quad u = -\cos\varphi,\ v = \sin\varphi \quad \text{at} \quad \rho = b. \tag{15}$$

Then Eq. (10) may be rewritten as follows

$$\cos^2\gamma\, q_p(\lambda) + \sin^2\gamma\, t_p(\lambda) - k + \lambda m_p = 0, \tag{16}$$

where

$$\begin{aligned} q_p(\lambda) &= -\int_0^{2\pi} \frac{\partial}{\partial \rho}\left(\frac{\partial^2 w}{\partial \rho^2} + \frac{1}{\rho}\frac{\partial w}{\partial \rho} + \frac{1}{\rho^2}\frac{\partial^2 w}{\partial \varphi^2}\right)\bigg|_{\rho=b} d\varphi, \\ t_p(\lambda) &= \int_0^{2\pi} \left[\left(\frac{\partial u}{\partial \rho} + \frac{\nu}{\rho}\left(u + \frac{\partial v}{\partial \varphi}\right)\right)\cos\varphi \right. \\ &\quad \left. + \frac{1-\nu}{2}\left(\frac{\partial v}{\partial \rho} - \frac{1}{\rho}\left(v - \frac{\partial u}{\partial \varphi}\right)\right)\sin\varphi\right]_{\rho=b} d\varphi. \end{aligned} \tag{17}$$

3 Asymptotic Approach

The second term in Eq. (5) does not permit to find an exact solution of the boundary-value problem (5), (7), (14)–(16). Its solution may be found in the form of series

$$\begin{aligned} w &= \sum_{j=0}^{\infty} \varepsilon^j w_j(\rho, \varphi), \quad u = \sum_{j=0}^{\infty} \varepsilon^j u_j(\rho, \varphi), \\ v &= \sum_{j=0}^{\infty} \varepsilon^j v_j(\rho, \varphi), \quad \lambda = \sum_{j=0}^{\infty} \varepsilon^{2j} \lambda_j. \end{aligned} \tag{18}$$

The substitution of Eqs. (17) into Eqs. (5), (7), (14)–(16) produces the sequence of the boundary-value problems with respect to w_j, u_j, v_j, λ_j. We consider these problems step by step.

In the zeroth order approximation, one has the homogeneous equations

$$\Delta^2 w_0 - \lambda_0 w_0 = 0 \tag{19}$$

$$\mathbf{L}_p \mathbf{X}_0^T = 0, \tag{20}$$

with the non-homogeneous boundary conditions (14) and (15), respectively. Here $\mathbf{X}_0 = (u_0, v_0)$ is the 2-vector.

The boundary-value problem (14), (19) has the following solution

$$w_0(\rho; \lambda_0) = A_1 J_0(\lambda_0^{1/4}\rho) + A_2 Y_0(\lambda_0^{1/4}\rho) + A_3 I_0(\lambda_0^{1/4}\rho) + A_4 K_0(\lambda_0^{1/4}\rho). \tag{21}$$

Here and below, J_k, Y_k are the kth-order Bessel functions of the first and second kinds respectively, J_k, Y_k are the modified kth-order Bessel functions of the first and second kinds respectively, and $A_j (j = 1, \dots, 4)$ are constants, which are found from the nonhomogeneous boundary conditions (14).

Remark 14.1 A solution of Eq. (19) may be found in the form of the function $w_0 = \chi_{n0}(\rho) \sin n\varphi$, where $n \geq 1$. However, it satisfies the boundary conditions (8) only at $w_p = 0$ and corresponds to the plate vibration forms, which do not stimulate oscillations of the adjoint prosthesis. These modes of the reconstructed TM with frequencies from the subset Ω_2 will be called here the "dead" ones.

The boundary-value problem (15), (20) has the simple solution

$$\begin{aligned} u_0 &= \left[c_{01}(1-3\nu)\rho^2 + \frac{c_{02}}{\rho^2} + c_{03}(3-\nu)^2 \ln\rho - c_{03}(1-\nu^2) - c_{04} \right] \cos\varphi, \\ v_0 &= \left[c_{01}(5+\nu)\rho^2 + \frac{c_{02}}{\rho^2} - c_{03}(3-\nu)^2 \ln\rho - 2c_{03}(1+\nu) + c_{04} \right] \sin\varphi, \end{aligned} \tag{22}$$

where constants c_{0i} are determined from the boundary conditions (15).

In the first order approximation, one has the following equations

$$\Delta^2 w_1 - \lambda_0 w_1 = \Delta_T w_0, \tag{23}$$

$$\mathbf{L}_p \mathbf{X}_1^T = 0, \tag{24}$$

where $\mathbf{X}_1 = (u_1, v_1)$. Equation (24) with the homogeneous boundary conditions

$$u_1 = v_1 = 0 \quad \text{at} \quad \rho = b, 1 \tag{25}$$

have the trivial solution $u_1 = 0$, $v_1 = 0$. We consider the non-homogeneous Eq. (23) with the homogeneous boundary conditions

$$w_1 = \frac{\mathrm{d}w_1}{\mathrm{d}\rho} = 0 \quad \text{at} \quad \rho = b, 1. \tag{26}$$

Taking Eq. (1) into account, Eq. (23) may be rewritten as follows

$$\Delta^2 w_1 - \lambda_0 w_1 = f_c(\rho; \lambda_0) \cos\varphi + f_s(\rho; \lambda_0) \sin\varphi, \tag{27}$$

where $f_c(\rho; \lambda_0)$, $f_s(\rho; \lambda_0)$ are known functions, which are not written out here.

Let $z_k(\rho)(k = 1, 2, \ldots)$ be an infinite orthonormal system of eigenfunctions of the boundary-value problem

$$\left(\frac{\mathrm{d}^2}{\mathrm{d}\rho^2} + \frac{1}{\rho}\frac{\mathrm{d}}{\mathrm{d}\rho} - \frac{1}{\rho^2}\right)^2 z - \beta^4 z = 0, \tag{28}$$

$$z = \frac{\mathrm{d}z}{\mathrm{d}\rho} = 0 \quad \text{at} \quad \rho = 1, \ b \tag{29}$$

and β_k be a corresponding sequence of eigenvalues. It is easily found that $\beta_1 = 6.830$, $\beta_2 = 11.286$, $\beta_3 = 15.764, \ldots$, and

$$z_k(\rho) = d_k^{-1} \left[J_1(\beta_k\rho) + C_2 Y_1(\beta_k\rho) + C_3 I_1(\beta_k\rho) + C_4 K_1(\beta_k\rho)\right], \tag{30}$$

$$d_k^2 = \int_b^1 \left[J_1(\beta_k\rho) + C_2 Y_1(\beta_k\rho) + C_3 I_1(\beta_k\rho) + C_4 K_1(\beta_k\rho)\right]^2 \rho \mathrm{d}\rho, \tag{31}$$

where C_2, C_3, C_4 are determined from the boundary conditions (29).

Then the solution of the boundary-value problem (26), (27) may be presented in the form of the series

$$w_1(\rho, \varphi; \lambda_0) = \sum_{k=1}^{\infty} \left[B_{ck}(\lambda_0) \cos\varphi + B_{sk}(\lambda_0) \sin\varphi\right], \tag{32}$$

$$B_{ck} = (\beta_k^4 - \lambda_0)^{-1} \int_b^1 f_c(\rho; \lambda_0) z_k(\rho) \rho \mathrm{d}\rho. \tag{33}$$

The functions B_{sk} are found from Eq. (33), where the subscript c is replaced by s. It is assumed here that $\lambda_0 \neq \beta_k^4$ for any $k = 1, 2, \ldots$.

In the second order approximation, one gets the non-homogeneous equations

$$\mathbf{L}_p \mathbf{X}_2^T = -\lambda_0 \tau \mathbf{X}_0 \tag{34}$$

$$\Delta^2 w_2 - \lambda_0 w_2 = \Delta_T w_1 + \lambda_2 w_0, \tag{35}$$

with the homogeneous boundary conditions (25), (26) for the vector $\mathbf{X}_2 = (u_2, v_2)$ and w_2.

The boundary-value problem (25), (34) has the following solution

$$\begin{aligned} u_2 &= \big[c_{21}(1-3\nu)\rho^2 + c_{22}\rho^{-2} + c_{23}(3-\nu)^2 \ln \rho - c_{23}(1-\nu^2) - c_{24} \\ &\quad + a_{c1}\rho^3 + a_{c2}\rho^{-1} + a_{c3}\rho \ln \rho + a_{c4}\rho\big] \cos \varphi, \\ v_2 &= \big[c_{21}(5+\nu)\rho^2 + c_{22}\rho^{-2} - c_{23}(3-\nu)^2 \ln \rho - 2c_{23}(1+\nu) + c_{24} \\ &\quad + a_{s1}\rho^3 + a_{s2}\rho^{-1} + a_{s3}\rho \ln \rho + a_{s4}\rho\big] \sin \varphi, \end{aligned} \tag{36}$$

where

$$\begin{aligned} a_{c1} &= \frac{2(-3+23\nu-10\nu^2)\tau\lambda_0 c_{01}}{45(1-\nu)}, \\ a_{c2} &= \frac{2\tau\lambda_0 c_{02}}{3(1-\nu)}, \\ a_{c3} &= \frac{2(-18+21\nu-8\nu^2+\nu^3)\tau\lambda_0 c_{03}}{3(1-\nu)}, \\ a_{c4} &= \frac{[(219-249\nu+97\nu^2-11\nu^3)c_{03}+(12-6\nu)c_{04}]\tau\lambda_0}{9(1-\nu)}, \\ a_{s1} &= \frac{(-81-4\nu+5\nu^2)\tau\lambda_0 c_{01}}{45(1-\nu)}, \\ a_{s2} &= \frac{\tau\lambda_0 c_{02}}{3}, \\ a_{s3} &= \frac{(45-39\nu+11\nu^2-\nu^3)\tau\lambda_0 c_{03}}{3(1-\nu)}, \\ a_{s4} &= \frac{[(-267_2 37\nu-65\nu^2+7\nu^3)c_{03}-(15-3\nu)c_{04}]\tau\lambda_0}{9(1-\nu^2)}. \end{aligned}$$

Here c_{0i} are the constants mentioned above (see Eq. (22), and $c_{2i} (i = 1, 2, 3, 4)$ are determined by substituting Eq. (36) into the boundary conditions (25), where the subscript 1 should be replaced by 2.

Let us consider the boundary-value problem (26), (35). When taking Eqs. (1) and (32) into account, Eq. (35) may be rewritten as follows

$$\begin{aligned} \Delta^2 w_2 - \lambda_0 w_2 = \{&G_{2c}(\rho;\lambda_0)\cos 2\varphi + G_{2s}(\rho;\lambda_0)\sin 2\varphi\} \\ &+ \lambda_2 w_0(\rho;\lambda_0) + G(\rho;\lambda_0), \end{aligned} \tag{37}$$

where G, G_{2c}, G_{2s} are known functions of ρ and λ_0.

The partial solution of Eq. (37) may be presented in the form

$$w_2 = w_2^\circ(\rho; \lambda_0, \lambda_2) + w_2^{(\varphi)}(\rho, \varphi; \lambda_0), \tag{38}$$

where w_2° is the partial solution of the equation

$$\Delta^2 w_2 - \lambda_0 w_2 = \lambda_2 w_0(\rho; \lambda_0) + G(\rho; \lambda_0), \tag{39}$$

and the function $w_2^{(\varphi)}$ satisfies the equation

$$\Delta^2 w_2 - \lambda_0 w_2 = \{G_{2c}(\rho; \lambda_0)\cos 2\varphi + G_{2s}(\rho; \lambda_0)\sin 2\varphi\}. \tag{40}$$

It should be noted that the function $w_2^{(\varphi)}$ gives a contribution only into the value of the correction $\varepsilon^4\lambda_4$ [see Eq. (17)]. Thus, when considering only the first three approximations, the function $w_2^{(\varphi)}$ may be omitted at this step (see Remark 14.2).

Let $y_n(\rho)(n = 1, 2, \ldots)$ be an infinite orthonormal system of eigenfunctions of the equation

$$\left(\frac{d^2}{d\rho^2} + \frac{1}{\rho}\frac{d}{d\rho}\right)^2 y - \zeta^4 y = 0 \tag{41}$$

with the homogeneous boundary conditions (29) for y, and ζ_n be a corresponding sequence of eigenvalues. Here $\zeta_1 = 6.734$, $\zeta_2 = 11.197$, $\zeta_3 = 15.690, \ldots$, and

$$y_n(\rho) = r_n^{-1}\left[J_0(\zeta_n\rho) + D_2 Y_0(\zeta_n\rho) + D_3 I_0(\zeta_n\rho) + D_4 K_0(\zeta_n\rho)\right], \tag{42}$$

$$r_n^2 = \int_b^1 \left[J_0(\zeta_n\rho) + D_2 Y_0(\zeta_n\rho) + D_3 I_0(\zeta_n\rho) + D_4 K_0(\zeta_n\rho)\right]^2 \rho d\rho, \tag{43}$$

where D_i are constants, which are found from the boundary conditions (29) for y.

Then the partial solution of the boundary-value problem (29), (39) will be as follows

$$w_2^\circ = \sum_{n=1}^{\infty} A_n y_n(\rho), \quad A_n(\lambda_0, \lambda_2) = \frac{\lambda_2\delta_n'(\lambda_0) + \delta_n''(\lambda_0)}{\zeta_n^4 - \lambda_0}, \tag{44}$$

where

$$\delta_n' = \int_b^1 w_0(\rho; \lambda_0) y_n(\rho)\rho d\rho, \quad \delta_n'' = \int_b^1 G(\rho; \lambda_0) y_n(\rho)\rho d\rho, \tag{45}$$

and it is assumed that $\lambda_0 \neq \zeta_n^4$.

Remark 14.2 The next steps for seeking functions w_j (at $j \geq 3$) is no longer valid because of the error of Eq. (5). When taking into account the error of the classical theory of Kirchhoff, the first three terms in the expansion (17) may be considered as the approximate solution of Eq. (5) if the inequality $\delta_p^2 h^{-2} \sin^2\gamma > h^3 a^{-3}$ holds.

Let us consider Eq. (16). In this equation, the functions $q_p(\lambda)$, $t_p(\lambda)$ may be estimated as

$$q_p = q_{p0} + \varepsilon q_{p1} + \varepsilon^2 q_{p2} + o(\varepsilon^2), \quad t_p = t_{p0} + \varepsilon t_{p1} + \varepsilon^2 t_{p2} + o(\varepsilon^2), \tag{46}$$

where

$$\begin{aligned} q_{p0}(\lambda_0) &= -\int_0^{2\pi} \frac{\partial}{\partial\rho}\left(\frac{\partial^2 w_0}{\partial\rho^2} + \frac{1}{\rho}\frac{\partial w_0}{\partial\rho} + \frac{1}{\rho^2}\frac{\partial^2 w_0}{\partial\varphi^2}\right)\bigg|_{\rho=b} d\varphi, \quad q_{p1} = 0, \\ q_{p2}(\lambda_0, \lambda_2) &= -\int_0^{2\pi} \frac{\partial}{\partial\rho}\left(\frac{\partial^2 w_2}{\partial\rho^2} + \frac{1}{\rho}\frac{\partial w_2}{\partial\rho} + \frac{1}{\rho^2}\frac{\partial^2 w_2}{\partial\varphi^2}\right)\bigg|_{\rho=b} d\varphi \\ &= -2\pi \sum_{n=1}^{\infty} \frac{\lambda_2 \delta_n'(\lambda_0) + \delta_n''(\lambda_0)}{\varsigma_n^4 - \lambda_0} \frac{\partial}{\partial\rho}\left(\frac{\partial^2 y_n}{\partial\rho^2} + \frac{1}{\rho}\frac{\partial y_n}{\partial\rho}\right)\bigg|_{\rho=b}, \\ t_{p0} &= \frac{8\pi b(\nu^2 - 1)}{(1-b^2)(1+\nu)^2 - (1+b^2)(3-\nu)^2 \ln b}, \quad t_{p1} = 0, \end{aligned} \tag{47}$$

$$\begin{aligned} t_{p2}(\lambda_0) = \int_0^{2\pi} \Bigg[&\left(\frac{\partial u_2}{\partial\rho} + \frac{\nu}{\rho}\left(u_2 + \frac{\partial v_2}{\partial\varphi}\right)\right)\cos\varphi \\ &+ \frac{1-\nu}{2}\left(\frac{\partial v_2}{\partial\rho} - \frac{1}{\rho}\left(v_2 - \frac{\partial u_2}{\partial\varphi}\right)\right)\sin\varphi\Bigg]_{\rho=b} d\varphi. \end{aligned}$$

Then, in the first-order approximation, one has the following transcendental equation

$$\cos^2\gamma\, q_{p0}(\lambda_0) + \sin^2\gamma\, t_{p0} - k + \lambda_0 m_p = 0 \tag{48}$$

with respect to the parameter λ_0.

Let $\lambda_0^{(\varsigma)}$ ($\varsigma = 1, 2, \ldots$) be roots of Eq. (48), where $\varsigma - 1$ means a number of fixed circles at the surface of the annular plate.

Equation (48) permits to find natural frequencies corresponding to the case when the prosthesis length increment $\Delta l = l - l_{min} = 0$. However, this case is never realized in surgical practice because of the very high risk of dislodgment of the prosthesis shaft from the perforation in the stapes footplate.

Considering the next approximation, one can find the correction

$$\lambda_2 = \lambda_2^{(\varsigma)} = \frac{2\pi \cos^2 \gamma \sum\limits_{n=1}^{\infty} \frac{\xi_n \delta_n''}{\zeta_n^4 - \lambda_0^{(\varsigma)}} + \sin^2 \gamma \, t_{p2}(\lambda_0)}{m_p - 2\pi \cos^2 \gamma \sum\limits_{n=1}^{\infty} \frac{\xi_n - \delta_n'}{\zeta_n^4 - \lambda_0^{(\varsigma)}}}, \tag{49}$$

$$\xi_n = \frac{\partial}{\partial \rho} \left(\frac{\partial^2 y_n}{\partial \rho^2} + \frac{1}{\rho} \frac{\partial y_n}{\partial \rho} \right) \Bigg|_{\rho=b}.$$

The parameter $\lambda_2^{(\varsigma)}$ depends on the prosthesis length increment $\Delta l > 0$ and the initial stresses (1) in the reconstructed TM.

4 Finite Element Model

If the contact point C of the prosthesis does not coincide with the TM center O (see Fig. 2a) and/or the initial stresses (1) are large, the constructed above solutions are not valid for a prediction of natural frequencies. For these cases we propose the finite element model via ANSYS software. Using ANSYS programming design language (APDL), the parametric finite element model with a regular mesh has been developed. The elaborated program code allows to change easily the size of finite elements and obtain the high quality mesh. The four-node SHELL181-type finite elements with six degrees of freedom at each node (translations in the direction of the x-, y- and the z- axes, and rotations about these axes) have been chosen for the description of the annular plate [15, 16]. In the common case, the external counter of the TM was considered to be non-circular and closed to the ellipse with the semiaxes a_1, $a_2 (a_1 < a_2)$, which are oriented along the x-, and y- axes, respectively (see Fig. 2a). The linear elastic isotropic material model for the cartilage transplant was used for the simulation. The prosthesis was assumed to be perfectly rigid body. The sparse direct solver based on a direct elimination of equations was used to define the initial stress-strain state of the reconstructed TM after inserting the prosthesis. To perform the modal analysis of the prestressed TM, the block Lanczos algorithm was applied. This algorithm includes additional compute kernels using blocks of vectors, sparse matrix multiplication using the assembled mass matrix, and repeated forward/backward solves using multiple right-hand side vectors [15]. The applied algorithm based on SHELL181-type elements permits to find the natural frequencies corresponding to both the flexural and tangential vibrations of the plate. In our computations in-plane vibrations are not considered. In addition, the modes corresponding to the flexural vibrations of the TM with the fixed prosthesis are also omitted.

5 Modes and Natural Frequencies

5.1 *Reconstructed Tympanic Membrane Represented by a Circular Plate*

When constructing the function w_1 and calculating the correction $\lambda_2^{(\varsigma)}$, the infinite series in (32) and (49) were replaced by truncated ones with K and N terms, respectively. Table 1 shows the normalized parameters $\lambda_2^{(\varsigma)}(\varsigma = 1, 2, 3)$ vs. $N = 1, 2, 3$ at the fixed values $K = 6$, $a = 5$ mm, $b_p = 1.5$ mm, $\Delta\, l_p = 0.01$ mm, $\gamma = 5\pi/36$, $E = 3.4\,\text{N/mm}^2$, $\nu = 0.4$, $h = 0.5$ mm, $m_p = 6.0$ mg, $T_{rw} = 10^{-5}$ H/m, $S_{rw} = 3.0\,\text{mm}^2$, $S_{ps} = 0.126\,\text{mm}^2$. For the parameters accepted above, one has $\varepsilon = 0.017$. Data from Table 1 testify a satisfactory convergence of series (49).

Tables 2 and 3 show the parameters $\lambda_0^{(\varsigma)}$, $\lambda_2^{(\varsigma)}$ (normalized by the number 187.952) and the first three natural frequencies $\omega^{(\varsigma)}$ ($\varsigma = 1, 2, 3$) versus the elastic modules E of the cartilage transplant and the prosthesis mass m_p, respectively. Calculations were performed for the geometrical and physical parameters taken above. Series (17) are asymptotic ones if

$$|\lambda_2^{(\varsigma)}/\lambda_0^{(\varsigma)}| = O(1) \quad \text{at} \quad \varepsilon \to 0. \tag{50}$$

It may be seen that the correction $\lambda_2^{(\varsigma)}$ shows weak dependence on the modules E and appreciable dependence on the mass m_p of the prosthesis. One can see also, that the asymptotic estimation (50) holds for all parameters under consideration. As expected, increasing the cartilage transplant stiffness or decreasing the prosthesis mass results in increase of the natural frequencies for all modes.

To verify the natural frequencies found by use of the asymptotic approach, the numerical finite element computations have been performed. Table 4 shows

Table 1 Corrections $\lambda_2^{(\varsigma)}$ versus a number of terms N in series (49)

N	$\lambda_2^{(1)}$	$\lambda_2^{(2)}$	$\lambda_2^{(3)}$
1	0.331	159.617	18.330
2	−2.193	142.902	740.067
3	−2.191	142.827	740.217
4	−2.191	142.827	740.217

Table 2 The dimensionless parameters $\lambda_0^{(\varsigma)}$, $\lambda_2^{(\varsigma)}$ and natural frequencies $\omega^{(\varsigma)}$ (Hz) versus the elastic modules E (H/mm^2) of the cartilage transplant

E	$\lambda_0^{(1)}$	$\lambda_2^{(1)}$	$\omega^{(1)}$	$\lambda_0^{(2)}$	$\lambda_2^{(2)}$	$\omega^{(2)}$	$\lambda_0^{(3)}$	$\lambda_2^{(3)}$	$\omega^{(3)}$
2.80	1.14	−2.26	139	15.74	−15.81	521	96.79	−47.91	1294
3.40	1.14	−2.59	153	15.74	−15.63	576	96.79	−47.35	1432

Table 3 The dimensionless parameters $\lambda_0^{(\varsigma)}$, $\lambda_2^{(\varsigma)}$ and natural frequencies $\omega^{(\varsigma)}$ (Hz) versus the prosthesis mass m_p (mg)

m_p	$\lambda_0^{(1)}$	$\lambda_2^{(1)}$	$\omega^{(1)}$	$\lambda_0^{(2)}$	$\lambda_2^{(2)}$	$\omega^{(2)}$	$\lambda_0^{(3)}$	$\lambda_2^{(3)}$	$\omega^{(3)}$
2.0	1.38	−3.71	103	51.61	82.34	632	303.10	386.71	1532
4.0	1.16	−2.76	95	46.41	112.10	599	280.73	547.74	1474
6.0	1.00	−2.19	88	43.19	142.80	578	268.60	740.20	1442

Table 4 Natural frequencies $\omega^{(\varsigma)}$ and $\omega_{FEM}^{(\varsigma)}$ (Hz) versus the cartilage transplant thickness h (mm)

h	$\omega^{(1)}$	$\omega_{FEM}^{(1)}$	$\omega^{(2)}$	$\omega_{FEM}^{(2)}$	$\omega^{(3)}$	$\omega_{FEM}^{(3)}$
0.3	105	87	325	303	824	840
0.5	154	108	576	494	1,432	1,397
0.8	229	175	953	915	2,333	1,964

frequencies $\omega^{(\varsigma)}$ and $\omega_{FEM}^{(\varsigma)}$ obtained by the asymptotic and finite element methods for different thicknesses of the cartilage transplant h. Here and below, the superscript *FEM* refers to magnitudes found by using FEM. Calculations have been done for $E = 3.4\,\text{N/mm}^2$, other parameters have been taken as above. Let $\Omega_{FEM}^{\varepsilon}$ be the subset of the natural frequencies, for which the amplitude ratio $A_p = w_p/W_{max}$ is a value of the order $O(\varepsilon)$, where w_p is the amplitude of oscillations of the prosthesis, and W_{max} is the maximum amplitude of the normal vibrations of the TM. The natural frequencies from $\Omega_{FEM}^{\varepsilon}$ correspond to modes of in-plane vibrations or flexural vibrations with the fixed prosthesis. Table 4 does not contain frequencies from the set $\Omega_{FEM}^{\varepsilon}$ for which $A_p \leq 0.05$. As it is seen, comparison of the frequencies found by the two different approaches demonstrates a satisfactory coincidence for sufficiently thin plates.

5.2 Reconstructed Tympanic Membrane Represented by an Elliptic Plate

As a rule, the prosthesis plate is placed between the TM center O and the point C_s [11, 12], which is the projection of the point S on the TM (see Fig. 2a). We have studied the influence of the prosthesis position on the natural frequencies and modes of the system. Under the FEM simulation, the external counter of the TM was considered close to the ellipse with the semiaxes $a_1 = 4$ mm and $a_2 = 5$ mm. In Table 5, the natural frequencies are presented for the three different cases: (a) the contact point C coincides with the center O; (b) the contact point C has the coordinates $x_C = 1.389$ mm, $y_C = -0.567$ mm; (c) the contact point with the coordinates $x_C = 2.315$ mm, $y_C = -0.945$ mm is far from the center O. The first, forth and

Table 5 Natural frequencies ω_{FEM}(Hz) and amplitude ratios A_p for different positions of the prosthesis at the reconstructed eardrum

Mode number	$\omega_{FEM}^{(a)}$	$A_p^{(a)}$	Mode number	$\omega_{FEM}^{(b)}$	$A_p^{(b)}$	Mode number	$\omega_{FEM}^{(c)}$	$A_p^{(c)}$
1	34	0.70	1	24	0.23	1	19	0.07
4	50	0.38	2	31	0.37	3	31	0.38
6	53	0.40	4	42	0.14	4	38	0.20
16	118	0.17	6	60	0.07	6	51	0.08
19	129	0.11	8	69	0.07	7	57	0.08
42	214	0.11	9	79	0.06	11	83	0.05

(a) $x_C = 0, y_C = 0$; (b) $x_C = 1.389$ mm, $y_C = -0.567$ mm; (c) $x_C = 2.315$ mm, $y_C = -0.945$ mm

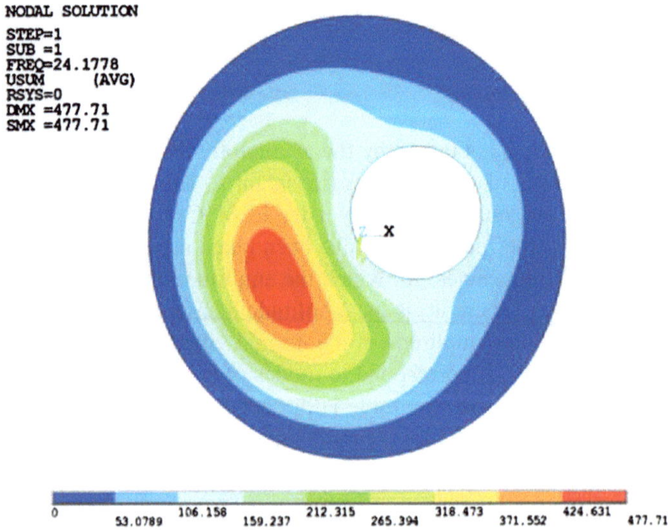

Fig. 4 Normal displacements of the reconstructed TM corresponding to the 1st mode of the RME with the natural frequency $\omega_{FEM}^{(b)} = 24$ Hz and the amplitude ratio $A_p^{(b)} = 0.23$

seventh columns show the mode numbers for the cases (a), (b) and (c), respectively. The natural frequencies corresponding to these cases are denoted by the superscripts (a), (b), (c). Frequencies from the set $\Omega_{FEM}^{\varepsilon}$, where $\varepsilon < 0.05$, are omitted here.

In Fig. 4, the TM displacements corresponding to the 1st mode with the natural frequency $\omega_{FEM}^{(b)} = 24$ Hz and fixed prosthesis are shown. The 2nd mode with the natural frequency $\omega_{FEM}^{(b)} = 31$ Hz and the amplitude ratio $A_p^{(b)} = 0.47$ is presented in Fig. 5, 3rd mode with the natural frequency $\omega_{FEM}^{(b)} = 33$ Hz and the amplitude ratio $A_p^{(b)} = 0.47$ is presented in Fig. 6. The TM parts shown in dark blue color are immobile or have very small displacements, and the ones marked by red have the maximum amplitude.

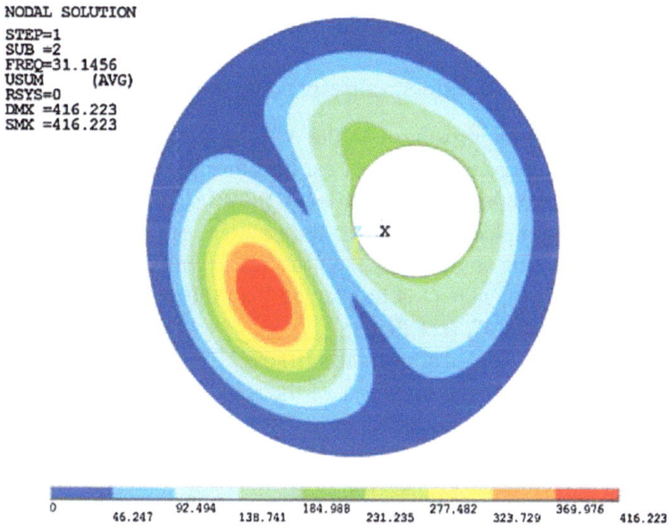

Fig. 5 Normal displacements of the reconstructed TM corresponding to the 2nd mode of the RME with the natural frequency $\omega_{FEM}^{(b)} = 31$ Hz and the amplitude ratio $A_p^{(b)} = 0.37$

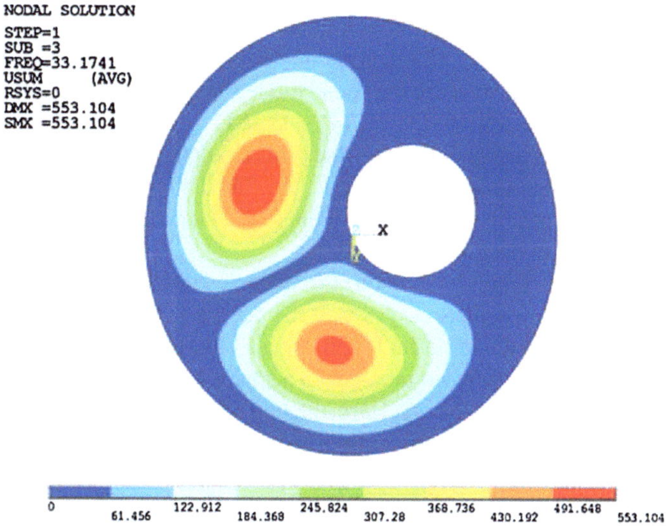

Fig. 6 Normal displacements of the reconstructed TM corresponding to the 3rd mode of the RME with the natural frequency $\omega_{FEM}^{(b)} = 33$ Hz and the fixed prothesis

It can be seen that shifting the prosthesis plate from the center O to the point C_s results in:

- decrease of the lowest natural frequencies of the RME;
- decrease of a number of the "dead modes" with frequencies from the subset $\Omega_{FEM}^{\varepsilon}$;

- increase of a number of the "functional modes", which "transmit" energy of the TM vibrations into the inner ear;
- decrease of the amplitude ratio A_p corresponding to the "functional modes".

Thus, when shifting the prosthesis to the edge of the reconstructed eardrum, its capacity to transmit energy into the inner ear becomes lower. However, it should be noted that the question of the optimal position of the prosthesis at the TM for this type of reconstruction is complex and includes the problem of the prosthesis stability in the middle ear cavity. It is obvious that the prosthesis (with an inclined shaft) placed in a vicinity of the TM center O is less stable than the prosthesis allocated in the neighborhood of the point C_s (see Fig. 2a).

In all calculations performed above, the stiffness $k^* = k^*_{rw}$ was assumed to depend on the TM stiffness and tensile force T_{rw} of the round window. So, for all geometrical and physical parameters accepted above, $k^* = k^*_{rw} = 2.51 \times 10^{-11}$ N/mm has too small magnitude. To ensure stability of the prosthesis at the TM and exclude its falling out from the hole in the stapes footplate, the modified version of the prosthesis like TORP (see Fig. 2b) with a conical spring attached to the shaft is proposed. Here, $k^*_{rw} \ll k^*_{sp}$. Therefore, it was assumed that $k^* \approx k^*_{sp}$ in all calculations. Table 6 shows the natural frequencies calculated for different values of the spring stiffness k^*_{sp} for the case when the contact point C has the coordinates $x_C = 1.389$ mm, $y_C = -0.567$ mm. The TM was represented by the elliptic plate with the semiaxes $a_1 = 4$ mm and $a_2 = 5$ mm. For all computations the prosthesis length increment $\Delta l = l - l_{min} = 1.0$ mm, and all other parameters were taken as above. In Table 6, gaps correspond to frequencies for the "dead modes" with very small parameters A_p, the modes with numbers $5, 10, 11, 12, 14 - 17, 20 - 27, \ldots$ are also "dead" and omitted here. As seen, increasing the stiffness k^*_{sp} results in decreasing the eigenfrequencies. However this influence is very weak. Calculations show that the natural frequencies are appreciably influenced by the parameter k^*_{sp} for very high stiffness k^*_{sp}, which may lead to buckling of the TM.

It is interesting to compare the natural frequencies found above with the lowest eigenfrequencies of the oscillating system of the middle ear in normal. The first eight frequencies of the normal middle ear found using FEM [17] are presented in Table 7. The calculations were performed taking into account the influence of all ligaments and muscles in the middle ear cavity. Here, the first five frequencies correspond to the modes for which eardrum is almost immovable and the most intensive oscillations are performed by the malleus-incus-stapes chain, and only the sixth and subsequent frequencies relate to modes with intensive vibrations of the tympanic membrane.

When comparing the data from Tables 2, 3, 4, 5, 6, and 7, one can conclude: the total reconstruction of the middle ear (including tympanoplasty and stapedotomy) results in appearance of a lot of modes with very low eigenfrequencies, some of these modes being characterized by the fixed prosthesis.

Table 6 Natural frequencies ω_{FEM}(Hz) and amplitude ratios A_p versus the stiffness k^*_{sp}

Mode number	$\omega^{(a)}_{FEM}$	$A^{(a)}_p$	$\omega^{(b)}_{FEM}$	$A^{(b)}_p$	$\omega^{(c)}_{FEM}$	$A^{(c)}_p$
1	24.18	0.227	24.09	0.216	23.02	0.137
2	31.15	0.367	31.10	0.364	30.92	0.342
3	–	0.000	–	0.000	31.68	0.088
4	42.36	0.141	42.26	0.140	41.35	0.136
6	59.69	0.067	59.64	0.066	59.15	0.057
7	–	0.00	68.26	0.055	65.96	0.075
8	69.01	0.068	68.81	0.057	–	0.000
9	78.51	0.059	78.50	0.057	–	0.000
13	98.14	0.063	98.16	0.061	97.41	0.051
18	–	0.000	–	0.000	125.30	0.053
19	127.04	0.061	126.81	0.060	–	0.000
28	–	0.000	–	0.000	163.08	0.050

(a) $k^*_{sp} = 2.51 \times 10^{-5}$ N/mm; (b) $k^*_{sp} = 2.51 \times 10^{-3}$ N/mm; (c) $k^*_{sp} = 2.51 \times 10^{-2}$ N/mm

Table 7 Natural frequencies of the normal middle ear [17]

Mode number	1	2	3	4	5	6	7	8
ω, Hz	91	172	228	271	483	647	857	1189

6 Conclusions

In this paper, the mathematical model of the reconstructed middle ear subjected to both tympanoplasty and stapedotomy has been proposed. The biomechanical oscillating system of the reconstructed middle ear consists of the almost circular annular plate made of the cartilage transplant and the conjugated titanium prosthesis with a pliable shaft. The end of the prosthesis shaft is inserted into the vestibule through a small perforation drilled out in the stapes footplate. It was assumed that the prosthesis can perform only the longitudinal oscillations along the perforation axis. With the view of ensuring stability of the prosthesis in the middle ear cavity, the modified version of the total prosthesis (like TORP) with a conical spring attached to its shaft has been proposed. To study initial deformations and flexural vibrations of the annular elastic plate, the governing equations based on the hypothesis of Kirchhoff and taking into account the initial membrane stresses and the in-plane displacements as well have been used. As the first step, the initial stresses in the plate have been found. Then, for cases when these stresses were small and the prosthesis plate was attached to the reconstructed eardrum in its center, the differential equations describing motion of the annular plate were solved by using the asymptotic approach. On the contrary, if the initial stresses in the TM were turned to be large or/and the contact points of the prosthesis with the plate were chosen to be far from its center, the finite element simulation was applied in our study.

The qualitative analysis of the constructed asymptotic solutions and numerical calculations have shown that the total reconstruction, including both tympanoplasty and stapedotomy, results in considerable decrease of the natural frequencies of the middle ear oscillating system and appearance of a set of the "dead modes" with the fixed prothesis. The RME eigenfrequencies depend strongly upon the parameters of the cartilage transplant, the mass of the prosthesis and the point of its placement at the reconstructed eardrum. In terms of the surgical procedure, the basic conclusions which could assist the clinician are as follows:

- to avoid appearance of the "dead modes"and guarantee the best transfer of the energy of external sound waves to the inner ear, the prosthesis should be adapted in its shape during surgery to individual middle-ear anatomy by adjusting the angle between the stapes stem and its plate so that the prosthesis plate was positioned as close to the center of the reconstructed TM as possible;
- variation of the thickness and stiffness of the cartilage transplant allows to correct slightly the eigenfrequency spectrum and approach some frequencies to those of the normal middle ear;
- the attachment of the conical spring to the prosthesis shaft can guarantee its stability in the middle ear cavity but does not influence on the natural frequencies appreciably.

References

1. Kobrak, G.: The Middle Ear. The university of Chicago Press, Chicago (1958)
2. Rosowski, J.J.: Mechanism of sound conduction in normal and diseased ears. In: Rosowski, J.J., Merchant, S.N. (eds.) The Function and Mechanics of Normal, Diseased and Reconstructed Middle Ear, pp. 137–145. Kugler Publications, The Hague, The Netherlands (2000)
3. Jahnke, K., Lieberum, B., Kühn, W.: Missing handle of malleus: reinforcement of the tympanic membrane. In: Hüttenbrink, K.B. (ed.) Middle Ear Mechanics in Research and Otosurgery, Department of Oto-Rhino-Laringology, pp. 197–199. Dresden University of Technology, Dresden (1997)
4. Wullstein, H.: Operationen zur Verbesserung des Gehöres. Georg Thieme Verlag, Stuttgart (1968)
5. Offergeld, C.F.E., Hüttenbrink, K.B., Zahnert, T., Hofmann, G.: Experimental investigations of osicular joint ankylosis. In: Rosowski, J.J., Merchant, S.N. (eds.) The Function and Mechanics of Normal, Diseased and Reconstructed Middle Ear, pp. 177–186. Kugler Publications, The Hague, The Netherlands (2000)
6. Hüttenbrink, K.B.: Mechanical aspects of middle ear reconstruction. In: Hüttenbrink, K.B. (ed.) Middle Ear Mechanics in Research and Otosurgery, Department of Oto-Rhino-Laringology, pp. 165–168. Dresden University of Technology, Dresden (1997)
7. Mikhasev, G.I., Slavashevich, I.L.: Eigenfrequencies of the middle ear oscillating systems after total reconstruction. Vestnik St Petersburg University Ser **1**(3), 107–116 (2012)
8. Mürbe, D., Zahnert, T., Bornitz, M., Hüttenbrink, K.B.: Acoustic properties of different cartilage reconstruction techniques of the tympanic membrane. Laryngoscope **112**, 1769–1776 (2002)

9. Zahnert, T., Hüttenbrink, K.B., Mürbe, D., Bornitz, M.: Experimental investigations of the use of cartilage in tympanic membrane reconstruction. Am. J. Otol. **21**, 322–328 (2000)
10. Gelfand, S.A.: Hearing: An Introduction to Psychological and Physiological coustics. Marcel Dekker Inc, New York and Basel (1981)
11. Mikhasev, G., Ermochenko, S., Bornitz, M.: On the strain-stress state of the reconstructed middle ear after inserting a malleusincus prosthesis. Math. Med. Biol. **27**, 289–312 (2010)
12. Koike, T., Wada, H., Kobayashi, T.: Analysis of the finite-element method of transfer function of reconstructed middle ears and their postoperative changes. In: Rosowski, J.J., Merchant, S.N. (eds.) The Function and Mechanics of Normal, Diseased and Reconstructed Middle Ear, pp. 309–320. Kugler Publications, The Hague, The Netherlands (2000)
13. Pozrikidis, C.: Hydrodynamics of a vibrating stapes prosthesis in stapedotomy. Eng. Anal. Boundary Elem. **31**, 1–9 (2007)
14. Mikhasev, G.I., Tovstik, P.E.: Localized Vibrations and Waves in Thin Shells. Asymptotic Methods, FIZMATLIT. Nauka, Moscow (2009)
15. Basov, K.A.: ANSYS: User's Guide. DMK Press, Moscow (2005)
16. Madenci, E., Guver, I.: The Finite Element Method and Application in Engineering Using ANSYS. Springer, Heidelberg (2006)
17. Beer, H.J., Bornitz, M., Hardke, H.J., Schmidt, R., Hofman, G., Vogel, V., Zarnet, T., Hüttenbrink, K.B.: Finite element modeling of the human eardrum and applications. In: Hüttenbrink, K.B. (ed.) Middle Ear Mechanics in Research and Otosurgery, Deptartmant of Oto-Rhino-Laringology, pp. 40–47. Dresden University of Technology, Dresden (1997)

A New Approach for Studying Nonlinear Dynamic Response of a Thin Fractionally Damped Plate with 2:1 and 2:1:1 Internal Resonances

Yury A. Rossikhin and Marina V. Shitikova

Abstract Dynamic behaviour of a nonlinear plate embedded in a fractional derivative viscoelastic medium and subjected to the conditions of the internal resonances two-to-one and one-to-one, as well as the internal combinational resonances has been studied by Rossikhin and Shitikova in [12, 13]. Nonlinear equations, the linear parts of which occur to be coupled, were solved by the method of multiple time scales. A new approach proposed in this paper allows one to uncouple the linear parts of equations of motion of the plate, while the same method, the method of multiple time scales, has been utilized for solving nonlinear equations. The new approach enables one to solve the problems of vibrations of thin bodies more efficiently.

1 Introduction

It is well known that the nonlinear vibrations of plates is an important area of applied mechanics, since plates are used as structural elements in many fields of industry and technology. Different methods: analytical [8, 12, 13, 19], numerical [2, 9] and experimental [2] could be employed to investigate the nonlinear vibrations of plates. Extensive review of recent research developments in the field could be found in [3, 4], and [18].

The study of free undamped [4, 9], as well as damped [5, 6, 12, 13] nonlinear systems is important because the dynamics characteristics of the system—defined by the amplitude-frequency relations and modes of vibrations—are determined [9]. Moreover, nonlinear vibrations could be accompanied by such a phenomenon as the internal resonance, resulting in multimode response with a strong interaction of the modes involved [7]. The internal resonance could be found within some combination of natural frequencies of one and the same type of vibrations. As an example, a 1:3

Y.A. Rossikhin (✉) · M.V. Shitikova
Research Center of Wave Dynamics, Voronezh State University of Architecture and Civil Engineering, Voronezh 394006, Russia
e-mail: YAR@vgasu.vrn.ru

M.V. Shitikova
e-mail: shitikova@vmail.ru

H. Altenbach and G.I. Mikhasev (eds.), *Shell and Membrane Theories in Mechanics and Biology*, Advanced Structured Materials 45, DOI 10.1007/978-3-319-02535-3_15

internal resonance was discovered in [9], when the fourth natural frequency of the out-of-plane vibrations was three times larger than the fundamental frequency of the out-of-plane vibrations.

Another type of the internal resonance was investigated in [12, 13], when one frequency of the in-plane vibrations was equal (a 1:1 internal resonance) or twice as larger (a 1:2 internal resonance) than some frequency of the out-of-plane vibrations. The combinational resonances of the additive and difference type were also investigated in [13].

In the given paper, nonlinear free vibrations of a thin plate described by three coupled nonlinear differential equations are considered when the plate is being under the conditions of the internal resonance resulting in the interaction of modes corresponding to the mutually orthogonal displacements. The displacement functions are determined in terms of eigenfunctions of linear vibrations. The procedure resulting in decoupling linear parts of equations is proposed with the further utilization of the method of multiple scales for solving nonlinear governing equations of motion, in so doing the amplitude functions are expanded into power series in terms of the small parameter and depend on different time scales. It is shown that the phenomenon of internal resonance can be very critical, since in a thin plate internal resonance of the type two-to-one, one-to-one, as well as additive and difference combinational resonances are always present.

2 Problem Formulation and the Method of Solution

Let us consider the dynamic behavior of a free supported non-linear thin rectangular plate, vibrations of which in a viscoelastic medium are described in the Cartesian system of coordinates by the following three differential equations written in the dimensionless form [12, 13]:

$$u_{xx} + \frac{1-\nu}{2}\beta_1^2 u_{yy} + \frac{1+\nu}{2}\beta_1 v_{xy} + w_x\left(w_{xx} + \frac{1-\nu}{2}\beta_1^2 w_{yy}\right) + \frac{1+\nu}{2}\beta_1^2 w_y w_{xy}$$
$$= \ddot{u} + \kappa_1 D^\gamma u, \tag{1}$$

$$\beta_1^2 v_{yy} + \frac{1-\nu}{2} v_{xx} + \frac{1+\nu}{2}\beta_1 u_{xy} + \beta_1 w_y\left(\beta_1^2 w_{yy} + \frac{1-\nu}{2} w_{xx}\right) + \frac{1+\nu}{2}\beta_1 w_x w_{xy}$$
$$= \ddot{v} + \kappa_2 D^\gamma v, \tag{2}$$

$$\frac{\beta_2^2}{12}(w_{xxxx} + 2\beta_1^2 w_{xxyy} + \beta_1^4 w_{yyyy}) - w_{xx}(u_x + \nu\beta_1 v_y) - w_x(u_{xx} + \nu\beta_1 v_{xy})$$
$$- \frac{1-\nu}{2}\beta_1[w_{xy}(\beta_1 u_y + v_x) + w_y(\beta_1 u_{xy} + v_{xx})]$$
$$- \beta_1^2[w_{yy}(\nu u_x + \beta_1 v_y) + w_y(\nu u_{xy} + \beta_1 v_{yy})]$$

$$-\frac{1-\nu}{2}\beta_1[w_{xy}(\beta_1 u_y + v_x) + w_x(\beta_1 u_{yy} + v_{xy})] = -\ddot{w} - \kappa_3 D^\gamma w, \quad (3)$$

subjected to the initial

$$u|_{t=0} = \dot{u}\Big|_{t=0} = 0, \quad v|_{t=0} = \dot{v}\Big|_{t=0} = 0, \quad w|_{t=0} = \dot{w}\Big|_{t=0} = 0, \quad (4)$$

as well as the boundary conditions

$$w|_{x=0} = w\Big|_{x=1} = 0, \quad v|_{x=0} = v\Big|_{x=1} = 0, \quad u_x\Big|_{x=0} = u_x\Big|_{x=1} = 0,$$
$$w_{xx}\Big|_{x=0} = w_{xx}\Big|_{x=1} = 0, \quad (5)$$

$$w|_{y=0} = w\Big|_{y=1} = 0, \quad u|_{y=0} = u\Big|_{y=1} = 0, \quad v_y\Big|_{y=0} = v_y\Big|_{y=1} = 0,$$
$$w_{yy}\Big|_{y=0} = w_{yy}\Big|_{y=1} = 0, \quad (6)$$

where $u = u(x, y, t)$, $v = v(x, y, t)$, and $w = w(x, y, t)$ are the displacements of points located in the plate's middle surface in the $x-$, $y-$, and $z-$ directions, respectively, ν is Poisson's ratio, $\beta_1 = a/b$ and $\beta_2 = h/a$ are the parameters defining the dimensions of the plate, a and b are the plate's dimensions along the $x-$ and $y-$ axes, respectively, h is the thickness, t is the time, an overdot denotes the time-derivative, lower indices label the derivatives with respect to the corresponding coordinates, κ_i $(i = 1, 2, 3)$ are damping coefficients, and D^γ is the Riemann-Liouville fractional derivative of the γ-order [14]

$$D^\gamma F = \frac{\mathrm{d}}{\mathrm{d}t}\int_0^t \frac{F(t-t')}{\Gamma(1-\gamma)t'^\gamma}\,\mathrm{d}t'. \quad (7)$$

It has been noted in [11, 12] that a fractional derivative is the immediate extension of an ordinary derivative. In fact, when $\gamma \to 1$, $D^\gamma x$ tends to $\dot{x}$, i.e., at $\gamma \to 1$ the fractional derivative goes over into the ordinary derivative, and the mathematical model of the viscoelastic shell under consideration transforms into the Kelvin-Voigt model, wherein the elastic element behaves nonlinearly, but the viscous element behaves linearly. When $\gamma \to 0$, the fractional derivative $D^\gamma x$ tends to $x(t)$. To put it otherwise, the introduction of the new fractional parameter along with the parameters κ_i allows one to change not only the magnitude of viscosity at the cost of an increase or decrease in the parameters κ_i, but also the character of viscosity at the sacrifice of variations in the fractional parameter.

In Eqs. (1)–(3), the dimensionless values are introduced similarly as it has been done in [12].

We seek the solution of Eqs. (1)–(3) in the form

$$u(x,y,t)=\sum_{m=1}^{\infty}\sum_{n=1}^{\infty}x_{1mn}(t)\eta_{1mn}(x,y),$$
$$v(x,y,t)=\sum_{m=1}^{\infty}\sum_{n=1}^{\infty}x_{2mn}(t)\eta_{2mn}(x,y), \tag{8}$$
$$w(x,y,t)=\sum_{m=1}^{\infty}\sum_{n=1}^{\infty}x_{3mn}(t)\eta_{3mn}(x,y),$$

where $x_{1mn}(t)$, $x_{2mn}(t)$, and $x_{3mn}(t)$ are the generalized displacements corresponding to the displacements in the plane of the plate and to its deflection, respectively, but the natural functions satisfying the boundary conditions (5) and (6) have the form

$$\eta_{1mn}(x,y)=\cos\pi mx\ \sin\pi ny,\quad \eta_{2mn}(x,y)=\sin\pi mx\ \cos\pi ny, \tag{9}$$
$$\eta_{3mn}(x,y)=\sin\pi mx\ \sin\pi ny,$$

and m and n are integers.

Linear undamped natural modes of flexural in-plane and out-of-plane vibrations of the plate are the solution of eigenvalue problem

$$-\ddot{x}_{1mn}\eta_{1mn}+x_{1mn}\left(\eta_{1mn,xx}+\frac{1-\nu}{2}\beta_1^2\eta_{1mn,yy}\right)+\frac{1+\nu}{2}\beta_1\eta_{2mn,xy}\,x_{2mn}=0, \tag{10}$$

$$-\ddot{x}_{2mn}\eta_{2mn}+x_{2mn}\left(\beta_1^2\eta_{2mn,yy}+\frac{1-\nu}{2}\eta_{2mn,xx}\right)+\frac{1+\nu}{2}\beta_1\eta_{1mn,xy}\,x_{1mn}=0, \tag{11}$$

$$\ddot{x}_{3mn}\eta_{3mn}+\frac{\beta_2^2}{12}\left(\eta_{3mn,xxxx}+2\beta_1^2\,\eta_{3mn,xxyy}+\beta_1^4\,\eta_{3mn,yyyy}\right)x_{3mn}=0. \tag{12}$$

The set of Eqs. (10) and (11) has the characteristic equation

$$\omega_{mn}^4-(S_{11}^{mn}+S_{22}^{mn})\omega_{mn}^2+S_{11}^{mn}S_{22}^{mn}-S_{12}^{mn}S_{21}^{mn}=0, \tag{13}$$

the roots of which are the dimensionless natural frequencies of the flexural in-plane vibrations of the plate

$$\omega_{1\,mn}^2=\pi^2(m^2+\beta_1^2n^2),\qquad \omega_{2\,mn}^2=\frac{1-\nu}{2}\,\omega_{1mn}^2, \tag{14}$$

where $S_{11}^{mn} = \pi^2 \left(m^2 + \frac{1-\nu}{2} \beta_1^2 n^2\right)$,

$$S_{12}^{mn} = S_{21}^{mn} = \frac{1+\nu}{2} \beta_1 \pi^2 mn, \quad S_{22}^{mn} = \pi^2 \left(\frac{1-\nu}{2} m^2 + \beta_1^2 n^2\right). \tag{15}$$

From Eq. (12), the natural frequency of the flexural out-of-plane vibrations of the plate could be obtained

$$\omega_{3\,mn}^2 = \frac{\pi^4 \beta_2^2}{12} (m^2 + \beta_1^2 n^2)^2. \tag{16}$$

Substituting (9) into Eqs. (1)–(3), multiplying (1), (2), and (3) by η_{1lk}, η_{2lk}, and η_{3lk}, respectively, integrating over x and y, and using the orthogonality conditions for linear modes within the domains of $0 \le x, y \le 1$, we are led to a coupled set of nonlinear ordinary differential equations of the second order in $x_{i\,mn}$ $(i = 1, 2, 3)$:

$$\ddot{x}_{\alpha\,mn} + \kappa_\alpha D^\gamma x_{\alpha\,mn} + S_{\alpha\beta}^{mn} x_{\beta\,mn} = -F_{\alpha\,mn} \quad (\alpha, \beta = 1, 2), \tag{17}$$

$$\ddot{x}_{3\,mn} + \kappa_3 D^\gamma x_{3\,mn} + \omega_{3mn}^2 x_{3\,mn} = -F_{3\,mn}, \tag{18}$$

where the summation is carried out over two repeated indices, the elements of the matrix S_{ij}^{mn} are defined as (15).

The nonlinear parts $F_{i\,mn}$ of Eqs. (17) have the form

$$F_{1\,mn} = 4 \sum_{m_1} \sum_{n_1} \sum_{m_2} \sum_{n_2} x_{3m_1n_1} x_{3m_2n_2} A_{1\,mn}^{m_1n_1m_2n_2}, \tag{19}$$

$$F_{2\,mn} = 4 \sum_{m_1} \sum_{n_1} \sum_{m_2} \sum_{n_2} x_{3m_1n_1} x_{3m_2n_2} A_{2\,mn}^{m_1n_1m_2n_2}, \tag{20}$$

$$F_{3\,mn} = -4 \sum_{m_1} \sum_{n_1} \sum_{m_2} \sum_{n_2} \left[x_{3m_1n_1} x_{1m_2n_2} C_{mn}^{m_1n_1m_2n_2} + x_{3m_1n_1} x_{2m_2n_2} D_{mn}^{m_1n_1m_2n_2}\right], \tag{21}$$

where

$$A_{1\,mn}^{m_1n_1m_2n_2} = \pi^3 m_1 \left(m_2^2 + \frac{1-\nu}{2} \beta_1^2 n_2^2\right) a_{1mn}^{m_1n_1m_2n_2} - \frac{1+\nu}{2} \pi^3 \beta_1^2 n_1 m_2 n_2 a_{2mn}^{m_1n_1m_2n_2},$$

$$A_{2\,mn}^{m_1n_1m_2n_2} = \pi^3 \beta_1 n_1 \left(\beta_1^2 n_2^2 + \frac{1-\nu}{2} m_2^2\right) a_{3mn}^{m_1n_1m_2n_2} - \frac{1+\nu}{2} \pi^3 \beta_1 m_1 m_2 n_2 a_{4mn}^{m_1n_1m_2n_2},$$

$$C_{mn}^{m_1n_1m_2n_2} = \pi^3 m_2 \left(m_1^2 + \nu\beta_1^2 n_1^2\right) a_{5mn}^{m_1n_1m_2n_2} + (1-\nu)\pi^3 \beta_1^2 m_1 n_1 n_2 a_{6mn}^{m_1n_1m_2n_2}$$
$$- \pi^3 m_1 \left(m_2^2 + \frac{1-\nu}{2} \beta_1^2 n_2^2\right) a_{7mn}^{m_1n_1m_2n_2} - \frac{1+\nu}{2} \beta_1^2 \pi^3 n_1 m_2 n_2 a_{8mn}^{m_1n_1m_2n_2},$$

$$D_{mn}^{m_1n_1m_2n_2} = \pi^3\beta_1 n_2\left(\nu m_1^2 + \beta_1^2 n_1^2\right) a_{5mn}^{m_1n_1m_2n_2} + (1-\nu)\beta_1\pi^3 m_1 n_1 m_2 a_{6mn}^{m_1n_1m_2n_2}$$
$$- \frac{1+\nu}{2}\beta_1\pi^3 m_1 m_2 n_2 a_{7mn}^{m_1n_1m_2n_2} - \pi^3\beta_1 n_1\left(\frac{1-\nu}{2} m_2^2 + \beta_1^2 n_2^2\right) a_{8mn}^{m_1n_1m_2n_2},$$

$$a_{1mn}^{m_1n_1m_2n_2} = \int_0^1\int_0^1 \cos\pi m_1 x\ \sin\pi n_1 y\ \sin\pi m_2 x\ \sin\pi n_2 y\ \cos\pi m x\ \sin\pi n y\ dxdy,$$

$$a_{2mn}^{m_1n_1m_2n_2} = \int_0^1\int_0^1 \sin\pi m_1 x\ \cos\pi n_1 y\ \cos\pi m_2 x\ \cos\pi n_2 y\ \cos\pi m x\ \sin\pi n y\ dxdy,$$

$$a_{3mn}^{m_1n_1m_2n_2} = \int_0^1\int_0^1 \sin\pi m_1 x\ \cos\pi n_1 y\ \sin\pi m_2 x\ \sin\pi n_2 y\ \sin\pi m x\ \cos\pi n y\ dxdy,$$

$$a_{4mn}^{m_1n_1m_2n_2} = \int_0^1\int_0^1 \cos\pi m_1 x\ \sin\pi n_1 y\ \cos\pi m_2 x\ \cos\pi n_2 y\ \sin\pi m x\ \cos\pi n y\ dxdy,$$

$$a_{5mn}^{m_1n_1m_2n_2} = \int_0^1\int_0^1 \sin\pi m_1 x\ \sin\pi n_1 y\ \sin\pi m_2 x\ \sin\pi n_2 y\ \sin\pi m x\ \sin\pi n y\ dxdy,$$

$$a_{6mn}^{m_1n_1m_2n_2} = \int_0^1\int_0^1 \cos\pi m_1 x\ \cos\pi n_1 y\ \cos\pi m_2 x\ \cos\pi n_2 y\ \sin\pi m x\ \sin\pi n y\ dxdy,$$

$$a_{7mn}^{m_1n_1m_2n_2} = \int_0^1\int_0^1 \cos\pi m_1 x\ \sin\pi n_1\ \cos\pi m_2 x\ \sin\pi n_2 y\ \sin\pi m x\ \sin\pi n y\ dxdy,$$

$$a_{8mn}^{m_1n_1m_2n_2} = \int_0^1\int_0^1 \sin\pi m_1 x\ \cos\pi n_1 y\ \sin\pi m_2 x\ \cos\pi n_2 y\ \sin\pi m x\ \sin\pi n y\ dxdy.$$

Since the second-rank tensor $S_{\alpha\beta}^{mn}$ is symmetric, then it has two real eigenvalues $\omega_{\alpha\, mn}$ defined in (14) which are in correspondence with two mutually orthogonal eigenvectors

$$\mathbf{L}_{mn}^{\mathrm{I}}\left\{L_{1\, mn}^{\mathrm{I}} = \frac{\pi m}{\omega_{1\, mn}},\quad L_{2\, mn}^{\mathrm{I}} = \frac{\pi\beta_1 n}{\omega_{1\, mn}}\right\}, \tag{22}$$

$$\mathbf{L}_{mn}^{\mathrm{II}}\left\{L_{1\, mn}^{\mathrm{II}} = \frac{\pi\beta_1 n}{\omega_{1\, mn}},\quad L_{2\, mn}^{\mathrm{II}} = -\frac{\pi m}{\omega_{1\, mn}}\right\}, \tag{23}$$

i.e.,

$$L_{\alpha\, mn}^{\mathrm{I}} L_{\alpha\, mn}^{\mathrm{I}} = L_{\alpha\, mn}^{\mathrm{II}} L_{\alpha\, mn}^{\mathrm{II}} = 1,\quad L_{\alpha\, mn}^{\mathrm{I}} L_{\alpha\, mn}^{\mathrm{II}} = 0. \tag{24}$$

Thus, the matrix $S^{mn}_{\alpha\beta}$ and the generalized displacements $x_{\alpha\, mn}$ entering in Eqs. (17) and (18) could be expanded in terms of the vectors (22) [15] as

$$S^{mn}_{\alpha\beta} = \omega^2_{1\, mn} L^{\mathrm{I}}_{\alpha\, mn} L^{\mathrm{I}}_{\beta\, mn} + \omega^2_{2\, mn} L^{\mathrm{II}}_{\alpha\, mn} L^{\mathrm{II}}_{\beta\, mn}, \tag{25}$$

$$x_{\alpha\, mn} = X_{1\, mn} L^{\mathrm{I}}_{\alpha\, mn} + X_{2\, mn} L^{\mathrm{II}}_{\alpha\, mn}. \tag{26}$$

Substituting (25) and (26) in Eqs. (17) and (18) and then multiplying (17) successively by $L^{\mathrm{I}}_{i\, mn}$ and $L^{\mathrm{II}}_{i\, mn}$ with due account for (24) we obtain the following three equations:

$$\ddot{X}_{1mn} + \kappa_1 D^{\gamma} X_{1mn} + \omega^2_{1mn} X_{1mn} = -\sum_{\alpha=1}^{2} F_{\alpha\, mn} L^{\mathrm{I}}_{\alpha\, mn}, \tag{27}$$

$$\ddot{X}_{2mn} + \kappa_2 D^{\gamma} X_{2mn} + \omega^2_{2mn} X_{2mn} = -\sum_{\alpha=1}^{2} F_{\alpha\, mn} L^{\mathrm{II}}_{\alpha\, mn}, \tag{28}$$

$$\ddot{X}_{3mn} + \kappa_3 D^{\gamma} X_{3mn} + \omega^2_{3mn} X_{3mn} = -F_{3\, mn}, \tag{29}$$

where $X_{3mn} = x_{3mn}$.

Assume hereafter that the vibration process occurs in such a way that only three natural modes corresponding to the generalized displacements $X_{1s_1s_2}$, $X_{2l_1l_2}$, and $X_{3k_1k_2}$ are excited and dominate over other natural modes. In this case, the right parts of Eqs. (27)–(29) are significantly simplified, and equations of free vibrations (27)–(29) take the form

$$\ddot{X}_{1s_1s_2} + \kappa_1 D^{\gamma} X_{1s_1s_2} + \omega^2_{1s_1s_2} X_{1s_1s_2} = -\zeta^{k_1k_2}_{1s_1s_2} X^2_{3k_1k_2}, \tag{30}$$

$$\ddot{X}_{2l_1l_2} + \kappa_2 D^{\gamma} X_{2l_1l_2} + \omega^2_{2l_1l_2} X_{2l_1l_2} = -\zeta^{k_1k_2}_{2l_1l_2} X^2_{3k_1k_2}, \tag{31}$$

$$\ddot{X}_{3k_1k_2} + \kappa_3 D^{\gamma} X_{3k_1k_2} + \omega^2_{3k_1k_2} X_{3k_1k_2} = -\zeta^{s_1s_2}_{13\, k_1k_2} X_{1s_1s_2} X_{3k_1k_2} - \zeta^{l_1l_2}_{23\, k_1k_2} X_{2l_1l_2} X_{3k_1k_2}, \tag{32}$$

where

$$\zeta^{k_1k_2}_{1s_1s_2} = 4\left(A^{k_1k_2k_1k_2}_{1\; s_1s_2} L^{\mathrm{I}}_{1s_1s_2} + A^{k_1k_2k_1k_2}_{2\; s_1s_2} L^{\mathrm{I}}_{2s_1s_2} \right), \tag{33}$$

$$\zeta^{k_1k_2}_{2l_1l_2} = 4\left(A^{k_1k_2k_1k_2}_{1\; l_1l_2} L^{\mathrm{II}}_{1l_1l_2} + A^{k_1k_2k_1k_2}_{2\; l_1l_2} L^{\mathrm{II}}_{2l_1l_2} \right), \tag{34}$$

$$\zeta^{s_1s_2}_{12k_1k_2} = 4\left(C^{k_1k_2s_1s_2}_{k_1k_2} L^{\mathrm{I}}_{1s_1s_2} + D^{k_1k_2s_1s_2}_{k_1k_2} L^{\mathrm{I}}_{2s_1s_2} \right), \tag{35}$$

$$\zeta^{l_1l_2}_{23k_1k_2} = 4\left(C^{k_1k_2l_1l_2}_{k_1k_2} L^{\mathrm{II}}_{1l_1l_2} + D^{k_1k_2l_1l_2}_{k_1k_2} L^{\mathrm{II}}_{2l_1l_2} \right), \tag{36}$$

$$A_{1\,s_1s_2}^{k_1k_2k_1k_2} = \pi^3 k_1 \left(k_1^2 + \frac{1-\nu}{2}\,\beta_1^2 k_2^2 \right) a_{1s_1s_2}^{k_1k_2k_1k_2} - \frac{1+\nu}{2}\,\pi^3 \beta_1^2 k_1 k_2^2 a_{2s_1s_2}^{k_1k_2k_1k_2}, \tag{37}$$

$$A_{2\,l_1l_2}^{k_1k_2k_1k_2} = \pi^3 \beta_1 k_2 \left(\beta_1^2 k_2^2 + \frac{1-\nu}{2}\,k_1^2 \right) a_{3l_1l_2}^{k_1k_2k_1k_2} - \frac{1+\nu}{2}\,\pi^3 \beta_1 k_1^2 k_2 a_{4l_1l_2}^{k_1k_2k_1k_2}, \tag{38}$$

$$\begin{aligned} C_{k_1k_2}^{k_1k_2s_1s_2} &= \pi^3 s_1 \left(k_1^2 + \nu \beta_1^2 k_2^2 \right) a_{5k_1k_2}^{k_1k_2s_1s_2} + (1-\nu)\pi^3 \beta_1^2 k_1 k_2 s_2 a_{6k_1k_2}^{k_1k_2s_1s_2} \\ &\quad - \pi^3 k_1 \left(s_1^2 + \frac{1-\nu}{2}\,\beta_1^2 s_2^2 \right) a_{7k_1k_2}^{k_1k_2s_1s_2} - \frac{1+\nu}{2}\,\beta_1^2 \pi^3 s_1 s_2 k_2 a_{8k_1k_2}^{k_1k_2s_1s_2}, \end{aligned} \tag{39}$$

$$\begin{aligned} D_{k_1k_2}^{k_1k_2l_1l_2} &= \pi^3 \beta_1 l_2 \left(\nu k_1^2 + \beta_1^2 k_2^2 \right) a_{5k_1k_2}^{k_1k_2l_1l_2} + (1-\nu)\beta_1 \pi^3 l_1 k_1 k_2 a_{6k_1k_2}^{k_1k_2l_1l_2} \\ &\quad - \frac{1+\nu}{2}\,\beta_1 \pi^3 k_1 l_1 l_2 a_{7k_1k_2}^{k_1k_2l_1l_2} - \pi^3 \beta_1 k_2 \left(\frac{1-\nu}{2}\,l_1^2 + \beta_1^2 l_2^2 \right) a_{8k_1k_2}^{k_1k_2l_1l_2}. \end{aligned} \tag{40}$$

From relationships (33)–(40) we could calculate all coefficients entering in Eqs. (30)–(32).

Omitting hereafter the subindices s_1s_2, k_1k_2, and l_1l_2 for ease of presentation, Eqs. (30)–(32) could be rewritten as

$$\ddot{X}_1 + \kappa_1 D^\gamma X_1 + \omega_1^2 X_1 + \zeta_1 X_3^2 = 0, \tag{41}$$

$$\ddot{X}_2 + \kappa_2 D^\gamma X_2 + \omega_2^2 X_2 + \zeta_2 X_3^2 = 0, \tag{42}$$

$$\ddot{X}_3 + \kappa_3 D^\gamma X_3 + \omega_3^2 X_3 + X_3 \left(\zeta_{13} X_1 + \zeta_{23} X_2 \right) = 0. \tag{43}$$

3 Method of Solution

An approximate solution of Eqs. (41)–(43) for small but finite amplitudes weakly varying with time can be represented by a third-order uniform expansion in terms of different time scales in the following form [7]:

$$X_i = \varepsilon X_{i1}(T_0, T_1, T_2 \ldots) + \varepsilon^2 X_{i2}(T_0, T_1, T_2 \ldots) + \varepsilon^3 X_{i3}(T_0, T_1, T_2 \ldots) + \ldots, \tag{44}$$

where $i = 1, 2, 3$, ε is a small dimensionless parameter of the same order of magnitude as the amplitudes, $T_n = \varepsilon^n t$ are new independent variables, among them: $T_0 = t$ is a fast scale characterizing motions with the natural frequencies, and $T_1 = \varepsilon t$ and $T_2 = \varepsilon^2 t$ are slow scales characterizing the modulation of the amplitudes and phases of the modes with nonlinearity.

Recall that the first and the second time-derivatives are defined, respectively, as follows

$$\frac{\mathrm{d}}{\mathrm{d}t} = D_0 + \varepsilon D_1 + \varepsilon^2 D_2 + \cdots, \quad \frac{\mathrm{d}^2}{\mathrm{d}t^2} = D_0^2 + 2\varepsilon D_0 D_1 + \varepsilon^2 \left(D_1^2 + 2 D_0 D_2 \right) + \cdots, \tag{45}$$

while the fractional-order time-derivative could be represented following [11] as

$$\left(\frac{\mathrm{d}}{\mathrm{d}t}\right)^{\gamma} = \left(D_0 + \varepsilon D_1 + \varepsilon^2 D_2 + \cdots\right)^{\gamma}$$
$$= D_0^{\gamma} + \varepsilon\gamma D_0^{\gamma-1} D_1 + \frac{1}{2}\varepsilon^2\gamma\left[(\gamma-1)D_0^{\gamma-2}D_1^2 + 2D_0^{\gamma-1}D_2\right] + \cdots, \tag{46}$$

where $D_n = \partial/\partial T_n$, and $D_0^{\gamma}, D_0^{\gamma-1}, D_0^{\gamma-2}, \ldots$ are the Riemann–Liouville fractional derivatives in time t defined in (7).

Considering that the viscosity is small, i.e.,

$$\kappa_i = \varepsilon^k \mu_i \tau_i^{\gamma},$$

where τ_i is the relaxation time of the ith generalized displacement, μ_i is a finite value, and the choice of k depends on the order of smallness of the viscosity coefficients κ_i, substituting (44)–(46) in Eqs. (41)–(43), after equating the coefficients at like powers of ε to zero, we are led to a set of recurrence equations to various orders:

- to order ε

$$D_0^2 X_{11} + \omega_1^2 X_{11} = 0, \tag{47}$$
$$D_0^2 X_{21} + \omega_2^2 X_{21} = 0, \tag{48}$$
$$D_0^2 X_{31} + \omega_3^2 X_{31} = 0; \tag{49}$$

- to order ε^2

$$D_0^2 X_{12} + \omega_1^2 X_{12} = -2D_0 D_1 X_{11} - \zeta_1 X_{31}^2 - \mu_1(2-k)\tau_1^{\gamma} D_0^{\gamma} X_{11}, \tag{50}$$
$$D_0^2 X_{22} + \omega_2^2 X_{22} = -2D_0 D_1 X_{21} - \zeta_2 X_{31}^2 - \mu_2(2-k)\tau_2^{\gamma} D_0^{\gamma} X_{21}, \tag{51}$$
$$D_0^2 X_{32} + \omega_3^2 X_{32} = -2D_0 D_1 X_{31} - \zeta_{13} X_{11} X_{31} - \zeta_{23} X_{21} X_{31} - \mu_3(2-k)\tau_3^{\gamma} D_0^{\gamma} X_{31}, \tag{52}$$

- to order ε^3

$$D_0^2 X_{13} + \omega_1^2 X_{13} = -2D_0 D_1 X_{12} - \left(D_1^2 + 2D_0 D_2\right) X_{11} - 2\zeta_1 X_{31} X_{32} - \mu_1(2-k)\tau_1^{\gamma}\left(D_0^{\gamma} X_{12} + D_0^{\gamma-1} D_1 X_{11}\right) - \mu_1(k-1)\tau_1^{\gamma} D_0^{\gamma} X_{11}, \tag{53}$$
$$D_0^2 X_{23} + \omega_2^2 X_{23} = -2D_0 D_1 X_{22} - \left(D_1^2 + 2D_0 D_2\right) X_{21} - 2\zeta_2 X_{31} X_{32} - \mu_2(2-k)\tau_2^{\gamma}\left(D_0^{\gamma} X_{22} + D_0^{\gamma-1} D_1 X_{21}\right) - \mu_2(k-1)\tau_2^{\gamma} D_0^{\gamma} X_{21}, \tag{54}$$
$$D_0^2 X_{33} + \omega_3^2 X_{33} = -2D_0 D_1 X_{32} - \left(D_1^2 + 2D_0 D_2\right) X_{31} - \zeta_{13}\left(X_{11} X_{32} + X_{12} X_{31}\right) - \zeta_{23}\left(X_{21} X_{32} + X_{22} X_{31}\right) - \mu_3(2-k)\tau_3^{\gamma}\left(D_0^{\gamma} X_{32} + D_0^{\gamma-1} D_1 X_{31}\right) - \mu_3(k-1)\tau_3^{\gamma} D_0^{\gamma} X_{31}. \tag{55}$$

In order to construct the uniformly valid solution, it is necessary on each step to use the solution from the preceding step and to eliminate secular terms during integration [7].

We shall seek the solution of Eqs. (47)–(49) in the form

$$X_{j1} = A_j(T_1, T_2)\exp(i\omega_j T_0) + \bar{A}_j(T_1, T_2)\exp(-i\omega_j T_0) \quad (j = 1, 2, 3), \tag{56}$$

where $A_j(T_1, T_2)$ $(j = 1, 2, 3)$ are unknown complex functions, and $\bar{A}_j(T_1, T_2)$ are the complex conjugates of $A_j(T_1, T_2)$.

To solve the sets of Eqs. (50)–(52) and (54)–(55), it is necessary to specify the action of the fractional derivative D_0^γ (7) on the functions X_{j1}, i.e., to calculate $D_0^\gamma e^{i\omega_j t}$. It has been shown in [17] that

$$D_0^\gamma e^{i\omega_j t} = (i\omega_j)^\gamma e^{i\omega_j t} + \frac{\sin \pi\gamma}{\pi} \int_0^\infty \frac{u^\gamma}{u + i\omega_j}\, e^{-ut} \mathrm{d}u. \tag{57}$$

The first term in (57) is equivalent to the action of the fractional derivative with the low limit tending to $-\infty$

$$D_0^\gamma x(t) = \frac{1}{\Gamma(1-\gamma)} \frac{\mathrm{d}}{\mathrm{d}t} \int_{-\infty}^{t} \frac{x(s)\mathrm{d}s}{(t-s)^\gamma}, \tag{58}$$

the application of which for the exponential function is reduced to

$$D_0^\gamma e^{i\omega_j t} = (i\omega_j)^\gamma e^{i\omega_j t}, \tag{59}$$

while the second term of (57), as it has been proved in [16], does not influence the solution constructed via the method of multiple time scales restricted to the zeroth- and first-order approximations.

In other words, even the utilization of exact formula (57) in the problem under consideration produces completely equivalent results given by the approximate formula (59) if the solution is constructed via the method of multiple time scales within the considered orders of approximation. Thus, in further analysis we will utilize formula (59).

For further analysis it is also a need to specify the order of weak damping. In this paper, we shall restrict ourselves by considering the case of viscosity of the order of ε, while the case of higher orders of smallness of viscosity is a matter of investigation in another paper.

Then at $k = 1$ Eqs. (50)–(52) are reduced to

$$D_0^2 X_{12} + \omega_1^2 X_{12} = -2D_0 D_1 X_{11} - \zeta_1 X_{31}^2 - \mu_1 \tau_1^\gamma D_0^\gamma X_{11}, \tag{60}$$

$$D_0^2 X_{22} + \omega_2^2 X_{22} = -2D_0 D_1 X_{21} - \zeta_2 X_{31}^2 - \mu_2 \tau_2^\gamma D_0^\gamma X_{21}, \tag{61}$$

$$D_0^2 X_{32} + \omega_3^2 X_{32} = -2D_0 D_1 X_{31} - X_{31}(\zeta_{13} X_{11} + \zeta_{23} X_{21}) - \mu_3 \tau_3^\gamma D_0^\gamma X_{31}. \quad (62)$$

Substituting (56) in the right-hand side of Eqs. (60)–(62) with due account for (59) yields

$$\begin{aligned}
D_0^2 X_{12} + \omega_1^2 X_{12} &= -2i\omega_1 D_1 A_1(T_1)\exp(i\omega_1 T_0) - \zeta_1 \left[A_3^2 \exp(2i\omega_3 T_0) + A_3 \bar{A}_3\right] \\
&\quad - \mu_1 \tau_1^\gamma A_1 (i\omega_1)^\gamma \exp(i\omega_1 T_0) + \text{cc}, \qquad (63)\\
D_0^2 X_{22} + \omega_2^2 X_{22} &= -2i\omega_2 D_1 A_2(T_1)\exp(i\omega_2 T_0) - \zeta_2 \left[A_3^2 \exp(2i\omega_3 T_0) + A_3 \bar{A}_3\right] \\
&\quad - \mu_2 \tau_2^\gamma A_2 (i\omega_2)^\gamma \exp(i\omega_2 T_0) + \text{cc}, \qquad (64)\\
D_0^2 X_{32} + \omega_3^2 X_{32} &= -2i\omega_3 D_1 A_3(T_1)\exp(i\omega_3 T_0) - \mu_3 \tau_3^\gamma A_3 (i\omega_3)^\gamma \exp(i\omega_3 T_0) \\
&\quad - \zeta_{13} \left\{A_1 A_3 \exp[i(\omega_1 + \omega_3)T_0] + A_1 \bar{A}_3 \exp[i(\omega_1 - \omega_3)T_0]\right\} \\
&\quad - \zeta_{23} \left\{A_2 A_3 \exp[i(\omega_2 + \omega_3)T_0] + A_2 \bar{A}_3 \exp[i(\omega_2 - \omega_3)T_0]\right\} + \text{cc}, \qquad (65)
\end{aligned}$$

where cc is the complex conjugate part to the preceding terms.

Reference to Eqs. (63)–(65) shows that the following types of the internal resonance could occur on this step:

1. the two-to-one internal resonance, when one natural frequency is twice the other natural frequency,

$$\omega_1 = 2\omega_3 \quad (\omega_2 \neq \omega_1, \quad \omega_2 \neq 2\omega_3), \qquad (66)$$

$$\omega_2 = 2\omega_3 \quad (\omega_1 \neq \omega_2, \quad \omega_1 \neq 2\omega_3); \qquad (67)$$

2. the one-to-one-to-two internal resonance, i.e.,

$$\omega_1 = \omega_2 = 2\omega_3, \quad \text{or} \quad 1:1:2. \qquad (68)$$

4 Governing Nonlinear Differential Equations Describing Amplitude-Phase Modulation for Different Types of the Internal Resonance of the Order of ε

To deduce the nonlinear differential equations describing the modulation of amplitudes and phases of the nonlinear plate under consideration, we should consider separately each type of the internal resonance which could occur with due account for weak damping of the order ε.

4.1 Internal Resonance 2:1

Let us consider the case (66), when $\omega_1 = 2\omega_3$, while $\omega_2 \neq \omega_1$ and $\omega_2 \neq 2\omega_3$. The case (67) could be treated similarly.

Then eliminating secular terms in Eqs. (63)–(65), we obtain the following solvability equations:

$$2i\omega_1 D_1 A_1(T_1) + \mu_1 (i\omega_1\tau_1)^{\gamma} A_1 + \zeta_1 A_3^2 = 0, \tag{69}$$

$$2i\omega_2 D_1 A_2(T_1) + \mu_2 (i\omega_2\tau_2)^{\gamma} A_2 = 0, \tag{70}$$

$$2i\omega_3 D_1 A_3(T_1) + \mu_3 (i\omega_3\tau_3)^{\gamma} A_3 + \zeta_{13} A_1 \bar{A}_3 = 0. \tag{71}$$

Reference to the set of Eqs. (69)–(71) shows that its second equation is independent of other two, while the first and third ones represent a set of two nonlinear equations. The similar situation was noted in [12] for a fractionally damped nonlinear plate in the case of the two-to-one internal resonance, when equations describing the plate in-plane motion are coupled.

Let us multiply Eqs. (69)–(71), respectively, by $\bar{A}_1$, $\bar{A}_2$, and $\bar{A}_3$ and find their complex conjugates. Adding every pair of the mutually adjoint equations with each other and subtracting one from another, and considering that

$$A_i = a_i e^{i\varphi_i} \quad (i = 1, 2, 3), \tag{72}$$

as a result we have

$$\left(a_1^2\right)^{\cdot} + s_1 a_1^2 + \omega_1^{-1}\zeta_1 a_1 a_3^2 \sin\delta = 0, \tag{73}$$

$$\left(a_3^2\right)^{\cdot} + s_3 a_3^2 - \omega_3^{-1}\zeta_{13} a_1 a_3^2 \sin\delta = 0, \tag{74}$$

$$\dot{\varphi}_1 - \frac{1}{2}\,\sigma_1 - \frac{1}{2}\,\zeta_1\omega_1^{-1} a_3^2 a_1^{-1} \cos\delta = 0, \tag{75}$$

$$\dot{\varphi}_3 - \frac{1}{2}\,\sigma_3 - \frac{1}{2}\,\zeta_{13}\omega_3^{-1} a_1 \cos\delta = 0, \tag{76}$$

$$\left(a_2^2\right)^{\cdot} + s_2 a_2^2 = 0, \tag{77}$$

$$\dot{\varphi}_2 - \frac{1}{2}\,\sigma_2 = 0, \tag{78}$$

where a dot denotes differentiation with respect to T_1, $\delta = 2\varphi_3 - \varphi_1$, and

$$s_i = \mu_i \tau_i^{\gamma} \omega_i^{\gamma-1} \sin\psi, \quad \sigma_i = \mu_i \tau_i^{\gamma} \omega_i^{\gamma-1} \cos\psi, \quad \psi = \frac{1}{2}\,\pi\gamma \quad (i = 1, 2, 3). \tag{79}$$

From Eqs. (73)–(76) we find

$$\frac{\frac{\omega_1}{\zeta_1}\left[\left(a_1^2\right)^{\cdot} + s_1 a_1^2\right]}{\frac{\omega_3}{\zeta_{13}}\left[\left(a_3^2\right)^{\cdot} + s_3 a_3^2\right]} = -1, \tag{80}$$

$$\dot{\delta} - \Sigma - \frac{1}{a_1}\left[\zeta_{13}\omega_3^{-1} a_1^2 - \frac{1}{2}\,\zeta_1\omega_1^{-1} a_3^2\right]\cos\delta = 0, \tag{81}$$

where $\Sigma = \sigma_3 - \frac{1}{2}\sigma_1$.

Introducing new functions $\xi_1(T_1)$ and $\xi_3(T_1)$, such that

$$a_1^2 = \frac{\zeta_1}{\omega_1}\,\xi_1(T_1)e^{-s_1T_1}, \quad a_3^2 = \frac{\zeta_{13}}{\omega_3}\,\xi_3(T_1)e^{-s_3T_1}, \tag{82}$$

and substituting (82) in (80) yield

$$\dot{\xi}_3 + \dot{\xi}_1\, e^{(s_3-s_1)T_1} = 0, \tag{83}$$

whence it follows

$$\xi_3 = -\left[\xi_1 e^{(s_3-s_1)T_1} - (s_3-s_1)\int\limits_0^{T_1} \xi_1 e^{(s_3-s_1)T_1}\mathrm{d}T_1\right] + E_0, \tag{84}$$

where

$$E_0 = \xi_{10} + \xi_{30}$$

is the initial energy of the plate defined by the initial conditions

$$\xi_1\Big|_{T_1=0} = \xi_{10}, \quad \xi_2\Big|_{T_1=0} = \xi_{20}, \quad \xi_3\Big|_{T_1=0} = \xi_{30}, \quad \delta\Big|_{T_1=0} = \delta_0. \tag{85}$$

From (73) and (74), we could find

$$a_1 = \frac{\omega_3}{\zeta_{13}}\left(2\,\frac{\dot{a}_3}{a_3} + s_3\right)\frac{1}{\sin\delta} = 2\,\frac{\omega_3}{\zeta_{13}}\,\frac{\dot{\xi}_3}{\xi_3}\,\frac{1}{\sin\delta}, \tag{86}$$

$$a_3^2 a_1^{-1} = -\frac{\omega_1}{\zeta_1}\,(2\dot{a}_1 + s_1 a_1)\,\frac{1}{\sin\delta} = 2\,\frac{\omega_1}{\zeta_1}\,\frac{\dot{\xi}_1}{\xi_1}\,\frac{1}{\sin\delta}. \tag{87}$$

Considering (86) and (87), Eq. (81) could be rewritten as

$$\tan\delta\left(\dot{\delta} - \Sigma\right) = \frac{1}{2}\,\frac{\dot{\xi}_1}{\xi_1} + \frac{\dot{\xi}_3}{\xi_3}, \tag{88}$$

integration of which yields

$$G_0 \exp\left(-\Sigma\int\limits_0^{T_1} \tan\delta\,\mathrm{d}T_1\right) = \sqrt{\xi_1}\,\xi_3\cos\delta, \tag{89}$$

where G_0 is a constant of integration to be found from the initial conditions (85).

Substituting (82) in (73) and (74), we have

$$\dot{\xi}_1 = -b\sqrt{\xi_1}\,\xi_3\,e^{\left(\frac{1}{2}s_1 - s_3\right)T_1}\sin\delta, \tag{90}$$

$$\dot{\xi}_3 = b\sqrt{\xi_1}\,\xi_3\,e^{-\frac{1}{2}s_1T_1}\sin\delta, \tag{91}$$

where

$$b = \frac{\zeta_{13}}{\omega_3}\sqrt{\frac{\zeta_1}{\omega_1}}.$$

The nonlinear set of Eqs. (84), (90), and (89) with the initial conditions (85) completely describe the vibrational process of the mechanical system being investigated under the condition of the internal resonance two-to-one and could be solved numerically.

4.1.1 Particular Case

In the particular case at $\gamma = 1$, $\sigma_1 = \sigma_3 = 0$, and hence $\Sigma = 0$. Assuming that $s_1 = s_3 = s$, from (84) we find

$$\xi_3 + \xi_1 = E_0, \tag{92}$$

whence it follows that

$$\xi_1 = E_0\,\xi, \quad \xi_3 = E_0(1-\xi), \tag{93}$$

or

$$a_1^2 = E_0\,\frac{\zeta_1}{\omega_1}\,\xi\exp(-sT_1), \quad a_3^2 = E_0\,\frac{\zeta_{13}}{\omega_1}(1-\xi)\exp(-sT_1). \tag{94}$$

Considering (94) and integrating (80) yield the law of energy dissipation

$$E = \frac{\omega_1}{\zeta_1}\,a_1^2 + \frac{\omega_3}{\zeta_{13}}\,a_3^2 = E_0 e^{-sT_1}, \tag{95}$$

where E is the energy of the system.

In the case under consideration, Eqs. (81) and (91), take, respectively, the form

$$\dot{\delta} = -\frac{1}{2}\,b\sqrt{E_0}\,\frac{1-3\xi}{\sqrt{\xi}}\,e^{-\frac{1}{2}sT_1}\cos\delta, \tag{96}$$

$$\dot{\xi} = -b\sqrt{E_0}\,\sqrt{\xi}(1-\xi)e^{-\frac{1}{2}sT_1}\sin\delta, \tag{97}$$

while (88) could be rewritten as

$$\dot{\delta}\tan\delta = \frac{1}{2}\frac{\dot{\xi}_1}{\xi_1} + \frac{\dot{\xi}_3}{\xi_3}. \tag{98}$$

Integrating (98) with due account for (93) yields

$$G(\xi, \delta) = \sqrt{\xi}(1-\xi)\cos\delta = G_0(\xi_0, \delta_0). \tag{99}$$

Eliminating the variable δ from (97) and (99), and integrating over T_1, we have

$$\int\limits_{\xi_0}^{\xi} \frac{d\xi}{\sqrt{\xi^3 - 2\xi^2 + \xi - G_0^2}} = \frac{2b\sqrt{E_0}}{s}\left(e^{-\frac{1}{2}sT_1} - 1\right). \tag{100}$$

The integral in the left-hand side of Eq. (100) can be transformed to an incomplete elliptic integral of the first kind and can be easily calculated using special tables [1].

For qualitative analysis of the solution let us introduce into consideration the phase fluid moving along the plane $\xi\delta$ in the channel of the finite width ($0 \le \xi \le 1$) and the infinite length ($-\infty < \delta < +\infty$) with the velocity $\mathbf{v}$ ($v_\xi = \dot{\xi}$ and $v_\delta = \dot{\delta}$) [10]. Each point with the coordinates $\xi\delta$ on the phase plane corresponds to certain magnitudes of the amplitudes a_1 and a_3 of two interacting modes of vibrations at the fixed instant, and to the phase difference of these modes relative to each other at the same instant.

In the phase plane $\xi\delta$, Eq. (99) defines the stream function $G(\xi, \delta)$ such that

$$v_\xi = \dot{\xi} = b\sqrt{E_0}\,\frac{\partial G}{\partial \delta}\,e^{-\frac{1}{2}sT_1}, \quad v_\delta = \dot{\delta} = -b\sqrt{E_0}\,\frac{\partial G}{\partial \xi}\,e^{-\frac{1}{2}sT_1}, \tag{101}$$

what is fulfilled along each streamline. In other words, the picture of the streamlines is unchanged with the time, but the field of the velocities constructed along the streamlines is time dependent in such a manner that at each point $\xi\delta$ of this field the direction of the velocity vector $\mathbf{v}$ remains constant, and its modulus decreases by the exponential law, resulting in quasi-steady vibrations of the viscoelastic plate.

Consequently, at $\gamma = 1$, the stream-function $G(\xi, \delta) = G_0(\xi_0, \delta_0)$ defined by (99) is the first integral of Eqs. (73)–(76), in parallel with the first integral (95), corresponding to the law of energy dissipation.

Comparison of the solution (89) with that in (99) shows that in the general case the phase fluid stream-lines disappear, and the phase fluid particles begin to describe intricate trajectories on the phase plane, i.e., quasi-stable motion of the plate goes over into the transient one.

4.1.2 Free Undamped Vibrations

In the absence of damping, i.e., when $\gamma = 0$, $\mu = 0$, and $s = 0$, the first integral takes the form

$$E = \frac{\omega_1}{\zeta_1}\, a_1^2 + \frac{\omega_3}{\zeta_{13}}\, a_3^2 = E_0, \tag{102}$$

what corresponds to the law of conservation of system's energy.

The second first integral (99) defines the stream function $G(\xi, \delta)$ such that

$$v_\xi = \dot{\xi} = b\sqrt{E_0}\,\frac{\partial G}{\partial \delta}, \quad v_\delta = \dot{\delta} = -b\sqrt{E_0}\,\frac{\partial G}{\partial \xi}, \tag{103}$$

which describes steady– state vibrations of an elastic plate.

Finally, Eq. (100) allowing one to find the function $\xi(T_1)$ and, thus to solve the problem under consideration, takes the form

$$\int\limits_{\xi_0}^{\xi} \frac{d\xi}{\sqrt{\xi^3 - 2\xi^2 + \xi - G_0^2}} = -b\sqrt{E_0}\, T_1 \tag{104}$$

Therefore, during free vibrations of the plate being under the conditions of the two-to-one internal resonance three regimes can be observed: stationary (absence of damping at $\gamma = 0$ and $\mu = 0$), quasi-stationary (damping is defined by an ordinary derivative at $\gamma = 1$), and transient (damping is defined by a fractional derivative at $0 < \gamma < 1$).

4.2 *Internal Resonance 1:1:2*

Let us consider the case (68), when $\omega_1 = \omega_2 = 2\omega_3$. Then eliminating secular terms in Eqs. (63)–(65), we obtain the following solvability equations:

$$2i\omega_1 D_1 A_1(T_1) + \mu_1\,(i\omega_1\tau_1)^\gamma\, A_1 + \zeta_1 A_3^2 = 0, \tag{105}$$

$$2i\omega_2 D_1 A_2(T_1) + \mu_2\,(i\omega_2\tau_2)^\gamma\, A_2 + \zeta_2 A_3^2 = 0, \tag{106}$$

$$2i\omega_3 D_1 A_3(T_1) + \mu_3\,(i\omega_3\tau_3)^\gamma\, A_3 + \zeta_{13} A_1 \bar{A}_3 + \zeta_{23} A_2 \bar{A}_3 = 0. \tag{107}$$

Reference to the set of Eqs. (105)–(107) shows that, in contrast to Eqs. (69)–(71) corresponding to the case (66) when $\omega_1 = 2\omega_3$, while $\omega_2 \neq \omega_1$ and $\omega_2 \neq 2\omega_3$, now we obtain the set of three coupled nonlinear equations.

Representing functions A_i entering in (105)–(107) in the polar form (72) and applying the same procedure as it has been carried out above for 2:1 internal resonance, we have

$$\left(a_1^2\right)^{\cdot} + s_1 a_1^2 + \omega_1^{-1} \zeta_1 a_1 a_3^2 \sin(2\varphi_3 - \varphi_1) = 0, \tag{108}$$

$$\left(a_2^2\right)^{\cdot} + s_2 a_2^2 + \omega_2^{-1} \zeta_2 a_2 a_3^2 \sin(2\varphi_3 - \varphi_2) = 0, \tag{109}$$

$$\begin{aligned}\left(a_3^2\right)^{\cdot} &+ s_3 a_3^2 - \omega_3^{-1} \zeta_{13} a_1 a_3^2 \sin(2\varphi_3 - \varphi_1)\\ &- \omega_3^{-1} \zeta_{23} a_2 a_3^2 \sin(2\varphi_3 - \varphi_2) = 0,\end{aligned} \tag{110}$$

$$\dot{\varphi}_1 - \frac{1}{2}\sigma_1 - \frac{1}{2}\zeta_1 \omega_1^{-1} a_3^2 a_1^{-1} \cos(2\varphi_3 - \varphi_1) = 0, \tag{111}$$

$$\dot{\varphi}_2 - \frac{1}{2}\sigma_2 - \frac{1}{2}\zeta_2 \omega_2^{-1} a_3^2 a_2^{-1} \cos(2\varphi_3 - \varphi_2) = 0, \tag{112}$$

$$\begin{aligned}\dot{\varphi}_3 &- \frac{1}{2}\sigma_3 - \frac{1}{2}\zeta_{13} \omega_3^{-1} a_1 \cos(2\varphi_3 - \varphi_1)\\ &- \frac{1}{2}\zeta_{23} \omega_3^{-1} a_2 \cos(2\varphi_3 - \varphi_2) = 0,\end{aligned} \tag{113}$$

The nonlinear set of Eqs. (108)–(113) with the initial conditions (85) completely describes the vibrational process of the mechanical system being investigated under the condition of the internal resonance 1:1:2 and could be solved numerically.

Introducing new functions $\xi_1(T_1)$, $\xi_2(T_1)$, and $\xi_3(T_1)$ such that

$$a_1^2 = \frac{\zeta_1 \omega_3}{\zeta_{13} \omega_1} \xi_1 e^{-s_1 T_1}, \quad a_2^2 = \frac{\zeta_2 \omega_3}{\zeta_{23} \omega_2} \xi_2 e^{-s_2 T_1}, \quad a_3^2 = \xi_3 e^{-s_3 T_1}, \tag{114}$$

and adding Eqs. (108)–(110) with due account for (114), we obtain

$$\dot{\xi}_1 e^{-s_1 T_1} + \dot{\xi}_2 e^{-s_2 T_1} + \dot{\xi}_3 e^{-s_3 T_1} = 0. \tag{115}$$

Equation (115) describes the law of energy variation for this case of the internal resonance.

5 Numerical Investigations

As example, let us carry out the qualitative analysis of the case of two-to-one internal resonance (66).

For this case, the stream-function $G(\xi, \delta)$ is constructed according to (99) and the stream-lines of the phase fluid in the phase plane $\xi - \delta$ are presented in Fig. 1. Magnitudes of G are indicated by digits near the curves which correspond to the stream-lines; the flow direction of the phase fluid elements are shown by arrows on the stream-lines.

Reference to Fig. 1 shows that the phase fluid flows within the circulation zones, which tend to be located around the perimeter of the rectangles bounded by the lines $\xi = 0$, $\xi = 1$, and $\delta = \pm(\pi/2) \pm 2\pi n$ $(n = 0, 1, 2, \ldots)$. As this takes place, the flow

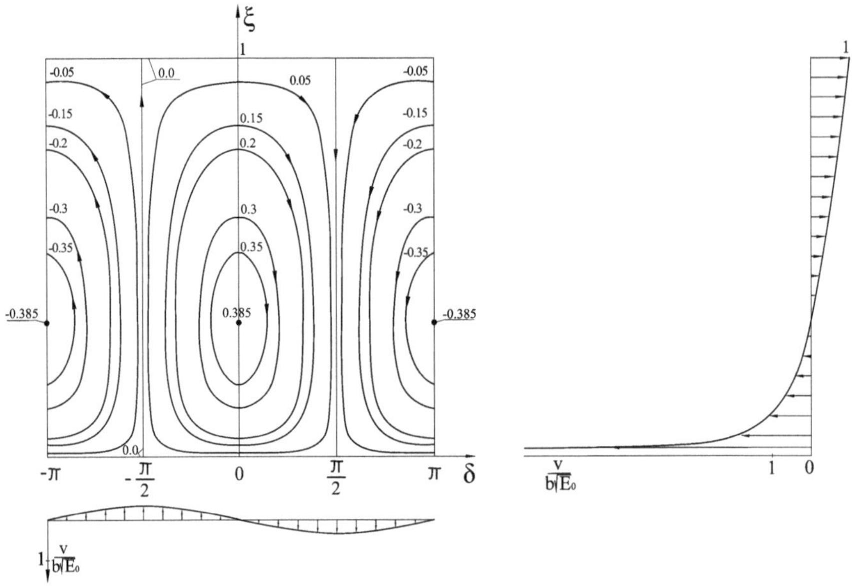

Fig. 1 Phase portrait for the case of 2:1 internal resonance

in each such rectangle becomes isolated. On all four rectangle sides $G = 0$ and inside it, the value G preserves its sign. The function G attains its extreme magnitudes at the points with the coordinates $\xi = \frac{1}{3}$, $\delta = \pm\pi n$ $(n = 0, 1, 2, \ldots)$.

Stream-lines give a pictorial estimate of the connection of G with all types of the energy– exchange mechanism. Thus, in the case of undamped vibrations, i.e., when the damping coefficient is equal to zero and $s = 0$, the points with the coordinates $\xi_0 = \frac{1}{3}$, $\delta_0 = \pm\pi n$ $(n = 0, 1, 2, \ldots)$ correspond to the stationary regime, since $\dot{\delta} = 0$ and $\dot{\xi} = 0$ according to (96) and (97). The stationary points $\xi_0 = \frac{1}{3}$, $\delta_0 = \pm\pi n$ are centers, as with a small deviation from a center, a phase element begins to move around the stationary point along a closed trajectory. Closed stream-lines correspond to the periodic change of both amplitudes and phases.

Along the lines $\delta = \pm(\pi/2) \pm 2\pi n$ $(n = 0, 1, 2, \ldots)$ pure amplitude modulated aperiodic motions are realized, since with an increase in time t from 0 to ∞ the value ξ increases from ξ_0 to 1 (along the line $\delta = -\pi/2$) or decreases from ξ_0 to 0 (along the line $\delta = \pi/2$), and from Eq. (104) it follows that

$$\xi = \left[\frac{1 + \sqrt{\xi_0} - (1 - \sqrt{\xi_0})\exp(-b\sqrt{E_0}\,T_1)}{1 + \sqrt{\xi_0} + (1 - \sqrt{\xi_0})\exp(-b\sqrt{E_0}\,T_1)}\right]^2, \delta(T_1) = \delta_0 = \frac{\pi}{2} \pm \pi n, n = 0, 1, 2, \ldots \tag{116}$$

Along the line $\xi = 1$ only phase modulated motions are realized, because when $\xi = \xi_0 = 1$ the amplitudes $a_3 = \text{const}$ and $a_1 = 0$, and from (96) and (97) we could find that

$$b\sqrt{E_0}\,T_1 = \ln \frac{\tan\left(\frac{\delta}{2}+\frac{\pi}{4}\right)}{\tan\left(\frac{\delta_0}{2}+\frac{\pi}{4}\right)}, \quad \xi(T_1) = \xi_0 = 1. \tag{117}$$

The transition of fluid elements from the points $\xi = 0, \delta = \pi/2 \pm 2\pi n$ to the points $\xi = 0, \quad \delta = -\pi/2 \pm 2\pi n$ proceeds instantly, because according to the distribution of the phase velocity along the section $\delta = 0$ (see Fig. 1) the magnitude of $\mathbf{v}$ tends to infinity as $\xi \to 0$. The distribution of the velocity along the vertical lines $\delta = \pm\pi n$ $(n = 0, 1, 2, \ldots)$ has the aperiodic character, while in the vicinity of the line $\xi = 1/3$ it possesses the periodic character.

At the presence of conventional viscosity, i.e., when $\gamma = 1$, the picture of the stream-lines remains unchanged, but the velocities of the phase fluid particles decay exponentially during their motion along the stream-lines. As this takes place, the stationary and boundary regimes (pure amplitude or pure phase modulations) become aperiodic regimes, that is the amplitudes and phases attenuate exponentially with time. In other words, according to (101), the field of the velocities constructed along the streamlines is time dependent in such a manner that at each point $\xi\delta$ of this field the direction of the velocity vector $\mathbf{v}$ remains constant, and its modulus decreases by the exponential law, resulting in quasi-steady vibrations of the viscoelastic plate.

Thus, along the lines $\delta = \pm(\pi/2) \pm 2\pi n$ $(n = 0, 1, 2, \ldots)$ in the presence of conventional viscosity the solution (116) should be rewritten as

$$\xi = \left[\frac{1 - c_0 \exp^{f(T_1)}}{1 + c_0 \exp^{f(T_1)}}\right]^2, \quad \delta(T_1) = \delta_0 = \frac{\pi}{2} \pm \pi n, \quad n = 0, 1, 2, \ldots \tag{118}$$

where

$$f(T_1) = \frac{2b\sqrt{E_0}}{s}\left(1 - e^{-\frac{1}{2}sT_1}\right), \quad c_0 = \frac{1-\sqrt{\xi_0}}{1+\sqrt{\xi_0}}.$$

The T_1-dependence of the values ξ and $1-\xi$ which are proportional to the square of the amplitudes a_3 and a_1, respectively, at $\delta_0 = 0$ and $\xi = 0.5$ is presented in Fig. 2, wherein the solid and dashed lines correspond, respectively, to the cases of damped and undamped vibrations. Reference to Fig. 2 shows the damping of the energy exchange between two subsystems. Moreover, the points of touching of solid curves, i.e., the points of tangency of the envelopes of vibrations $(\xi_i = 1 - \xi_i, T_{1\,i})$, allow one to determine the value characterizing the damping of energy during vibrations via the formula

$$\frac{\ln(a_{i+1}/a_i)}{T_{1\,(i-1)} - T_{1\,i}},$$

where a_i are defined by (94).

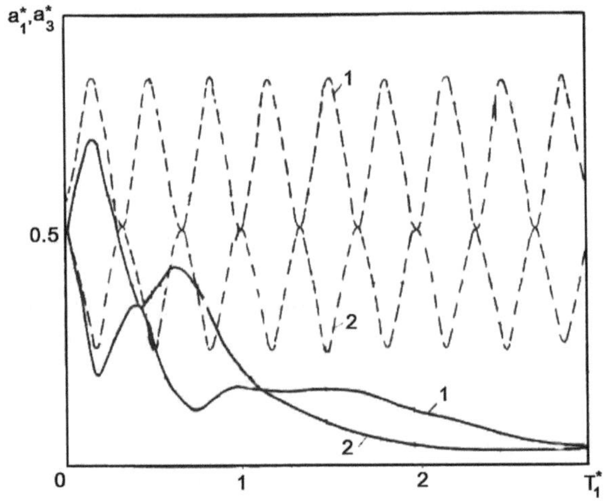

Fig. 2 The T_1^*-dependence of the dimensionless amplitudes of vibrations for the case of 2:1 internal resonance

In the general case, when $0 < \gamma < 1$, the phase fluid stream-lines disappear, and the phase fluid particles begin to describe intricate trajectories on the phase plane, i.e., quasi-stable motion of the shell goes over into the transient one.

6 Conclusion

Free damped vibrations of a non-linear plate in a fractional derivative viscoelastic medium have been investigated. Nonlinear vibrations are described by coupled three equations with respect to the three orthogonal displacements.

The proposed analytical approach for investigating the damped vibrations of the nonlinear plate subjected to the conditions of the internal resonance has been possible owing to the new procedure suggested in this paper, resulting in decoupling linear parts of equations with the further utilization of the method of multiple scales for solving nonlinear governing equations of motion.

It has been shown that the phenomenon of internal resonance could be very critical, since in the thin plate under consideration the internal resonance is always present. Moreover, its type depends on the order of smallness of the viscosity involved into consideration. Thus, at the ε-order, damped vibrations occur within the two-to-one and one-to-one-to-two internal resonance. Other types of the internal resonance, such as one-to-one, one-to-one-to-one, and combinational resonances of the additive and difference types could be found at ε^2-order, i.e., the type of the resonance depends on the order of smallness of the fractional derivative entering in the equations of motion of the plate. That is why all possible cases of the internal resonances of the order of ε have been examined in the present paper, while the internal resonances of the order of ε^2 will be considered in detail in another paper.

For each type of the resonance, the nonlinear sets of resolving equations in terms of amplitudes and phase differences have been obtained. It has been shown that for the two-to-one internal resonance there exist such particular cases when it is possible to obtain two first integrals, namely: the energy integral and the stream-function, what allows one to reduce the problem to the calculation of elliptic integrals.

The influence of viscosity on the energy exchange mechanism has been analyzed. It has been shown that each mode is characterized by its damping coefficient connected with the natural frequency by the exponential relationship with a negative fractional exponent.

Acknowledgments The research described in this publication has been supported by the Russian Ministry of High Education and Science. (No 7.22.2014/K)

References

1. Abramowitz, M., Stegun, I. (eds.): Handbook of Mathematical Functions with Formulas, Graphs, and Tables, Applied Mathematical Services, vol. 55. National Bureau of Standards U.S.A, Washington (1964)
2. Amabili, M.: Nonlinear vibrations of rectangular plates with different boundary conditions: theory and experiments. Comput. Struct. **82**, 2587–2605 (2004)
3. Amabili, M.: Nonlinear Vibrations and Stability of Shells and Plates. Cambridge University Press, New York (2008)
4. Breslavsky, I., Amabili, M., Legrand, M.: Physically and geometrically non-linear vibrations of thin rectangular plates. Int. J. Non-Linear Mech. **58**, 30–40 (2014)
5. Khodzhaev, D., Eshmatov, B.: Nonlinear vibrations of a viscoelastic plate with concentrated masses. J. Appl. Mech. Tech. Phys. **48**(6), 905–914 (2007)
6. Kim, T., Kim, J.: Nonlinear vibration of viscoelastic laminated composite plates. Int. J. Solids Struct. **39**(10), 2857–2870 (2002)
7. Nayfeh, A.: Perturbation Methods. Wiley, New York (1973)
8. Rashidi, M., Shooshtari, A., Beg, O.: Homotopy perturbation study of nonlinear vibration of von Kármán rectangular plates. Comput. Struct. **106–107**, 46–55 (2002)
9. Ribeiro, P., Petyt, M.: Non-linear free vibration of isotropic plates with internal resonance. Int. J. Non-Linear Mech. **35**(2), 263–278 (2000)
10. Rossikhin, Y., Shitikova, M.: Analysis of nonlinear free vibrations of suspension bridges. J. Sound Vibr. **186**, 369–393 (1995)
11. Rossikhin, Y., Shitikova, M.: Application of fractional calculus for analysis of nonlinear damped vibrations of suspension bridges. ASCE J. Eng. Mech. **124**, 1029–1036 (1998)
12. Rossikhin, Y., Shitikova, M.: Free damped nonlinear vibrations of a viscoelastic plate under the two-to-one internal resonance. Mater. Sci. Forum **440–441**, 29–36 (2003)
13. Rossikhin, Y., Shitikova, M.: Analysis of free non-linear vibrations of a viscoelastic plate under the conditions of different internal resonances. Int. J. Non-Linear Mech. **41**, 313–325 (2006)
14. Rossikhin, Y., Shitikova, M.: Application of fractional calculus for dynamic problems of solid mechanics: Novel trends and recent results. Appl. Mech. Reviews, **63**(1):010,801-1-010,801-52. (2010)
15. Rossikhin, Y., Shitikova, M.: Analysis of damped vibrations of thin bodies embedded into a fractional derivative viscoelastic medium. J. Mech. Behav. Mater. **21**(5–6), 155–159 (2012a)
16. Rossikhin, Y., Shitikova, M.: On fallacies in the decision between the Caputo and Riemann-Liouville fractional derivatives for the analysis of the dynamic response of a nonlinear viscoelastic oscillator. Mech. Research Commun. **45**, 22–27 (2012b)

17. Rossikhin, Y., Shitikova, M., Shcheglova, T.: Forced vibrations of a nonlinear oscillator with weak fractional damping. J. Mech. Mat. Struct. **4**(9), 1619–1636 (2009)
18. Sathyamoorthy, M.: Nonlinear vibrations of plates: an update of recent research developments. Appl. Mech. Review **49**(10), 55–62 (1996)
19. Shooshtari, A., Khadem, S.: A multiple scale method solution for the nonlinear vibration of rectangular plates. Sci. Iranica **14**, 64–71 (2007)

On Stability of Inhomogeneous Elastic Cylinder of Micropolar Material

Denis N. Sheydakov

Abstract The present research is dedicated to the buckling analysis of nonlinearly elastic rods made of porous materials. In the framework of a general stability theory for three-dimensional bodies, we have studied the stability of a circular micropolar rod subject to axial compression and external pressure. It is assumed that the elastic properties of the rod vary along the radius. Applying linearization the neutral equilibrium equations are derived, which describe the perturbed state of a rod. These linearized equations have been solved numerically for a few commonly used porous materials. The critical curves and corresponding buckling modes have been found, and the stability regions have been constructed in the plane of loading parameters. Using these results, we have studied the influence of elastic properties as well as the rod size on the loss of stability. Special attention has been given to the analysis of how the pattern of variation for elastic properties of material affects the stability of a micropolar rod.

1 Introduction

Due to the increasing number of new construction materials, the problem of stability analysis for bodies with a microstructure becomes important. Examples of such new materials are porous materials. Engineering structures made of porous materials, especially metal and polymer foams, are widely used in modern industries with airspace or automotive applications among others [2–4, 6, 10]. They combine low weight, high specific strength and excellent possibilities to absorb energy. As a rule, these constructions have a functionally graded structure (for example, a porous core covered by hard and stiff shell). This is necessary for corrosion or thermal protection, and optimization of mechanical properties in the process of loading.

D.N. Sheydakov (✉)
South Scientific Center of Russian Academy of Sciences, Chekhova Ave. 41,
344006 Rostov-on-Don, Russia
e-mail: sheidakov@mail.ru

H. Altenbach and G.I. Mikhasev (eds.), *Shell and Membrane Theories in Mechanics and Biology*, Advanced Structured Materials 45, DOI 10.1007/978-3-319-02535-3_16

2 Initial Strain State of Inhomogeneous Rod

We consider the inhomogeneous elastic circular rod of radius r_0 and length l, and made of metal or polymer foam. Due to the microstructure influence, the behavior of foams cannot be adequately described within the framework of the classical theory of elasticity. One approach to the modeling of the porous elastic body is to use the model of micropolar continuum, or Cosserat continuum [1, 5, 9, 12, 16, 23], i.e. medium with couple stresses and rotational degrees of freedom. Then, in the case of axial compression of the rod under external hydrostatic pressure, the radius-vector $\mathbf{R}$ and proper orthogonal tensor of microrotation H, which determine the position and rotation of a micropolar medium particle, are given by the following relations [15, 21, 24]:

$$\begin{aligned} &R = f(r), \qquad \Phi = \varphi, \qquad Z = \alpha z \\ &0 \leq r \leq r_0, \qquad 0 \leq \varphi \leq 2\pi, \qquad 0 \leq z \leq l \end{aligned} \tag{1}$$

$$\begin{aligned} \mathbf{R} &= f\,(r)\,\mathbf{e}_R + \alpha z\,\mathbf{e}_Z \\ \mathsf{H} &= \mathbf{e}_r \otimes \mathbf{e}_R + \mathbf{e}_\varphi \otimes \mathbf{e}_\Phi + \mathbf{e}_z \otimes \mathbf{e}_Z \end{aligned} \tag{2}$$

Here r, φ, z are the cylindrical coordinates in the reference configuration (Lagrangian coordinates), R, Φ, Z are the Eulerian cylindrical coordinates, $\left\{\mathbf{e}_r, \mathbf{e}_\varphi, \mathbf{e}_z\right\}$ and $\{\mathbf{e}_R, \mathbf{e}_\Phi, \mathbf{e}_Z\}$ are orthonormal vector bases of Lagrangian and Eulerian coordinates, respectively, α is the compression ratio along the axis of the rod, $f(r)$ is the unknown function which characterizes the radial deformation of the rod.

According to the expressions (1), (2), the deformation gradient C is (hereinafter the $'$ denotes the derivative with respect to r):

$$\mathsf{C} = \operatorname{Grad}\mathbf{R} = f'\mathbf{e}_r \otimes \mathbf{e}_R + \frac{f}{r}\mathbf{e}_\varphi \otimes \mathbf{e}_\Phi + \alpha\mathbf{e}_z \otimes \mathbf{e}_Z \tag{3}$$

where Grad is gradient operator in the Lagrangian coordinates:

$$\operatorname{Grad}\mathbf{R} = \mathbf{e}_r\frac{\partial\mathbf{R}}{\partial r} + \frac{1}{r}\mathbf{e}_\varphi\frac{\partial\mathbf{R}}{\partial\varphi} + \mathbf{e}_z\frac{\partial\mathbf{R}}{\partial z}$$

It follows from relations (2), (3) that the wryness tensor L is equal to zero [7, 17, 19, 20]

$$\mathsf{L} \times \mathsf{I} = -\left(\operatorname{grad}\mathsf{H}\right) \cdot \mathsf{H}^{\mathrm{T}} = \mathbf{0}$$

(I is the unit tensor) and the stretch tensor Y is expressed as follows

$$\mathsf{Y} = f'\mathbf{e}_r \otimes \mathbf{e}_r + \frac{f}{r}\mathbf{e}_\varphi \otimes \mathbf{e}_\varphi + \alpha\mathbf{e}_z \otimes \mathbf{e}_z \tag{4}$$

We assume that the elastic properties of the rod vary along the radius, and is described by the model of physically linear micropolar material, the specific strain energy of which is a quadratic form of the tensors $\mathbf{Y}-\mathbf{I}$ and $\mathbf{L}$ [8, 13]:

$$\begin{aligned}W(\mathbf{Y},\mathbf{L}) = {} & \tfrac{1}{2}\lambda(r)\mathrm{tr}^2(\mathbf{Y}-\mathbf{I}) + \tfrac{1}{2}[\mu(r)+\kappa(r)]\,\mathrm{tr}\left[(\mathbf{Y}-\mathbf{I})\cdot(\mathbf{Y}-\mathbf{I})^{\mathrm{T}}\right] \\ & +\frac{1}{2}\mu(r)\mathrm{tr}\,(\mathbf{Y}-\mathbf{I})^2 + \frac{1}{2}\gamma_1(r)\mathrm{tr}^2\mathbf{L} \\ & +\frac{1}{2}\gamma_2(r)\mathrm{tr}\left(\mathbf{L}\cdot\mathbf{L}^{\mathrm{T}}\right) + \frac{1}{2}\gamma_3(r)\mathrm{tr}\,\mathbf{L}^2\end{aligned} \qquad (5)$$

Here $\lambda(r)$, $\mu(r)$ are functions describing the change in the Lame parameters, $\kappa(r)$, $\gamma_1(r)$, $\gamma_2(r)$, $\gamma_3(r)$ are micropolar elastic parameters changing with the radial coordinate.

It follows from expressions (2), (4), and (5) that the Piola-type couple stress tensor $\mathbf{G}$ is equal to zero for the deformation of axial compression of the circular rod under external pressure

$$\mathbf{G} = \frac{\partial W}{\partial \mathbf{L}}\cdot\mathbf{H} = \left[\gamma_1(\mathrm{tr}\,\mathbf{L})\,\mathbf{I} + \gamma_2\mathbf{L} + \gamma_3\mathbf{L}^{\mathrm{T}}\right]\cdot\mathbf{H} = \mathbf{0}$$

and Piola-type stress tensor $\mathbf{D}$ is

$$\begin{aligned}\mathbf{D} &= \frac{\partial W}{\partial \mathbf{Y}}\cdot\mathbf{H} = \left[\lambda\mathrm{tr}\,(\mathbf{Y}-\mathbf{I})\,\mathbf{I} + \mu\left(\mathbf{Y}^{\mathrm{T}}-\mathbf{I}\right) + (\mu+\kappa)(\mathbf{Y}-\mathbf{I})\right]\cdot\mathbf{H} \\ &= \left[\lambda s + \chi\left(f'-1\right)\right]\mathbf{e}_r\otimes\mathbf{e}_R + \left[\lambda s + \chi\left(\frac{f}{r}-1\right)\right]\mathbf{e}_\varphi\otimes\mathbf{e}_\Phi \\ &\quad + \left[\lambda s + \chi\left(\alpha-1\right)\right]\mathbf{e}_z\otimes\mathbf{e}_Z; \\ s &= f' + \frac{f}{r} + \alpha - 3, \qquad \chi = 2\mu+\kappa\end{aligned} \qquad (6)$$

The equilibrium equations of nonlinear micropolar elasticity in the absence of mass forces and moments are written as follows [8, 24]

$$\mathrm{Div}\mathbf{D} = \mathbf{0}, \qquad \mathrm{Div}\mathbf{G} + \left(\mathbf{C}^{\mathrm{T}}\cdot\mathbf{D}\right)_{\times} = \mathbf{0} \qquad (7)$$

where Div is the divergence in the Lagrangian coordinates. The symbol $_\times$ represents the vector invariant of a second-order tensor. The boundary conditions

$$\mathbf{e}_r\cdot\mathbf{D}|_{r=r_0} = -pJ\mathbf{e}_r\cdot\mathbf{C}^{-\mathrm{T}}, \qquad \mathbf{e}_r\cdot\mathbf{G}|_{r=r_0} = \mathbf{0}; \qquad J = \det\mathbf{C} \qquad (8)$$

express the action of external hydrostatic pressure p (referred to the unit area of the deformed configuration) on the lateral surface of the rod ($r = r_0$). Using the relations

(3), (6), the boundary-value problem (7), (8) is written as follows:

$$
\begin{aligned}
&f'' + \left(\frac{\lambda' + \chi'}{\lambda + \chi} + \frac{1}{r}\right) f' + \left(\frac{\lambda'}{\lambda + \chi} - \frac{1}{r}\right) \frac{f}{r} + \frac{(\alpha - 3)\lambda' - \chi'}{\lambda + \chi} = 0 \\
&[\lambda(r_0) + \chi(r_0)]\, f'(r_0) + [\lambda(r_0) + \alpha p] \frac{f(r_0)}{r_0} + (\alpha - 3)\lambda(r_0) - \chi(r_0) = 0
\end{aligned}
\tag{9}
$$

By solving this problem, we find the unknown function $f(r)$. In most cases, this is done numerically. The analytical solution can be obtained only for some simple cases of the material inhomogeneity.

3 Perturbed State of Inhomogeneous Rod

We assume that in addition to the above-described state of equilibrium for the inhomogeneous rod, there is an infinitely close equilibrium state under the same external loads, which is determined by the radius vector $\mathbf{R} + \eta \mathbf{v}$ and microrotation tensor $\mathbf{H} - \eta \mathbf{H} \times \boldsymbol{\omega}$. Here η is a small parameter, $\mathbf{v}$ is vector of additional displacements, $\boldsymbol{\omega}$ is a linear incremental rotation vector, which characterizes the small rotation of the micropolar medium particles, measured from the initial strain state.

The perturbed state of equilibrium for the micropolar medium is described by the equations [8, 11, 18]:

$$
\operatorname{Div} \mathbf{D}^\bullet = \mathbf{0}, \qquad \operatorname{Div} \mathbf{G}^\bullet + \left[\operatorname{Grad} \mathbf{v}^{\mathrm{T}} \cdot \mathbf{D} + \mathbf{C}^{\mathrm{T}} \cdot \mathbf{D}^\bullet\right]_\times = \mathbf{0} \tag{10}
$$

where $\mathbf{D}^\bullet$ and $\mathbf{G}^\bullet$ are the linearized Piola-type stress and couple stress tensors. In the case of physically linear micropolar material (5), the following relations are valid for these tensors [8, 22]:

$$
\begin{aligned}
\mathbf{D}^\bullet = \left(\tfrac{\partial W}{\partial \mathbf{Y}}\right)^\bullet \cdot \mathbf{H} + \tfrac{\partial W}{\partial \mathbf{Y}} \cdot \mathbf{H}^\bullet &= \left(\lambda \left(\operatorname{tr} \mathbf{Y}^\bullet\right) \mathbf{I} + (\mu + \kappa)\, \mathbf{Y}^\bullet + \mu \mathbf{Y}^{\bullet \mathrm{T}}\right) \cdot \mathbf{H} \\
&- \left(\lambda \operatorname{tr} (\mathbf{Y} - \mathbf{I})\, \mathbf{I} + \mu \left(\mathbf{Y}^{\mathrm{T}} - \mathbf{I}\right) + (\mu + \kappa)\, (\mathbf{Y} - \mathbf{I})\right) \cdot \mathbf{H} \times \boldsymbol{\omega}
\end{aligned}
\tag{11}
$$

$$
\begin{aligned}
\mathbf{G}^\bullet = \left(\tfrac{\partial W}{\partial \mathbf{L}}\right)^\bullet \cdot \mathbf{H} + \tfrac{\partial W}{\partial \mathbf{L}} \cdot \mathbf{H}^\bullet &= \left(\gamma_1 \left(\operatorname{tr} \mathbf{L}^\bullet\right) \mathbf{I} + \gamma_2 \mathbf{L}^\bullet + \gamma_3 \mathbf{L}^{\bullet \mathrm{T}}\right) \cdot \mathbf{H} \\
&- \left(\gamma_1 \left(\operatorname{tr} \mathbf{L}\right) \mathbf{I} + \gamma_2 \mathbf{L} + \gamma_3 \mathbf{L}^{\mathrm{T}}\right) \cdot \mathbf{H} \times \boldsymbol{\omega}
\end{aligned}
\tag{12}
$$

$$
\mathbf{Y}^\bullet = (\operatorname{Grad} \mathbf{v} + \mathbf{C} \times \boldsymbol{\omega}) \cdot \mathbf{H}^{\mathrm{T}}, \qquad \mathbf{L}^\bullet = \operatorname{Grad} \boldsymbol{\omega} \cdot \mathbf{H}^{\mathrm{T}}
$$

Here $\mathbf{Y}^\bullet$ is the linearized stretch tensor, $\mathbf{L}^\bullet$ is the linearized wryness tensor.

The linearized boundary conditions on the lateral surface of the rod ($r = r_0$) are written as follows [15, 21]

$$\mathbf{e}_r \cdot \mathbf{D}^{\bullet}\big|_{r=r_0} = -pJ\mathbf{e}_r \cdot \mathbf{C}^{-\mathrm{T}} \cdot \left[(\operatorname{div}\mathbf{v})\mathbf{I} - \operatorname{grad}\mathbf{v}^{\mathrm{T}}\right], \qquad \mathbf{e}_r \cdot \mathbf{G}^{\bullet}\big|_{r=r_0} = \mathbf{0} \quad (13)$$

where div and grad are the divergence and gradient in the Eulerian coordinates.

We assume that there is no friction at the ends of the rod ($z = 0, l$) and constant normal displacements are given. This leads to the following linearized end conditions:

$$\begin{aligned} \mathbf{e}_z \cdot \mathbf{D}^{\bullet} \cdot \mathbf{e}_R\big|_{z=0,l} &= \mathbf{e}_z \cdot \mathbf{D}^{\bullet} \cdot \mathbf{e}_{\Phi}\big|_{z=0,l} = \mathbf{e}_z \cdot \mathbf{v}|_{z=0,l} = 0 \\ \mathbf{e}_z \cdot \mathbf{G}^{\bullet} \cdot \mathbf{e}_Z\big|_{z=0,l} &= \mathbf{e}_r \cdot \boldsymbol{\omega}|_{z=0,l} = \mathbf{e}_{\varphi} \cdot \boldsymbol{\omega}|_{z=0,l} = 0 \end{aligned} \quad (14)$$

The vector of additional displacements $\mathbf{v}$ and vector of incremental rotation $\boldsymbol{\omega}$ in the basis of Eulerian cylindrical coordinates are written as

$$\mathbf{v} = v_R\mathbf{e}_R + v_{\Phi}\mathbf{e}_{\Phi} + v_Z\mathbf{e}_Z, \qquad \boldsymbol{\omega} = \omega_R\mathbf{e}_R + \omega_{\Phi}\mathbf{e}_{\Phi} + \omega_Z\mathbf{e}_Z \quad (15)$$

To solve the linearized boundary-value problem (10), (13), (14) for a system of six partial differential equations, we use the following substitution

$$\begin{aligned} v_R &= V_R(r)\cos n\varphi \cos\beta z, & \omega_R &= \Omega_R(r)\sin n\varphi \sin\beta z \\ v_{\Phi} &= V_{\Phi}(r)\sin n\varphi \cos\beta z, & \omega_{\Phi} &= \Omega_{\Phi}(r)\cos n\varphi \sin\beta z \\ v_Z &= V_Z(r)\cos n\varphi \sin\beta z, & \omega_Z &= \Omega_Z(r)\sin n\varphi \cos\beta z \end{aligned} \quad (16)$$

$$\beta = \pi m/l, \qquad m, n = 0, 1, 2, \ldots$$

which leads to the separation of variables φ, z in this problem and allows to satisfy the linearized end conditions (14). As a result, the stability analysis of the inhomogeneous circular micropolar rod is reduced to solving a linear homogeneous boundary-value problem for a system of six ordinary differential equations (a detailed scalar formulation of this problem can be found in the Appendix). For its solvability, it is necessary to formulate an additional six conditions at $r = 0$, which can be obtained by requiring the boundedness of unknown functions V_R, V_{Φ}, V_Z, Ω_R, Ω_{Φ}, Ω_Z and their derivatives [25]:

$$\begin{aligned} n = 0:\ & V_R(0) = V_{\Phi}(0) = V_Z'(0) = 0, \quad \Omega_R(0) = \Omega_{\Phi}(0) = \Omega_Z'(0) = 0 \\ n = 1:\ & V_R'(0) = V_{\Phi}'(0) = V_Z(0) = 0, \quad \Omega_R'(0) = \Omega_{\Phi}'(0) = \Omega_Z(0) = 0 \end{aligned} \quad (17)$$

4 Numerical Results

In the present paper, we have carried out the stability analysis for the inhomogeneous rod made of dense polyurethane foam. The micropolar elastic parameters for this material have been previously identified by [13, 14] through a series of experiments in which the size effect was measured during torsion or bending of small rods.

It was assumed that the pattern of change in the elastic parameters λ, μ, κ along the radius is the same, while the micropolar elastic parameters $\gamma_1, \gamma_2, \gamma_3$ are constant:

$$\begin{array}{lll} \lambda(r) = \lambda_0 \xi(r), & \mu(r) = \mu_0 \xi(r), & \kappa(r) = \kappa_0 \xi(r) \\ \lambda_0 = 797.3 \ \text{MPa}, & \mu_0 = 99.67 \ \text{MPa}, & \kappa_0 = 8.67 \ \text{MPa} \\ \gamma_1 = -26.65 \ \text{N}, & \gamma_2 = 45.3 \ \text{N}, & \gamma_3 = 34.65 \ \text{N} \end{array}$$

Several laws describing the change in the elastic properties of the material have been considered:

$$\text{EXP1} : \xi(r) = \frac{9,999 + 90,001^{r/r_0}}{10,000}, \qquad \text{SIN1} : \xi(r) = 9 \sin\left(\frac{\pi r}{2r_0} - \frac{\pi}{2}\right) + 10$$

$$\text{EXP2} : \xi(r) = \frac{9 + 91^{r/r_0}}{10}, \qquad \text{SIN2} : \xi(r) = \frac{9}{2} \sin\left(\frac{\pi r}{r_0} - \frac{\pi}{2}\right) + \frac{11}{2}$$

For all listed laws the rod is more rigid on the lateral surface and less rigid in the center: the values of the elastic parameters λ, μ, κ at $r = r_0$ are ten times higher than their values at $r = 0$ (see Fig. 1).

By numerical solution [26] of the linearized boundary-value problem described in Sect. 3 we have found the critical curves, corresponding to the various buckling modes of the inhomogeneous micropolar rod. Based on the analysis of these curves, the stability regions are constructed in the plane of loading parameters (axial compression and external pressure) for rods of various sizes. For convenience, the following dimensionless parameters are introduced:

$$\delta = 1 - \alpha, \quad p^* = p/\mu_0, \quad r_0^* = r_0/l_b$$

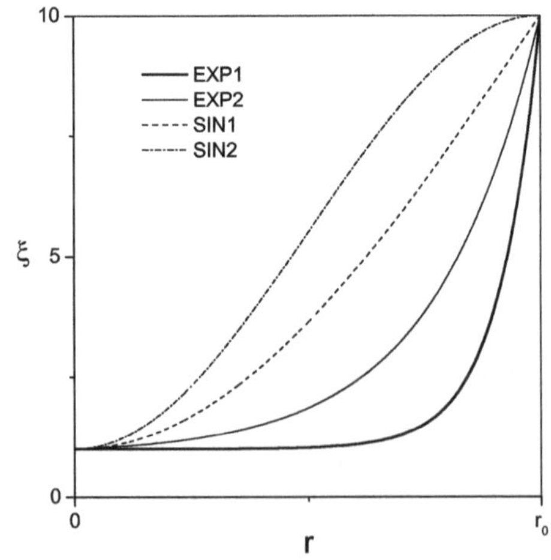

Fig. 1 Various laws $\xi(r)$ describing the change in the elastic parameters of the rod along the radius

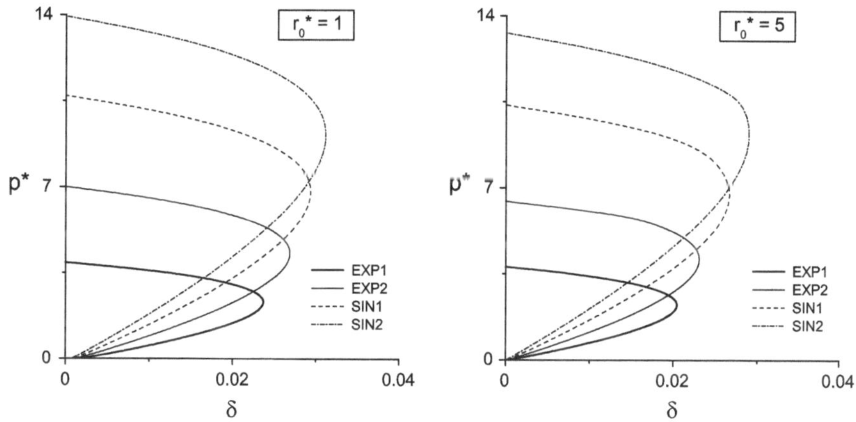

Fig. 2 Boundaries of the stability regions for the inhomogeneous rods of different sizes

where $l_b = \sqrt{\gamma_2/2\,(2\mu_0 + \kappa_0)}$ is the characteristic length for bending [13]. The ratio of the length of the undeformed rod to its diameter is 20 ($l = 40r_0$) for all presented results.

In Fig. 2 the stability regions (boundaries of the stability regions) are presented for various inhomogeneous rods of two sizes: $r_0^* = 1$ and $r_0^* = 5$. It follows from graphs that the micropolar rod becomes more stable with a decrease in size. This size effect is due to a microstructure influence and not observed for classical (non-polar) elastic bodies. It is quite significant for small rods ($r_0^* < 5$), while negligible for large ones.

According to the obtained results, a relatively small external pressure has a stabilizing effect on the deformation of axial compression. Specifically, we can always find some optimal value of the external pressure at which a compressed rod is most stable (the critical value of relative axial compression is maximum). This optimal value depends on the material properties and rod geometry. At the same time, the influence of high external pressure is negative, i.e. it destabilizes the deformation of circular rod (see Fig. 2).

Through comparison of the results of stability analysis for the exponential (EXP1, EXP2) and trigonometric (SIN1, SIN2) laws of change in the elastic parameters along the radius, we have determined that the micropolar rods with the exponential pattern of inhomogeneity are more stable with respect to the axial compression, but less stable with respect to the external pressure. This is more evident for smaller rods ($r_0^* < 5$).

5 Conclusion

In the framework of bifurcation approach, we have studied the stability of an inhomogeneous circular rod of micropolar material. For the physically linear material, a system of linearized equilibrium equations was derived, which describes the behavior of the inhomogeneous rod in a perturbed state. In the case of a circular rod made of dense polyurethane foam the stability regions were constructed in the plane of loading parameters for different laws describing the change in the elastic properties of material along the rod radius. Based on these results, in particular, we have determined that the micropolar rods with the exponential pattern of inhomogeneity are more stable with respect to the axial compression, but less stable with respect to the external pressure. Additionally, it was found that a relatively small external pressure has a stabilizing influence on the deformation of axial compression.

Appendix: Scalar Formulation of the Linearized Boundary Value Problem

With respect to the representations (2), (3), (15), the expressions for the linearized stretch tensor $\mathbf{Y}^{\bullet}$ and wryness tensor $\mathbf{L}^{\bullet}$ have the form:

$$\begin{aligned}
\mathbf{Y}^{\bullet} &= \left(\frac{\partial v_{\Phi}}{\partial r} - f'\omega_Z\right)\mathbf{e}_r\otimes\mathbf{e}_{\varphi} + \frac{1}{r}\left(\frac{\partial v_R}{\partial \varphi} - v_{\Phi} + f\omega_Z\right)\mathbf{e}_{\varphi}\otimes\mathbf{e}_r \\
&+\left(\frac{\partial v_Z}{\partial r} + f'\omega_{\Phi}\right)\mathbf{e}_r\otimes\mathbf{e}_z + \left(\frac{\partial v_R}{\partial z} - \alpha\omega_{\Phi}\right)\mathbf{e}_z\otimes\mathbf{e}_r \\
&+\frac{1}{r}\left(\frac{\partial v_Z}{\partial \varphi} - f\omega_R\right)\mathbf{e}_{\varphi}\otimes\mathbf{e}_z + \left(\frac{\partial v_{\Phi}}{\partial z} + \alpha\omega_R\right)\mathbf{e}_z\otimes\mathbf{e}_{\varphi} \\
&+\frac{\partial v_R}{\partial r}\mathbf{e}_r\otimes\mathbf{e}_r + \frac{1}{r}\left(\frac{\partial v_{\Phi}}{\partial \varphi} + v_R\right)\mathbf{e}_{\varphi}\otimes\mathbf{e}_{\varphi} + \frac{\partial v_Z}{\partial z}\mathbf{e}_z\otimes\mathbf{e}_z
\end{aligned} \tag{18}$$

$$\begin{aligned}
\mathbf{L}^{\bullet} &= \frac{\partial \omega_R}{\partial r}\mathbf{e}_r\otimes\mathbf{e}_r + \frac{1}{r}\left(\frac{\partial \omega_{\Phi}}{\partial \varphi} + \omega_R\right)\mathbf{e}_{\varphi}\otimes\mathbf{e}_{\varphi} + \frac{\partial \omega_Z}{\partial z}\mathbf{e}_z\otimes\mathbf{e}_z \\
&+\frac{\partial \omega_{\Phi}}{\partial r}\mathbf{e}_r\otimes\mathbf{e}_{\varphi} + \frac{1}{r}\left(\frac{\partial \omega_R}{\partial \varphi} - \omega_{\Phi}\right)\mathbf{e}_{\varphi}\otimes\mathbf{e}_r + \frac{\partial \omega_Z}{\partial r}\mathbf{e}_r\otimes\mathbf{e}_z \\
&+\frac{\partial \omega_R}{\partial z}\mathbf{e}_z\otimes\mathbf{e}_r + \frac{1}{r}\frac{\partial \omega_Z}{\partial \varphi}\mathbf{e}_{\varphi}\otimes\mathbf{e}_z + \frac{\partial \omega_{\Phi}}{\partial z}\mathbf{e}_z\otimes\mathbf{e}_{\varphi}
\end{aligned} \tag{19}$$

According to relations (2), (4), (11), (12), (15), (18) (19), the components of the linearized Piola-type stress tensor $\mathbf{D}^{\bullet}$ and couple stress tensor $\mathbf{G}^{\bullet}$ are written as follows:

$$\begin{aligned}
\mathbf{e}_r\cdot\mathbf{D}^{\bullet}\cdot\mathbf{e}_R &= (\lambda+\chi)\frac{\partial v_R}{\partial r} + \frac{\lambda}{r}\left(\frac{\partial v_{\Phi}}{\partial \varphi} + v_R\right) + \lambda\frac{\partial v_Z}{\partial z} \\
\mathbf{e}_r\cdot\mathbf{D}^{\bullet}\cdot\mathbf{e}_{\Phi} &= (\mu+\kappa)\frac{\partial v_{\Phi}}{\partial r} + \frac{\mu}{r}\left(\frac{\partial v_R}{\partial \varphi} - v_{\Phi}\right) + \left[\mu\left(f' + \frac{f}{r}\right) + \lambda s - \chi\right]\omega_Z \\
\mathbf{e}_r\cdot\mathbf{D}^{\bullet}\cdot\mathbf{e}_Z &= (\mu+\kappa)\frac{\partial v_Z}{\partial r} + \mu\frac{\partial v_R}{\partial z} - \left[\mu\left(f'+\alpha\right) + \lambda s - \chi\right]\omega_{\Phi}
\end{aligned}$$

$$
\begin{aligned}
\mathbf{e}_\varphi \cdot \mathbf{D}^\bullet \cdot \mathbf{e}_R &= \frac{\mu+\kappa}{r}\left(\frac{\partial v_R}{\partial \varphi} - v_\Phi\right) + \mu\frac{\partial v_\Phi}{\partial r} - \left[\mu\left(f' + \frac{f}{r}\right) + \lambda s - \chi\right]\omega_Z \\
\mathbf{e}_\varphi \cdot \mathbf{D}^\bullet \cdot \mathbf{e}_\Phi &= \lambda\frac{\partial v_R}{\partial r} + \frac{\lambda+\chi}{r}\left(\frac{\partial v_\Phi}{\partial \varphi} + v_R\right) + \lambda\frac{\partial v_Z}{\partial z} \\
\mathbf{e}_\varphi \cdot \mathbf{D}^\bullet \cdot \mathbf{e}_Z &= \frac{\mu+\kappa}{r}\frac{\partial v_Z}{\partial \varphi} + \mu\frac{\partial v_\Phi}{\partial z} + \left[\mu\left(\frac{f}{r} + \alpha\right) + \lambda s - \chi\right]\omega_R \\
\mathbf{e}_z \cdot \mathbf{D}^\bullet \cdot \mathbf{e}_R &= (\mu+\kappa)\frac{\partial v_R}{\partial z} + \mu\frac{\partial v_Z}{\partial r} + \left[\mu\left(f' + \alpha\right) + \lambda s - \chi\right]\omega_\Phi \\
\mathbf{e}_z \cdot \mathbf{D}^\bullet \cdot \mathbf{e}_\Phi &= (\mu+\kappa)\frac{\partial v_\Phi}{\partial z} + \frac{\mu}{r}\frac{\partial v_Z}{\partial \varphi} - \left[\mu\left(\frac{f}{r} + \alpha\right) + \lambda s - \chi\right]\omega_R \\
\mathbf{e}_z \cdot \mathbf{D}^\bullet \cdot \mathbf{e}_Z &= \lambda\frac{\partial v_R}{\partial r} + \frac{\lambda}{r}\left(\frac{\partial v_\Phi}{\partial \varphi} + v_R\right) + (\lambda+\chi)\frac{\partial v_Z}{\partial z}
\end{aligned}
\tag{20}
$$

$$
\begin{aligned}
\mathbf{e}_r \cdot \mathbf{G}^\bullet \cdot \mathbf{e}_R &= (\gamma_1+\gamma_2+\gamma_3)\frac{\partial \omega_R}{\partial r} + \frac{\gamma_1}{r}\left(\frac{\partial \omega_\Phi}{\partial \varphi} + \omega_R\right) + \gamma_1\frac{\partial \omega_Z}{\partial z} \\
\mathbf{e}_r \cdot \mathbf{G}^\bullet \cdot \mathbf{e}_\Phi &= \gamma_2\frac{\partial \omega_\Phi}{\partial r} + \frac{\gamma_3}{r}\left(\frac{\partial \omega_R}{\partial \varphi} - \omega_\Phi\right) \\
\mathbf{e}_r \cdot \mathbf{G}^\bullet \cdot \mathbf{e}_Z &= \gamma_2\frac{\partial \omega_Z}{\partial r} + \gamma_3\frac{\partial \omega_R}{\partial z} \\
\mathbf{e}_\varphi \cdot \mathbf{G}^\bullet \cdot \mathbf{e}_R &= \frac{\gamma_2}{r}\left(\frac{\partial \omega_R}{\partial \varphi} - \omega_\Phi\right) + \gamma_3\frac{\partial \omega_\Phi}{\partial r} \\
\mathbf{e}_\varphi \cdot \mathbf{G}^\bullet \cdot \mathbf{e}_\Phi &= \gamma_1\frac{\partial \omega_R}{\partial r} + \frac{\gamma_1+\gamma_2+\gamma_3}{r}\left(\frac{\partial \omega_\Phi}{\partial \varphi} + \omega_R\right) + \gamma_1\frac{\partial \omega_Z}{\partial z} \\
\mathbf{e}_\varphi \cdot \mathbf{G}^\bullet \cdot \mathbf{e}_Z &= \frac{\gamma_2}{r}\frac{\partial \omega_Z}{\partial \varphi} + \gamma_3\frac{\partial \omega_\Phi}{\partial z} \\
\mathbf{e}_z \cdot \mathbf{G}^\bullet \cdot \mathbf{e}_R &= \gamma_2\frac{\partial \omega_R}{\partial z} + \gamma_3\frac{\partial \omega_Z}{\partial r} \\
\mathbf{e}_z \cdot \mathbf{G}^\bullet \cdot \mathbf{e}_\Phi &= \gamma_2\frac{\partial \omega_\Phi}{\partial z} + \frac{\gamma_3}{r}\frac{\partial \omega_Z}{\partial \varphi} \\
\mathbf{e}_z \cdot \mathbf{G}^\bullet \cdot \mathbf{e}_Z &= \gamma_1\frac{\partial \omega_R}{\partial r} + \frac{\gamma_1}{r}\left(\frac{\partial \omega_\Phi}{\partial \varphi} + \omega_R\right) + (\gamma_1+\gamma_2+\gamma_3)\frac{\partial \omega_Z}{\partial z}
\end{aligned}
\tag{21}
$$

By taking into account the expressions (3), (6), (15), (16), (20), (21), we derive a scalar form of the linearized equilibrium equations (10)

$$
\begin{aligned}
&(\lambda+\chi)\,V_R'' + \left(\lambda' + \chi' + \frac{\lambda+\chi}{r}\right)V_R' - \frac{\lambda - \lambda' r + (\mu+\kappa)\,\zeta + \mu}{r^2}V_R \\
&+ \frac{n\,(\lambda+\mu)}{r}V_\Phi' - \frac{n\left(\lambda - \lambda' r + 3\mu + 2\kappa\right)}{r^2}V_\Phi + \beta\,(\lambda+\mu)\,V_Z' \\
&+ \beta\lambda' V_Z + \beta B_2 \Omega_\Phi - \frac{n B_3}{r}\Omega_Z = 0
\end{aligned}
$$

$$
(\mu+\kappa)\,V_\Phi'' - \frac{n\,(\lambda+\mu)}{r}V_R' - \frac{n\left(\lambda + \mu' r + 3\mu + 2\kappa\right)}{r^2}V_R
$$

$$+\left(\mu' + \kappa' + \frac{\mu+\kappa}{r}\right) V'_\Phi - \frac{(\lambda+\mu) n^2 + \mu' r + (\mu+\kappa)\zeta}{r^2} V_\Phi$$
$$- \frac{n\beta(\lambda+\mu)}{r} V_Z - \beta B_1 \Omega_R + B_3 \Omega'_Z + B'_3 \Omega_Z = 0$$

$$(\mu+\kappa) V''_Z - \beta(\lambda+\mu) V'_R - \frac{\beta(\lambda + \mu' r + \mu)}{r} V_R - \frac{n\beta(\lambda+\mu)}{r} V_\Phi$$
$$+\left(\mu' + \kappa' + \frac{\mu+\kappa}{r}\right) V'_Z - \left[(\lambda+\mu)\beta^2 + \frac{\zeta-1}{r^2}(\mu+\kappa)\right] V_Z$$
$$+\frac{nB_1}{r}\Omega_R - B_2 \Omega'_\Phi - \left(B'_2 + \frac{B_2}{r}\right)\Omega_\Phi = 0 \tag{22}$$

$$\gamma \Omega''_R + \left(\gamma' + \frac{\gamma}{r}\right)\Omega'_R - \left[\frac{\gamma - \gamma_1' r + \gamma_2(\zeta-1)}{r^2} - \left(\frac{f}{r} + \alpha\right) B_1\right]\Omega_R$$
$$- \frac{n(\gamma-\gamma_2)}{r}\Omega'_\Phi + \frac{n(\gamma - \gamma_1' r + \gamma_2)}{r^2}\Omega_\Phi - \beta(\gamma-\gamma_2)\Omega'_Z$$
$$- \beta\gamma_1' \Omega_Z - \beta B_1 V_\Phi + \frac{nB_1}{r} V_Z = 0$$

$$\gamma_2 \Omega''_\Phi + \frac{n(\gamma-\gamma_2)}{r}\Omega'_R + \frac{n(\gamma + \gamma_3' r + \gamma_2)}{r^2}\Omega_R + \left(\gamma_2' + \frac{\gamma_2}{r}\right)\Omega'_\Phi$$
$$- \left[\frac{(\gamma-\gamma_2) n^2 + \gamma_3' r + \gamma_2 \zeta}{r^2} - (f' + \alpha) B_2\right]\Omega_\Phi$$
$$- \frac{n\beta(\gamma-\gamma_2)}{r}\Omega_Z + \beta B_2 V_R + B_2 V'_Z = 0$$

$$\gamma_2 \Omega''_Z + \beta(\gamma-\gamma_2)\Omega'_R + \frac{\beta(\gamma + \gamma_3' r - \gamma_2)}{r}\Omega_R - \frac{n\beta(\gamma-\gamma_2)}{r}\Omega_\Phi$$
$$+ \left(\gamma_2' + \frac{\gamma_2}{r}\right)\Omega'_Z - \left[\gamma\beta^2 + \frac{n^2}{r^2}\gamma_2 - B_3\left(f' + \frac{f}{r}\right)\right]\Omega_Z$$
$$- \frac{nB_3}{r} V_R - B_3 V'_\Phi - \frac{B_3}{r} V_\Phi = 0$$

Here the following notations are used:

$$\gamma = \gamma_1 + \gamma_2 + \gamma_3, \qquad \zeta = n^2 + r^2\beta^2 + 1$$

$$\begin{aligned} B_1 &= \mu\left(\frac{f}{r} + \alpha\right) + \lambda s - \chi, \\ B_2 &= \mu(f' + \alpha) + \lambda s - \chi, \\ B_3 &= \mu\left(f' + \frac{f}{r}\right) + \lambda s - \chi \end{aligned}$$

The scalar representation of the linearized boundary conditions (13) on the lateral surface of the rod is written as:

$$
\begin{aligned}
&[\lambda(r_0)+\chi(r_0)]\,V_R'(r_0)+\frac{\lambda(r_0)+\alpha p}{r_0}\,[V_R(r_0)+nV_\Phi(r_0)]\\
&\quad+\beta\left[\lambda(r_0)+p\frac{f(r_0)}{r_0}\right]V_Z(r_0)=0\\
&[\mu(r_0)+\kappa(r_0)]\,V_\Phi'(r_0)+\frac{\alpha p-\mu(r_0)}{r_0}\,[nV_R(r_0)+V_\Phi(r_0)]+B_3(r_0)\Omega_Z(r_0)=0\\
&[\mu(r_0)+\kappa(r_0)]\,V_Z'(r_0)+\beta\left[p\frac{f(r_0)}{r_0}-\mu(r_0)\right]V_R(r_0)-B_2(r_0)\Omega_\Phi(r_0)=0\\
&\gamma(r_0)\Omega_R'(r_0)+\frac{\gamma_1(r_0)}{r_0}\,[\Omega_R(r_0)-n\Omega_\Phi(r_0)]-\beta\gamma_1(r_0)\Omega_Z(r_0)=0\\
&\frac{\gamma_3(r_0)}{r_0}\,[n\Omega_R(r_0)-\Omega_\Phi(r_0)]+\gamma_2(r_0)\Omega_\Phi'(r_0)=0\\
&\beta\gamma_3(r_0)\Omega_R(r_0)+\gamma_2(r_0)\Omega_Z'(r_0)=0
\end{aligned}
\tag{23}
$$

Acknowledgments This work was supported by the Russian Foundation for Basic Research (grants 12-01-00038-a and 12-01-00811-a).

References

1. Altenbach, J., Altenbach, H., Eremeyev, V.A.: On generalized Cosserat-type theories of plates and shells: a short review and bibliography. Arch. Appl. Mech. **80**, 73–92 (2010)
2. Ashby, M.F., Evans, A.G., Fleck, N.A., Gibson, L.J., Hutchinson, J.W., Wadley, H.N.G.: Metal Foams: A Design Guide. Butterworth-Heinemann, Boston (2000)
3. Banhart, J.: Manufacturing routes for metallic foams. J. Minerals **52**(12), 22–27 (2000)
4. Banhart, M.F.A.J., Fleck, N.A. (eds.): Metal Foams and Porous Metal Structures. MIT, Bremen (1999)
5. Cosserat, E., Cosserat, F.: Theorie des Corps Deformables. Hermann et Fils, Paris (1909)
6. Degischer, H.P., Kriszt, B. (eds.): Handbook of Cellular metals. Production, Processing, Applications. Wiley-VCH, Weinheim (2002)
7. Eremeyev, V.A., Pietraszkiewicz, W.: Material symmetry group of the non-linear polar-elastic continuum. Int. J. Solids Struct. **49**(14), 1993–2005 (2012)
8. Eremeyev, V.A., Zubov, L.M.: On the stability of elastic bodies with couple-stresses. Mech. Solids **29**(3), 172–181 (1994)
9. Eringen, A.C.: Microcontinuum Field Theory. I. Foundations and Solids. Springer, New York (1999)
10. Gibson, L.J., Ashby, M.F.: Cellular Solids: Structure and Properties, 2nd edn. Cambridge University Press, Cambridge (1997)
11. Green, A.E., Adkins, J.E.: Large Elastic Deformations and Non-Linear Continuum Mechanics. Clarendon Press, Oxford (1960)
12. Kafadar, C.B., Eringen, A.C.: Micropolar media - I. The classical theory. Int. J. Eng. Sci **9**, 271–305 (1971)
13. Lakes, R.: Experimental methods for study of Cosserat elastic solids and other generalized elastic continua. In: Mühlhaus, H. (ed) Continuum models for materials with micro-structure, pp. 1–22. Wiley, New York (1995)

14. Lakes, R.S.: Experimental microelasticity of two porous solids. Int. J. Solids Struct. **22**, 55–63 (1986)
15. Lurie, A.I.: Non-linear Theory of Elasticity. North-Holland, Amsterdam (1990)
16. Maugin, G.A.: On the structure of the theory of polar elasticity. Phil. Trans. Royal Soc. Lond. A **356**, 1367–1395 (1998)
17. Nikitin, E., Zubov, L.M.: Conservation laws and conjugate solutions in the elasticity of simple materials and materials with couple stress. J. Elast. **51**, 1–22 (1998)
18. Ogden, R.W.: Non-Linear Elastic Deformations. Dover, Mineola (1997)
19. Pietraszkiewicz, W., Eremeyev, V.A.: On natural strain measures of the non-linear micropolar continuum. Int. J. Solids Struct. **46**, 774–787 (2009a)
20. Pietraszkiewicz, W., Eremeyev, V.A.: On vectorially parameterized natural strain measures of the non-linear cosserat continuum. Int. J. Solids Struct. **46**(11–12), 2477–2480 (2009b)
21. Sheydakov, D.N.: Stability of elastic cylindrical micropolar shells subject to combined loads. In: Pietraszkiewicz, W., Kreja, I. (eds.) Shell Structures: Theory and Applications, vol. 2, pp. 141–144. CRC Press, Boca Raton (2010)
22. Sheydakov, D.N.: Buckling of elastic composite rod of micropolar material subject to combined loads. In: Altenbach, H., Erofeev, V., Maugin, G. (eds.) Mechanics of Generalized Continua–From Micromechanical Basics to Engineering Applications, Advanced Structured Materials, vol.7, pp. 255–271. Springer, Berlin (2011)
23. Toupin, R.A.: Theories of elasticity with couple-stress. Arch. Rat. Mech. Anal. **17**, 85–112 (1964)
24. Zubov, L.M.: Nonlinear Theory of Dislocations and Disclinations in Elastic Bodies. Springer, Berlin (1997)
25. Zubov, L.M., Sheidakov, D.N.: The effect of torsion on the stability of an elastic cylinder under tension. PMM J. Appl. Math. Mech. **69**(1), 49–56 (2005)
26. Zubov, L.M., Sheidakov, D.N.: Instability of a hollow elastic cylinder under tension, torsion, and inflation. Trans. ASME J. Appl. Mech. **75**(1), 011,002-1–011,002-6 (2008)

A New Approach for Studying Nonlinear Dynamic Response of a Thin Fractionally Damped Cylindrical Shell with Internal Resonances of the Order of ε

Marina V. Shitikova and Yury A. Rossikhin

Abstract Non-linear vibrations of a cylindrical shell embedded in a fractional derivative viscoelastic medium and subjected to the different conditions of the internal resonance of the order of ε, where ε is a small value, are investigated. The displacement functions are determined in terms of eigenfunctions of linear vibrations. The procedure resulting in decoupling linear parts of equations is proposed with the further utilization of the method of multiple scales for solving nonlinear governing equations of motion, in so doing the amplitude functions are expanded into power series in terms of the small parameter and depend on different time scales. The influence of viscosity on the energy exchange mechanism is analyzed. It is shown that each mode is characterized by its damping coefficient connected with the natural frequency by the exponential relationship with a negative fractional exponent. Comparison of the results obtained in this paper for the nonlinear shallow cylindrical shell in the cases of the internal resonance of the order of ε with those for a nonlinear plate, the motion of which is described also by three coupled nonlinear equations in terms of three displacements, reveals the fact that the shell equations could produce much more diversified variety of internal resonances, including combinational resonances of the additive and difference types, than the plate equations.

1 Introduction

Beginning with the paper by Witt and Gorelik [1], whose authors were among the first to show theoretically and experimentally the two-to-one internal resonance with the energy exchange from one subsystem to another using the simplest two-degree-of-freedom mechanical system as an example, interest of researchers to the problems of the internal resonance in mechanical systems does not relax. It will

M.V. Shitikova (✉) · Y.A. Rossikhin
Research Center of Wave Dynamics, Voronezh State University of Architecture and Civil Engineering, Voronezh 394006, Russia
e-mail: mvs@vgasu.vrn.ru

Y.A. Rossikhin
e-mail: yar@vgasu.vrn.ru

H. Altenbach and G.I. Mikhasev (eds.), *Shell and Membrane Theories in Mechanics and Biology*, Advanced Structured Materials 45, DOI 10.1007/978-3-319-02535-3_17

suffice to mention the state-of-the-art article [2] and the monograph [3], wherein the extensive review of literature in the field of internal resonances in different mechanical systems is presented. Different types of the internal resonance: one-to-one, two-to-one, as well as a variety of combinational resonances, when three and more natural modes interact, have been discussed. The enumerated internal resonances were investigated in mechanical systems with more than one degree-of-freedom, as well as in strings, beams, plates, and shells.

It has been emphasized by many researchers [4, 5] that the phenomenon of internal resonances can be very critical especially for circular cylindrical shells. Thus, the nonlinear vibrations of infinitely long circular cylindrical shells under the conditions of the two-to-one internal resonance were studied in [6] via the method of multiple time scales using the simple plane strain theory of shells. Parametrically excited vibrations of infinitely long cylindrical shells and nonlinear forced vibrations of a simply supported, circular cylindrical shell filled with an incompressible, inviscid, quiescent and dense fluid were investigated, respectively, in [5, 7] and [4] using Donnell's nonlinear shallow-shell theory. The flexural deformation is usually expanded by using the linear shell eigenmodes, in so doing the flexural response involves several nodal diameters and one or two longitudinal half-waves. Internal resonances of different types have been also analyzed in [8, 9].

The extensive review of studies on shallow shells nonlinear vibrations could be found in the state-of-the-art articles by [10–12], as well as in recent papers [13, 14]. In spite of the fact that many studies have been carried out on large amplitude vibrations of circular cylindrical shells and many different approaches to the problem have been used, we agree with [4] that this research area is still far from being well understood.

The problem of free, as well as forced nonlinear vibrations of cylindrical shells can be considered from different positions depending on the shell geometry, in so doing the nonlinear displacement field is approximated by a finite sum of global interpolating functions. However the choice of appropriate modal expansions is fundamental to guaranteeing accuracy of the results for large-amplitude vibrations [15]. Thus, for example, different expansions involving from 14 to 48 generalized co-ordinates, associated to natural modes of simply supported shells, have been discussed in [16].

A comparison of five shell theories for large-amplitude vibrations of circular cylindrical shells which are generally applied to geometrically non-linear problems that use only three variables, which are the three middle surface displacements, has been carried out in [16]. More complicated shell theories, suitable for moderately thick laminated shells exist, and they use five independent variables, three displacements and two rotations [17, 18] or even six variables if thickness variation is taken into account [19]. However, it has been emphasized in [16] that the majority of researchers utilize the Donnell–Mushtari–Vlasov type equations.

In recent years much attention is given to damping features of mechanical systems subjected to the conditions of different internal resonances. Damping properties of non-linear systems are described mainly by the first-order time-derivative of a generalized displacement [3]. However, as it has been shown by [20], who analyzed free damped vibrations of suspension combined system under the conditions of the one-to-one internal resonance, for good fit of the theoretical investigations with the experimental results it is better to describe the damping features of non-linear mechanical systems in terms of fractional time-derivatives of the generalized displacements [21]. The analysis of non-linear vibrations of a two-degree-of-freedom mechanical system, the damping features of which are described by a fractional derivative, has shown [22] that in the case when the system is under the conditions of the two-to-one or one-to-one internal resonance, viscosity may have a twofold effect on the system: a destabilizing influence producing unsteady energy exchange, and a stabilizing influence resulting in damping of the energy exchange mechanism. The same phenomenon was noted when considering non-linear vibrations of a fractionally damped plate under the conditions of the two-to-one [23] or one-to-one [24] internal resonance.

In the present paper, non-linear free damped vibrations of a thin cylindrical viscoelastic shell, the damping properties of which are described by the Riemann-Liouville fractional derivatives, are investigated. The dynamic behaviour of the shell is described by a set of three coupled non-linear differential equations with due account for the fact that the shell is being under the conditions of the internal resonance resulting in the interaction of modes corresponding to the mutually orthogonal displacements. The displacement functions are determined in terms of eigenfunctions of linear vibrations. The procedure resulting in decoupling linear parts of equations is proposed with the further utilization of the method of multiple scales for solving nonlinear governing equations of motion, in so doing the amplitude functions are expanded into power series in terms of the small parameter and depend on different time scales. It is shown that the phenomenon of internal resonance can be very critical, since in a circular cylindrical shell internal resonance of the type two-to-one, one-to-one, three-to-one, as well as additive and difference combinational resonances are always present.

In the present paper, all possible cases of the internal resonances of the order of ε, where ε is a small value, are examined, while the internal resonances of the higher orders should be considered in detail separately. The nonlinear sets of resolving equations in terms of amplitudes and phase differences have been obtained for each investigated type of the internal resonance. It is shown that for many types of the internal resonances there exist such particular cases when it is possible to obtain two first integrals, namely: the energy integral and the stream-function, what allows one to reduce the problem to the calculation of elliptic integrals.

Comparison of the results obtained in this paper for the nonlinear shallow cylindrical shell in the cases of the internal resonance of the order of ε with those for a nonlinear plate, the motion of which is described also by three coupled nonlinear equations in terms of three displacements [25], reveals the fact that the shell equations could produce much more diversified variety of internal resonances than the plate equations.

2 Problem Formulation and Governing Equations

Let us consider the dynamic behaviour of a free supported non-linear elastic circular cylindrical shell of radius R and length l, vibrations of which in a viscoelastic medium are described by the Donnell–Mushtari–Vlasov equations with respect to the three displacements [26] in the cylindrical system of coordinates written in a dimensionless form [27]:

$$u_{xx} + \frac{1-\nu}{2}\,\beta_1^2 u_{\varphi\varphi} + \frac{1+\nu}{2}\,\beta_1 v_{x\varphi} - \nu\beta_1 w_x + w_x\left(w_{xx} + \frac{1-\nu}{2}\,\beta_1^2 w_{\varphi\varphi}\right) + \frac{1+\nu}{2}\,\beta_1^2 w_\varphi w_{x\varphi} = \ddot{u} + \kappa_1 D^\gamma u, \tag{1}$$

$$\beta_1^2 v_{\varphi\varphi} + \frac{1-\nu}{2}\,v_{xx} + \frac{1+\nu}{2}\,\beta_1 u_{x\varphi} - \beta_1^2 w_\varphi + \beta_1 w_\varphi\left(\beta_1^2 w_{\varphi\varphi} + \frac{1-\nu}{2}\,w_{xx}\right) + \frac{1+\nu}{2}\,\beta_1 w_x w_{x\varphi} = \ddot{v} + \kappa_2 D^\gamma v, \tag{2}$$

$$\begin{aligned}
&\frac{\beta_2^2}{12}\left(w_{xxxx} + 2\beta_1^2 w_{xx\varphi\varphi} + \beta_1^4 w_{\varphi\varphi\varphi\varphi}\right) - \nu\beta_1 u_x - \beta_1^2 v_\varphi + \beta_1^2 w + \frac{1}{2}\,\nu\beta_1 (w_x)^2 \\
&+ \frac{1}{2}\,\beta_1^3 (w_\varphi)^2 - w_x\left(u_{xx} + \frac{1-\nu}{2}\,\beta_1^2 u_{\varphi\varphi} + \frac{1+\nu}{2}\,\beta_1 v_{x\varphi}\right) \\
&- \beta_1 w_\varphi\left(\beta_1^2 v_{\varphi\varphi} + \frac{1-\nu}{2}\,v_{xx} + \frac{1+\nu}{2}\,\beta_1 u_{x\varphi}\right) - w_{xx}(u_x + \nu\beta_1 v_\varphi - \nu\beta_1 w) \\
&- \beta_1^2 w_{\varphi\varphi}(\nu u_x + \beta_1 v_\varphi - \beta_1 w) - (1-\nu)\beta_1 w_{x\varphi}(\beta_1 u_\varphi + v_x) = -\ddot{w} - \kappa_3 D^\gamma w
\end{aligned} \tag{3}$$

In Eqs. (1)–(3), the dimensionless values are introduced similarly as it has been done in [27], and the x-axis is directed along the axis of the cylinder, φ is the polar angle in the plane perpendicular to the x-axis, $u = u(x, \varphi, t)$, $v = v(x, \varphi, t)$, and $w = w(x, \varphi, t)$ are the displacements of points located in the shell's middle surface in three mutually orthogonal directions x, φ, r, r is the polar radius, h is the thickness, ρ is the density, E and ν are the elastic modulus and Poisson's ratio, respectively, t is the time, $\beta_1 = l/R$ and $\beta_2 = h/l$ are the parameters defining the dimensions of the shell, lower indices label the derivatives with respect to the corresponding coordinates, κ_i $(i = 1, 2, 3)$ are damping coefficients, and D^γ is the Riemann–Liouville fractional derivative of the γ-order [21].

Equations (1)–(3) are subjected to the initial

$$u|_{t=0} = \dot{u}\Big|_{t=0} = 0, \quad v|_{t=0} = \dot{v}\Big|_{t=0} = 0, \quad w|_{t=0} = \dot{w}\Big|_{t=0} = 0, \tag{4}$$

and boundary conditions

$$w\big|_{x=0} = w\big|_{x=1} = 0, \quad v|_{x=0} = v\big|_{x=1} = 0, \quad u_x\big|_{x=0} = u_x\big|_{x=1} = 0,$$
$$w_{xx}\big|_{x=0} = w_{xx}\big|_{x=1} = 0. \tag{5}$$

Since the set of Eqs. (1)–(3) admits the solution of the Navier type, then the displacements could be represented in the form

$$u(x, \varphi, t) = \sum_{m=1}^{\infty}\sum_{n=1}^{\infty} x_{1\,mn}(t)\eta_{1\,mn}(x, \varphi), \tag{6}$$

$$v(x, \varphi, t) = \sum_{m=1}^{\infty}\sum_{n=1}^{\infty} x_{2\,mn}(t)\eta_{2\,mn}(x, \varphi), \tag{7}$$

$$w(x, \varphi, t) = \sum_{m=1}^{\infty}\sum_{n=1}^{\infty} x_{3\,mn}(t)\eta_{3\,mn}(x, \varphi), \tag{8}$$

where m and n are integers, $x_{i\,mn}(t)$ are the generalized displacements, and $\eta_{i\,mn}(x, \varphi)$ $(i = 1, 2, 3)$ are the eigenfunctions satisfying the boundary conditions (5):

$$\eta_{i\,mn}(x, \varphi) = \begin{cases} \cos \pi m x \sin n\varphi, \\ \sin \pi m x \cos n\varphi, \\ \sin \pi m x \sin n\varphi, \end{cases} \quad (0 \le x \le 1\ ,\ 0 \le \varphi \le 2\pi) \tag{9}$$

Substituting (6)–(9) into Eqs. (1)–(3), multiplying (1), (2), and (3) by η_{1lk}, η_{2lk}, and η_{3lk}, respectively, integrating over x and φ, and using the orthogonality conditions for linear modes within the domains of $0 \le x \le 1$ and $0 \le \varphi \le 2\pi$, we are led to a coupled set of nonlinear ordinary differential equations of the second order in $x_{i\,mn}$ $(i = 1, 2, 3)$:

$$\ddot{x}_{i\,mn} + \kappa_i D^{\gamma} x_{i\,mn} + S_{ij}^{mn} x_{j\,mn} = -F_{i\,mn} \quad (i, j = 1, 2, 3), \tag{10}$$

where the summation is carried out over two repeated indices, the elements of the matrix S_{ij}^{mn} are defined as follows:

$$S_{11}^{mn} = \left(\pi^2 m^2 + \frac{1-\nu}{2}\,\beta_1^2 n^2\right), \quad S_{12}^{mn} = S_{21}^{mn} = \frac{1+\nu}{2}\,\beta_1 \pi m n,$$
$$S_{13}^{mn} = S_{31}^{mn} = \nu\beta_1 \pi m, \quad S_{23}^{mn} = S_{32}^{mn} = \beta_1^2 n, \tag{11}$$
$$S_{22}^{mn} = \left(\frac{1-\nu}{2}\,\pi^2 m^2 + \beta_1^2 n^2\right), \quad S_{33}^{mn} = \frac{\beta_2^2}{12}\left(\pi^2 m^2 + \beta_1^2 n^2\right)^2 + \beta_1^2.$$

The nonlinear parts $F_{i\,mn}$ of Eq. (10) have the form

$$F_{1mn} = 2\sum_{m_1}\sum_{n_1}\sum_{m_2}\sum_{n_2} x_{3m_1n_1}x_{3m_2n_2}A_{mn}^{m_1n_1m_2n_2}, \tag{12}$$

$$F_{2mn} = \frac{2}{\pi}\sum_{m_1}\sum_{n_1}\sum_{m_2}\sum_{n_2} x_{3m_1n_1}x_{3m_2n_2}B_{mn}^{m_1n_1m_2n_2}, \tag{13}$$

$$F_{3mn} = -\frac{2}{\pi}\sum_{m_1}\sum_{n_1}\sum_{m_2}\sum_{n_2}\left[x_{3m_1n_1}x_{1m_2n_2}C_{mn}^{m_1n_1m_2n_2}\right.$$
$$\left. + x_{3m_1n_1}x_{2m_2n_2}D_{mn}^{m_1n_1m_2n_2} + x_{3m_1n_1}x_{3m_2n_2}E_{mn}^{m_1n_1m_2n_2}\right], \tag{14}$$

where

$$A_{mn}^{m_1n_1m_2n_2} = m_1\left(\pi^2m_2^2 + \frac{1-\nu}{2}\,\beta_1^2n_2^2\right)a_{1mn}^{m_1n_1m_2n_2} - \frac{1+\nu}{2}\,\beta_1^2n_1m_2n_2a_{2mn}^{m_1n_1m_2n_2},$$

$$B_{mn}^{m_1n_1m_2n_2} = \beta_1n_1\left(\beta_1^2n_2^2 + \frac{1-\nu}{2}\,\pi^2m_2^2\right)a_{3mn}^{m_1n_1m_2n_2}$$
$$-\frac{1+\nu}{2}\,\beta_1\pi^2m_1m_2n_2a_{4mn}^{m_1n_1m_2n_2},$$

$$C_{mn}^{m_1n_1m_2n_2} = \pi m_2\left(\pi^2m_1^2 + \nu\beta_1^2n_1^2\right)a_{5mn}^{m_1n_1m_2n_2} + (1-\nu)\beta_1^2\pi m_1n_1n_2a_{6mn}^{m_1n_1m_2n_2}$$
$$-\pi m_1\left(\pi^2m_2^2 + \frac{1-\nu}{2}\,\beta_1^2n_2^2\right)a_{7mn}^{m_1n_1m_2n_2}$$
$$-\frac{1+\nu}{2}\,\beta_1^2\pi n_1m_2n_2a_{8mn}^{m_1n_1m_2n_2},$$

$$D_{mn}^{m_1n_1m_2n_2} = \beta_1n_2\left(\nu\pi^2m_1^2 + \beta_1^2n_1^2\right)a_{5mn}^{m_1n_1m_2n_2} + (1-\nu)\beta_1\pi^2m_1n_1m_2a_{6mn}^{m_1n_1m_2n_2}$$
$$-\frac{1+\nu}{2}\,\beta_1\pi^2m_1m_2n_2a_{7mn}^{m_1n_1m_2n_2}$$
$$-\beta_1n_1\left(\frac{1-\nu}{2}\,\pi^2m_2^2 + \beta_1^2n_2^2\right)a_{8mn}^{m_1n_1m_2n_2},$$

$$E_{mn}^{m_1n_1m_2n_2} = \beta_1\left(\nu\pi^2m_2^2 + \beta_1^2n_2^2\right)a_{5mn}^{m_1n_1m_2n_2}$$
$$-\frac{\beta_1}{2}\left(\nu\pi^2m_1m_2a_{9mn}^{m_1n_1m_2n_2} + \beta_1^2n_1n_2a_{10mn}^{m_1n_1m_2n_2}\right),$$

and the coefficients $a_{k\,mn}^{m_1n_1m_2n_2}$ $(k = 1, 2, \ldots, 10)$ depending on the combinations of sine and cosine functions entering into the eigenfunctions (9) are presented in Appendix A.

Since the third-order tensor S_{ij}^{mn} is symmetric, then it has three real eigenvalues $\Omega_{i\,mn}$ $(i = 1, 2, 3)$ which are in correspondence with three mutually orthogonal eigenvectors

$$\mathbf{L}_{mn}^{\mathrm{I}}\left\{L_{i\,mn}^{\mathrm{I}}\right\}, \quad \mathbf{L}_{mn}^{\mathrm{II}}\left\{L_{i\,mn}^{\mathrm{II}}\right\}, \quad \mathbf{L}_{mn}^{\mathrm{III}}\left\{L_{i\,mn}^{\mathrm{III}}\right\}. \tag{15}$$

Thus, we could generalize a new approach suggested in [25] for a nonlinear plate for the case of a nonlinear cylindrical shell, that is we could apply the procedure resulting in decoupling linear parts of equations proposed in [25] with the further utilization of the method of multiple scales for solving nonlinear governing equations of motion. For this purpose, we expand the matrix S_{ij}^{mn} and the generalized displacements $x_{i\,mn}$ in terms of the vectors (15) as

$$S_{ij}^{mn} = \Omega_{1\,mn}^2 L_{i\,mn}^{\mathrm{I}} L_{j\,mn}^{\mathrm{I}} + \Omega_{2\,mn}^2 L_{i\,mn}^{\mathrm{II}} L_{j\,mn}^{\mathrm{II}} + \Omega_{3\,mn}^2 L_{i\,mn}^{\mathrm{III}} L_{j\,mn}^{\mathrm{III}}, \tag{16}$$

$$x_{i\,mn} = X_{1\,mn} L_{i\,mn}^{\mathrm{I}} + X_{2\,mn} L_{i\,mn}^{\mathrm{II}} + X_{3\,mn} L_{i\,mn}^{\mathrm{III}}. \tag{17}$$

Substituting (16) and (17) in Eq. (10) and then multiplying them successively by $L_{i\,mn}^{\mathrm{I}}$, $L_{i\,mn}^{\mathrm{II}}$, and $L_{i\,mn}^{\mathrm{III}}$ with due account for

$$\begin{aligned} L_{i\,mn}^{\mathrm{I}} L_{i\,mn}^{\mathrm{II}} &= L_{i\,mn}^{\mathrm{I}} L_{i\,mn}^{\mathrm{III}} = L_{i\,mn}^{\mathrm{II}} L_{i\,mn}^{\mathrm{III}} = 0, \\ L_{i\,mn}^{\mathrm{I}} L_{i\,mn}^{\mathrm{I}} &= L_{i\,mn}^{\mathrm{II}} L_{i\,mn}^{\mathrm{II}} = L_{i\,mn}^{\mathrm{III}} L_{i\,mn}^{\mathrm{III}} = 1, \end{aligned} \tag{18}$$

we obtain the following three equations:

$$\ddot{X}_{1mn} + \kappa_1 D^\gamma X_{1mn} + \Omega_{1mn}^2 X_{1mn} = -\sum_{i=1}^{3} F_{i\,mn} L_{i\,mn}^{\mathrm{I}}, \tag{19}$$

$$\ddot{X}_{2mn} + \kappa_2 D^\gamma X_{2mn} + \Omega_{2mn}^2 X_{2mn} = -\sum_{i=1}^{3} F_{i\,mn} L_{i\,mn}^{\mathrm{II}}, \tag{20}$$

$$\ddot{X}_{3mn} + \kappa_3 D^\gamma X_{3mn} + \Omega_{3mn}^2 X_{3mn} = -\sum_{i=1}^{3} F_{i\,mn} L_{i\,mn}^{\mathrm{III}}. \tag{21}$$

Assume hereafter that the vibration process occurs in such a way that only three natural modes corresponding to the generalized displacements $X_{1s_1s_2}$, $X_{2k_1k_2}$, and $X_{3l_1l_2}$ are excited and dominate over other natural modes. In this case, the right parts of Eqs. (19)–(21) are significantly simplified, and equations of free vibrations (19)–(21) take the form

$$\begin{aligned} &\ddot{X}_{1s_1s_2} + \kappa_1 D^\gamma X_{1s_1s_2} + \Omega_{1s_1s_2}^2 X_{1s_1s_2} + a_{11s_1s_2}^{\mathrm{I}} X_{1s_1s_2}^2 + a_{22s_1s_2}^{\mathrm{I}} X_{2k_1k_2}^2 + a_{33s_1s_2}^{\mathrm{I}} X_{3l_1l_2}^2 \\ &\quad + a_{12s_1s_2}^{\mathrm{I}} X_{1s_1s_2} X_{2k_1k_2} + a_{13s_1s_2}^{\mathrm{I}} X_{1s_1s_2} X_{3l_1l_2} + a_{23s_1s_2}^{\mathrm{I}} X_{2k_1k_2} X_{3l_1l_2} = 0, \end{aligned} \tag{22}$$

$$\begin{aligned} &\ddot{X}_{2k_1k_2} + \kappa_2 D^\gamma X_{2k_1k_2} + \Omega_{2k_1k_2}^2 X_{2k_1k_2} + a_{11k_1k_2}^{\mathrm{II}} X_{1s_1s_2}^2 + a_{22k_1k_2}^{\mathrm{II}} X_{2k_1k_2}^2 + a_{33k_1k_2}^{\mathrm{II}} X_{3l_1l_2}^2 \\ &\quad + a_{12k_1k_2}^{\mathrm{II}} X_{1s_1s_2} X_{2k_1k_2} + a_{13k_1k_2}^{\mathrm{II}} X_{1s_1s_2} X_{3l_1l_2} + a_{23k_1k_2}^{\mathrm{II}} X_{2k_1k_2} X_{3l_1l_2} = 0, \end{aligned} \tag{23}$$

$$\ddot{X}_{3l_1l_2} + \kappa_3 D^\gamma X_{3l_1l_2} + \Omega^2_{3l_1l_2} X_{3l_1l_2} + a^{\mathrm{III}}_{11l_1l_2} X^2_{1s_1s_2} + a^{\mathrm{III}}_{22l_1l_2} X^2_{2k_1k_2} + a^{\mathrm{III}}_{33l_1l_2} X^2_{3l_1l_2}$$
$$+ a^{\mathrm{III}}_{12l_1l_2} X_{1s_1s_2} X_{2k_1k_2} + a^{\mathrm{III}}_{13l_1l_2} X_{1s_1s_2} X_{3l_1l_2} + a^{\mathrm{III}}_{23l_1l_2} X_{2k_1k_2} X_{3l_1l_2} = 0, \quad (24)$$

where

$$a^\alpha_{11mn} = \frac{2}{\pi} \left\{ \left(L^{\mathrm{I}}_{3s_1s_2} \right)^2 \left(L^\alpha_{1mn} A^{s_1s_2s_1s_2}_{mn} + L^\alpha_{2mn} B^{s_1s_2s_1s_2}_{mn} + L^\alpha_{3mn} E^{s_1s_2s_1s_2}_{mn} \right) \right.$$
$$\left. + \left(L^I_{1s_1s_2} C^{s_1s_2s_1s_2}_{mn} + L^I_{2s_1s_2} D^{s_1s_2s_1s_2}_{mn} \right) L^{\mathrm{I}}_{3s_1s_2} L^\alpha_{3mn} \right\}, \quad (25)$$

$$a^\alpha_{22mn} = \frac{2}{\pi} \left\{ \left(L^{\mathrm{II}}_{3k_1k_2} \right)^2 \left(L^\alpha_{1mn} A^{k_1k_2k_1k_2}_{mn} + L^\alpha_{2mn} B^{k_1k_2k_1k_2}_{mn} + L^\alpha_{3mn} E^{k_1k_2k_1k_2}_{mn} \right) \right.$$
$$\left. + \left(L^{II}_{1k_1k_2} C^{k_1k_2k_1k_2}_{mn} + L^{II}_{2k_1k_2} D^{k_1k_2k_1k_2}_{mn} \right) L^{\mathrm{II}}_{3k_1k_2} L^\alpha_{3mn} \right\}, \quad (26)$$

$$a^\alpha_{33mn} = \frac{2}{\pi} \left\{ \left(L^{\mathrm{III}}_{3l_1l_2} \right)^2 \left(L^\alpha_{1mn} A^{l_1l_2l_1l_2}_{mn} + L^\alpha_{2mn} B^{l_1l_2l_1l_2}_{mn} + L^\alpha_{3mn} E^{l_1l_2l_1l_2}_{mn} \right) \right.$$
$$\left. + \left(L^{III}_{1l_1l_2} C^{l_1l_2l_1l_2}_{mn} + L^{III}_{2l_1l_2} D^{l_1l_2l_1l_2}_{mn} \right) L^{\mathrm{III}}_{3l_1l_2} L^\alpha_{3mn} \right\}, \quad (27)$$

$$a^\alpha_{13mn} = \frac{2}{\pi} \left\{ L^{\mathrm{I}}_{3s_1s_2} L^{\mathrm{III}}_{3l_1l_2} \left[\left(A^{s_1s_2l_1l_2}_{mn} + A^{l_1l_2s_1s_2}_{mn} \right) L^\alpha_{1mn} + \left(B^{s_1s_2l_1l_2}_{mn} + B^{l_1l_2s_1s_2}_{mn} \right) L^\alpha_{2mn} \right. \right.$$
$$\left. + \left(E^{s_1s_2l_1l_2}_{mn} + E^{l_1l_2s_1s_2}_{mn} \right) L^\alpha_{3mn} \right] + \left[L^{\mathrm{I}}_{3s_1s_2} L^{\mathrm{III}}_{1l_1l_2} C^{s_1s_2l_1l_2}_{mn} + L^{\mathrm{III}}_{3l_1l_2} L^{\mathrm{I}}_{1s_1s_2} C^{l_1l_2s_1s_2}_{mn} \right.$$
$$\left. \left. + L^{\mathrm{I}}_{3s_1s_2} L^{\mathrm{III}}_{2l_1l_2} D^{s_1s_2l_1l_2}_{mn} + L^{\mathrm{III}}_{3l_1l_2} L^{\mathrm{I}}_{2s_1s_2} D^{l_1l_2s_1s_2}_{mn} \right] L^\alpha_{3mn} \right\}, \quad (28)$$

$$a^\alpha_{23mn} = \frac{2}{\pi} \left\{ L^{\mathrm{II}}_{3k_1k_2} L^{\mathrm{III}}_{3l_1l_2} \left[\left(A^{k_1k_2l_1l_2}_{mn} + A^{l_1l_2k_1k_2}_{mn} \right) L^\alpha_{1mn} + \left(B^{k_1k_2l_1l_2}_{mn} + B^{l_1l_2k_1k_2}_{mn} \right) L^\alpha_{2mn} \right. \right.$$
$$\left. + \left(E^{k_1k_2l_1l_2}_{mn} + E^{l_1l_2k_1k_2}_{mn} \right) L^\alpha_{3mn} \right] + \left[L^{\mathrm{II}}_{3k_1k_2} L^{\mathrm{III}}_{1l_1l_2} C^{k_1k_2l_1l_2}_{mn} + L^{\mathrm{III}}_{3l_1l_2} L^{\mathrm{II}}_{1k_1k_2} C^{l_1l_2k_1k_2}_{mn} \right.$$
$$\left. \left. + L^{\mathrm{II}}_{3k_1k_2} L^{\mathrm{III}}_{2l_1l_2} D^{k_1k_2l_1l_2}_{mn} + L^{\mathrm{III}}_{3l_1l_2} L^{\mathrm{II}}_{2k_1k_2} D^{l_1l_2k_1k_2}_{mn} \right] L^\alpha_{3mn} \right\}, \quad (29)$$

$$a^\alpha_{12mn} = \frac{2}{\pi} \left\{ L^{\mathrm{I}}_{3s_1s_2} L^{\mathrm{II}}_{3k_1k_2} \left[\left(A^{s_1s_2k_1k_2}_{mn} + A^{k_1k_2s_1s_2}_{mn} \right) L^\alpha_{1mn} + \left(B^{s_1s_2k_1k_2}_{mn} + B^{k_1k_2s_1s_2}_{mn} \right) L^\alpha_{2mn} \right. \right.$$
$$\left. + \left(E^{s_1s_2k_1k_2}_{mn} + E^{k_1k_2s_1s_2}_{mn} \right) L^\alpha_{3mn} \right] + \left[L^{\mathrm{I}}_{3s_1s_2} L^{\mathrm{II}}_{1k_1k_2} C^{s_1s_2k_1k_2}_{mn} + L^{\mathrm{II}}_{3k_1k_2} L^{\mathrm{I}}_{1s_1s_2} C^{k_1k_2s_1s_2}_{mn} \right.$$
$$\left. \left. + L^{\mathrm{I}}_{3s_1s_2} L^{\mathrm{II}}_{2k_1k_2} D^{s_1s_2k_1k_2}_{mn} + L^{\mathrm{II}}_{3k_1k_2} L^{\mathrm{I}}_{2s_1s_2} D^{k_1k_2s_1s_2}_{mn} \right] L^\alpha_{3mn} \right\}. \quad (30)$$

From relationships (25)–(30) we could calculate all coefficients entering in Eq. (22) at $\alpha = \mathrm{I}$, $m = s_1$ and $n = s_2$, in Eq. (23) at $\alpha = \mathrm{II}$, $m = k_1$ and $n = k_2$, and in Eq. (24) at $\alpha = \mathrm{III}$, $m = l_1$ and $n = l_2$.

Omitting hereafter the subindices s_1s_2, k_1k_2, and l_1l_2 for ease of presentation, Eqs. (22)–(24) could be rewritten as

$$\ddot{X}_1 + \kappa_1 D^\gamma X_1 + \Omega_1^2 X_1 + a^{\mathrm{I}}_{11} X_1^2 + a^{\mathrm{I}}_{22} X_2^2 + a^{\mathrm{I}}_{33} X_3^2 + a^{\mathrm{I}}_{12} X_1 X_2$$
$$+ a^{\mathrm{I}}_{13} X_1 X_3 + a^{\mathrm{I}}_{23} X_2 X_3 = 0, \quad (31)$$
$$\ddot{X}_2 + \kappa_2 D^\gamma X_2 + \Omega_2^2 X_2 + a^{\mathrm{II}}_{11} X_1^2 + a^{\mathrm{II}}_{22} X_2^2 + a^{\mathrm{II}}_{33} X_3^2 + a^{\mathrm{II}}_{12} X_1 X_2$$
$$+ a^{\mathrm{II}}_{13} X_1 X_3 + a^{\mathrm{II}}_{23} X_2 X_3 = 0, \quad (32)$$
$$\ddot{X}_3 + \kappa_3 D^\gamma X_3 + \Omega_3^2 X_3 + a^{\mathrm{III}}_{11} X_1^2 + a^{\mathrm{III}}_{22} X_2^2 + a^{\mathrm{III}}_{33} X_3^2 + a^{\mathrm{III}}_{12} X_1 X_2$$
$$+ a^{\mathrm{III}}_{13} X_1 X_3 + a^{\mathrm{III}}_{23} X_2 X_3 = 0. \quad (33)$$

3 Method of Solution

An approximate solution of Eqs. (31)–(33) for small but finite amplitudes weakly varying with time can be represented by a two-order uniform expansion in terms of different time scales in the following form [28]:

$$X_i = \varepsilon X_{i1}(T_0, T_1 \dots) + \varepsilon^2 X_{i2}(T_0, T_1 \dots) + \cdots , \tag{34}$$

where $i = 1, 2, 3$, ε is a small dimensionless parameter of the same order of magnitude as the amplitudes, $T_n = \varepsilon^n t$ are new independent variables, among them: $T_0 = t$ is a fast scale characterizing motions with the natural frequencies, and $T_1 = \varepsilon t$ is a slow scale characterizing the modulation of the amplitudes and phases of the modes with nonlinearity.

Following the procedure described in [25], in doing so limiting ourselves by the case of small viscosity of the order of ε, i.e.,

$$\kappa_i = \varepsilon \mu_i \tau_i^{\gamma},$$

where τ_i is the relaxation time of the ith generalized displacement, μ_i is a finite value, and substituting (34) in Eqs. (31)–(33), after equating the coefficients at like powers of ε to zero, we are led to a set of recurrence equations to various orders:

- to order ε

$$D_0^2 X_{11} + \Omega_1^2 X_{11} = 0, \tag{35}$$
$$D_0^2 X_{21} + \Omega_2^2 X_{21} = 0, \tag{36}$$
$$D_0^2 X_{31} + \Omega_3^2 X_{31} = 0; \tag{37}$$

- to order ε^2

$$D_0^2 X_{12} + \Omega_1^2 X_{12} = -\ 2D_0 D_1 X_{11} - a_{11}^{\mathrm{I}} X_{11}^2 - a_{22}^{\mathrm{I}} X_{21}^2 - a_{33}^{\mathrm{I}} X_{31}^2 - a_{12}^{\mathrm{I}} X_{11} X_{21} - a_{13}^{\mathrm{I}} X_{11} X_{31} - a_{23}^{\mathrm{I}} X_{21} X_{31} - \mu_1 \tau_1^{\gamma} D_0^{\gamma} X_{11}, \tag{38}$$

$$D_0^2 X_{22} + \Omega_2^2 X_{22} = -\ 2D_0 D_1 X_{21} - a_{11}^{\mathrm{II}} X_{11}^2 - a_{22}^{\mathrm{II}} X_{21}^2 - a_{33}^{\mathrm{II}} X_{31}^2 - a_{12}^{\mathrm{II}} X_{11} X_{21} - a_{13}^{\mathrm{II}} X_{11} X_{31} - a_{23}^{\mathrm{II}} X_{21} X_{31} - \mu_2 \tau_2^{\gamma} D_0^{\gamma} X_{21}, \tag{39}$$

$$D_0^2 X_{32} + \Omega_3^2 X_{32} = -\ 2D_0 D_1 X_{31} - a_{11}^{\mathrm{III}} X_{11}^2 - a_{22}^{\mathrm{III}} X_{21}^2 - a_{33}^{\mathrm{III}} X_{31}^2 - a_{12}^{\mathrm{III}} X_{11} X_{21} - a_{13}^{\mathrm{III}} X_{11} X_{31} - a_{23}^{\mathrm{III}} X_{21} X_{31} - \mu_3 \tau_3^{\gamma} D_0^{\gamma} X_{31}. \tag{40}$$

We shall seek the solution of Eqs. (35)–(37) in the form

$$X_{j1} = A_j(T_1, T_2) \exp(i\Omega_j T_0) + \bar{A}_j(T_1, T_2) \exp(-i\Omega_j T_0) \quad (j = 1, 2, 3), \tag{41}$$

where $A_j(T_1, T_2)$ $(j = 1, 2, 3)$ are unknown complex functions, and $\bar{A}_j(T_1, T_2)$ are the complex conjugates of $A_j(T_1, T_2)$.

Substituting (41) in the right-hand side of Eqs. (38)–(40) with due account for the approximate formula $D_0^\gamma e^{i\Omega_j t} = (i\Omega_j)^\gamma e^{i\Omega_j t}$ [25, 29] yields

$$
\begin{aligned}
D_0^2 X_{12} + \Omega_1^2 X_{12} = & -2i\Omega_1 D_1 A_1(T_1)\exp(i\Omega_1 T_0) - a_{11}^{\mathrm{I}}\left[A_1^2\exp(2i\Omega_1 T_0) + A_1\bar{A}_1\right] \\
& - a_{22}^{\mathrm{I}}\left[A_2^2\exp(2i\Omega_2 T_0) + A_2\bar{A}_2\right] - a_{33}^{\mathrm{I}}\left[A_3^2\exp(2i\Omega_3 T_0) + A_3\bar{A}_3\right] \\
& - a_{12}^{\mathrm{I}}\left\{A_1A_2\exp\left[i(\Omega_1+\Omega_2)T_0\right] + A_1\bar{A}_2\exp\left[i(\Omega_1-\Omega_2)T_0\right]\right\} \\
& - a_{13}^{\mathrm{I}}\left\{A_1A_3\exp\left[i(\Omega_1+\Omega_3)T_0\right] + A_1\bar{A}_3\exp\left[i(\Omega_1-\Omega_3)T_0\right]\right\} \\
& - a_{23}^{\mathrm{I}}\left\{A_2A_3\exp\left[i(\Omega_2+\Omega_3)T_0\right] + A_2\bar{A}_3\exp\left[i(\Omega_2-\Omega_3)T_0\right]\right\} \\
& - \mu_1\tau_1^\gamma A_1\,(i\Omega_1)^\gamma\exp(i\Omega_1 T_0) + cc,
\end{aligned} \tag{42}
$$

$$
\begin{aligned}
D_0^2 X_{22} + \Omega_2^2 X_{22} = & -2i\Omega_2 D_1 A_2(T_1)\exp(i\Omega_2 T_0) - a_{11}^{\mathrm{II}}\left[A_1^2\exp(2i\Omega_1 T_0) + A_1\bar{A}_1\right] \\
& - a_{22}^{\mathrm{II}}\left[A_2^2\exp(2i\Omega_2 T_0) + A_2\bar{A}_2\right] - a_{33}^{\mathrm{II}}\left[A_3^2\exp(2i\Omega_3 T_0) + A_3\bar{A}_3\right] \\
& - a_{12}^{\mathrm{II}}\left\{A_1A_2\exp\left[i(\Omega_1+\Omega_2)T_0\right] + A_1\bar{A}_2\exp\left[i(\Omega_1-\Omega_2)T_0\right]\right\} \\
& - a_{13}^{\mathrm{II}}\left\{A_1A_3\exp\left[i(\Omega_1+\Omega_3)T_0\right] + A_1\bar{A}_3\exp\left[i(\Omega_1-\Omega_3)T_0\right]\right\} \\
& - a_{23}^{\mathrm{II}}\left\{A_2A_3\exp\left[i(\Omega_2+\Omega_3)T_0\right] + A_2\bar{A}_3\exp\left[i(\Omega_2-\Omega_3)T_0\right]\right\} \\
& - \mu_2\tau_2^\gamma A_2\,(i\Omega_2)^\gamma\exp(i\Omega_2 T_0) + cc,
\end{aligned} \tag{43}
$$

$$
\begin{aligned}
D_0^2 X_{32} + \Omega_3^2 X_{32} = & -2i\Omega_3 D_1 A_3(T_1)\exp(i\Omega_3 T_0) - a_{11}^{\mathrm{III}}\left[A_1^2\exp(2i\Omega_1 T_0) + A_1\bar{A}_1\right] \\
& - a_{22}^{\mathrm{III}}\left[A_2^2\exp(2i\Omega_2 T_0) + A_2\bar{A}_2\right] - a_{33}^{\mathrm{III}}\left[A_3^2\exp(2i\Omega_3 T_0) + A_3\bar{A}_3\right] \\
& - a_{12}^{\mathrm{III}}\left\{A_1A_2\exp\left[i(\Omega_1+\Omega_2)T_0\right] + A_1\bar{A}_2\exp\left[i(\Omega_1-\Omega_2)T_0\right]\right\} \\
& - a_{13}^{\mathrm{III}}\left\{A_1A_3\exp\left[i(\Omega_1+\Omega_3)T_0\right] + A_1\bar{A}_3\exp\left[i(\Omega_1-\Omega_3)T_0\right]\right\} \\
& - a_{23}^{\mathrm{III}}\left\{A_2A_3\exp\left[i(\Omega_2+\Omega_3)T_0\right] + A_2\bar{A}_3\exp\left[i(\Omega_2-\Omega_3)T_0\right]\right\} \\
& - \mu_3\tau_3^\gamma A_3\,(i\Omega_3)^\gamma\exp(i\Omega_3 T_0) + cc,
\end{aligned} \tag{44}
$$

where cc is the complex conjugate part to the preceding terms.

Reference to Eqs. (42)–(44) shows that the following types of the internal resonance could occur on this step:

1. the two-to-one internal resonance, when one natural frequency is twice the other natural frequency,

$$\Omega_1 = 2\Omega_2 \quad (\Omega_3 \neq \Omega_1, \quad \Omega_3 \neq 2\Omega_2), \quad \text{or} \quad \Omega_2 = 2\Omega_1, \tag{45}$$

$$\Omega_1 = 2\Omega_3 \quad (\Omega_2 \neq \Omega_1, \quad \Omega_2 \neq 2\Omega_3), \quad \text{or} \quad \Omega_3 = 2\Omega_1, \tag{46}$$

$$\Omega_2 = 2\Omega_3 \quad (\Omega_1 \neq \Omega_2, \quad \Omega_1 \neq 2\Omega_3), \quad \text{or} \quad \Omega_3 = 2\Omega_2; \tag{47}$$

2. the one-to-one-to-two or one-to-two-to-two internal resonance, i.e.,

$$\Omega_1 = \Omega_2 = 2\Omega_3, \quad \text{or} \quad 1:1:2, \tag{48}$$

$$\Omega_1 = 2\Omega_2 = \Omega_3, \quad \text{or} \quad 1:2:1, \tag{49}$$

$$2\Omega_1 = \Omega_2 = \Omega_3, \quad \text{or} \quad 2:1:1, \tag{50}$$

$$\Omega_1 = 2\Omega_2 = 2\Omega_3, \quad \text{or} \quad 1:2:2, \tag{51}$$

$$2\Omega_1 = \Omega_2 = 2\Omega_3, \quad \text{or} \quad 2:1:2, \tag{52}$$

$$2\Omega_1 = 2\Omega_2 = \Omega_3, \quad \text{or} \quad 2:2:1; \tag{53}$$

3. the combinational resonance of the additive-difference type, i.e.,

$$\Omega_1 = \Omega_2 + \Omega_3, \quad \text{or} \quad \Omega_2 = \Omega_1 - \Omega_3, \quad \text{or} \quad \Omega_3 = \Omega_1 - \Omega_2, \tag{54}$$

$$\Omega_1 = \Omega_2 - \Omega_3, \quad \text{or} \quad \Omega_2 = \Omega_1 + \Omega_3, \quad \text{or} \quad \Omega_3 = \Omega_2 - \Omega_1, \tag{55}$$

$$\Omega_1 = \Omega_3 - \Omega_2, \quad \text{or} \quad \Omega_2 = \Omega_3 - \Omega_1, \quad \text{or} \quad \Omega_3 = \Omega_1 + \Omega_2. \tag{56}$$

Comparison of the types of the internal resonances of the order of ε (45)–(56) just found for the nonlinear shallow cylindrical shell with those for a nonlinear plate, the motion of which is described also by three coupled nonlinear equations in terms of three displacements, presented in a companion paper [25] reveals the fact that the shell equations could produce much more diversified variety of internal resonances than the plate equations. The one-to-two-to-two internal resonances (51)–(53) and the combinational resonances of the additive-difference type (54)–(56) are characteristic only for cylindrical shells.

4 Governing Nonlinear Differential Equations Describing Amplitude-Phase Modulation for Different Types of the Internal Resonance of the Order of ε

To deduce the nonlinear differential equations describing the modulation of amplitudes and phases of the cylindrical shell under consideration, we should consider separately each type of the internal resonance which could occur with due account for weak damping of the order ε.

4.1 Internal Resonance 2:1

Let us consider the case (46), when $\Omega_3 = 2\Omega_1$, while $\Omega_2 \neq \Omega_3$ and $\Omega_2 \neq 2\Omega_1$. Then eliminating secular terms in Eqs. (42)–(44), we obtain the following solvability equations:

$$2i\Omega_1 D_1 A_1(T_1) + \mu_1 \tau_1^{\gamma} A_1 \, (i\Omega_1)^{\gamma} + a_{13}^{\mathrm{I}} \bar{A}_1 A_3 = 0, \tag{57}$$

$$2i\Omega_2 D_1 A_2(T_1) + \mu_2 \tau_2^{\gamma} A_2 \, (i\Omega_2)^{\gamma} = 0, \tag{58}$$

$$2i\Omega_3 D_1 A_3(T_1) + \mu_3 \tau_3^{\gamma} A_3 \, (i\Omega_3)^{\gamma} + a_{11}^{\mathrm{III}} A_1^2 = 0. \tag{59}$$

Reference to the set of Eqs. (57)–(59) shows that its second equation is independent of other two, while the first and third ones represent a set of two nonlinear equations.

Let us multiply Eqs. (57)–(59), respectively, by $\bar{A}_1$, $\bar{A}_2$, and $\bar{A}_3$ and find their complex conjugates. Adding every pair of the mutually adjoint equations with each other and subtracting one from another, and considering that

$$A_i = a_i e^{i\varphi_i} \quad (i = 1, 2, 3), \tag{60}$$

as a result we have

$$\left(a_1^2\right)^{\cdot} + s_1 a_1^2 = \Omega_1^{-1} a_{13}^{\mathrm{I}} a_1^2 a_3 \sin\delta, \tag{61}$$

$$\left(a_3^2\right)^{\cdot} + s_3 a_3^2 = -\Omega_3^{-1} a_{11}^{\mathrm{III}} a_1^2 a_3 \sin\delta, \tag{62}$$

$$\dot{\varphi}_1 - \frac{1}{2}\sigma_1 - \frac{1}{2} a_{13}^{I} \Omega_1^{-1} a_3 \cos\delta = 0, \tag{63}$$

$$\dot{\varphi}_3 - \frac{1}{2}\sigma_3 - \frac{1}{2} a_{11}^{\mathrm{III}} \Omega_3^{-1} a_1^2 a_3^{-1} \cos\delta = 0, \tag{64}$$

$$\left(a_2^2\right)^{\cdot} + s_2 a_2^2 = 0, \tag{65}$$

$$\dot{\varphi}_2 - \frac{1}{2}\sigma_2 = 0, \tag{66}$$

where a dot denotes differentiation with respect to T_1, $\delta = 2\varphi_1 - \varphi_3$, and

$$s_i = \mu_i \tau_i^{\gamma} \Omega_i^{\gamma-1} \sin\psi, \quad \sigma_i = \mu_i \tau_i^{\gamma} \Omega_i^{\gamma-1} \cos\psi, \quad \psi = \frac{1}{2}\pi\gamma \quad (i = 1, 2, 3). \tag{67}$$

The similar equations have been obtained in [25] for a fractionally damped nonlinear plate in the case of the two-to-one internal resonance. Thus, the detailed solution describing the dynamic response of the cylindrical shell with the two-to-one internal resonance could be found in [25] and is not repeated in this paper. Therefore, during free vibrations of the cylindrical shell being under the conditions of the two-to-one internal resonance three regimes could be observed: stationary (absence of damping at $\gamma = 0$ and $\mu = 0$), quasi-stationary (damping is defined by an ordinary derivative at $\gamma = 1$), and transient (damping is defined by a fractional derivative at $0 < \gamma < 1$).

4.2 *Combinational Additive-Difference Internal Resonance of the First Order*

Now let us consider the case of the combinational additive-difference internal resonance. As an example, we will analyze the case (54), when $\Omega_1 = \Omega_2 + \Omega_3$. Other possible cases, (55) and (56), could be treated similarly.

Eliminating secular terms in Eqs. (42)–(44) for the case under consideration and applying the same procedure as it has been carried out for the two-to-one internal resonance, we obtain the following solvability equations:

$$\left(a_1^2\right)^{\cdot} + s_1 a_1^2 = \Omega_1^{-1} a_{23}^{\mathrm{I}} a_1 a_2 a_3 \sin\delta, \tag{68}$$

$$\left(a_2^2\right)^{\cdot} + s_2 a_2^2 = -\Omega_2^{-1} a_{13}^{\mathrm{II}} a_1 a_2 a_3 \sin\delta, \tag{69}$$

$$\left(a_3^2\right)^{\cdot} + s_3 a_3^2 = -\Omega_3^{-1} a_{12}^{\mathrm{III}} a_1 a_2 a_3 \sin\delta, \tag{70}$$

$$\dot{\varphi}_1 - \frac{1}{2}\sigma_1 - \frac{1}{2}\frac{a_{23}^{\mathrm{I}}}{\Omega_1}\frac{a_2 a_3}{a_1}\cos\delta = 0, \tag{71}$$

$$\dot{\varphi}_2 - \frac{1}{2}\sigma_2 - \frac{1}{2}\frac{a_{13}^{\mathrm{II}}}{\Omega_2}\frac{a_1 a_3}{a_2}\cos\delta = 0, \tag{72}$$

$$\dot{\varphi}_3 - \frac{1}{2}\sigma_3 - \frac{1}{2}\frac{a_{12}^{\mathrm{III}}}{\Omega_3}\frac{a_1 a_2}{a_3}\cos\delta = 0, \tag{73}$$

where the phase difference has the form $\delta = \varphi_1 - (\varphi_2 + \varphi_3)$.

Introducing new functions $\xi_1(T_1)$, $\xi_2(T_1)$, and $\xi_3(T_1)$, such that

$$a_1^2 = \frac{a_{23}^{\mathrm{I}}}{\Omega_1}\xi_1 \exp(-s_1 T_1), \quad a_2^2 = \frac{a_{13}^{\mathrm{II}}}{\Omega_2}\xi_2 \exp(-s_2 T_1), \quad a_3^2 = \frac{a_{12}^{\mathrm{III}}}{\Omega_3}\xi_3 \exp(-s_3 T_1), \tag{74}$$

and substituting (74) in the left-hand sides of (68)–(70) yield

$$\dot{\xi}_1 \exp(-s_1 T_1) = a_1 a_2 a_3 \sin\delta, \tag{75}$$

$$\dot{\xi}_2 \exp(-s_2 T_1) = -a_1 a_2 a_3 \sin\delta, \tag{76}$$

$$\dot{\xi}_3 \exp(-s_3 T_1) = -a_1 a_2 a_3 \sin\delta, \tag{77}$$

the summation of which results in the following:

$$2\dot{\xi}_1 \exp(-s_1 T_1) + \dot{\xi}_2 \exp(-s_2 T_1) + \dot{\xi}_3 \exp(-s_3 T_1) = 0. \tag{78}$$

From Eqs. (75)–(77) we could find that

$$\frac{a_2 a_3}{a_1} = \frac{\dot{\xi}_1}{\xi_1}\frac{\Omega_1}{a_{23}^{\mathrm{I}}}\frac{1}{\sin\delta}, \tag{79}$$

$$\frac{a_1 a_3}{a_2} = -\frac{\dot{\xi}_2}{\xi_2}\frac{\Omega_2}{a_{13}^{\mathrm{II}}}\frac{1}{\sin\delta}, \tag{80}$$

$$\frac{a_1 a_2}{a_3} = -\frac{\dot{\xi}_3}{\xi_3}\frac{\Omega_3}{a_{12}^{\mathrm{III}}}\frac{1}{\sin\delta}, \tag{81}$$

while Eqs. (71)–(73) could be reduced to

$$\dot{\delta} = \Sigma + \frac{1}{2}\frac{a_{23}^{\mathrm{I}}}{\Omega_1}\frac{a_2 a_3}{a_1}\cos\delta - \frac{1}{2}\frac{a_{13}^{\mathrm{II}}}{\Omega_2}\frac{a_1 a_3}{a_2}\cos\delta - \frac{1}{2}\frac{a_{12}^{\mathrm{III}}}{\Omega_3}\frac{a_1 a_2}{a_3}\cos\delta, \tag{82}$$

where $\Sigma = \frac{1}{2}(\sigma_1 - \sigma_2 - \sigma_3)$.

Substituting (79)–(81) in (82) yields

$$\dot{\delta} - \Sigma = \frac{1}{2}\left(\frac{\dot{\xi}_1}{\xi_1} + \frac{\dot{\xi}_2}{\xi_2} + \frac{\dot{\xi}_3}{\xi_3}\right)\cot\delta. \tag{83}$$

The first integral of (83) could be written in the form

$$G_0 \exp\left(-\Sigma\int_0^{T_1}\tan\delta \mathrm{d}T_1\right) = \sqrt{\xi_1}\sqrt{\xi_2}\sqrt{\xi_3}\cos\delta, \tag{84}$$

where G_0 is a constant of integration to be found from the initial conditions

$$\xi_1\Big|_{T_1=0} = \xi_{10}, \quad \xi_2\Big|_{T_1=0} = \xi_{20}, \quad \xi_3\Big|_{T_1=0} = \xi_{30}, \quad \delta\Big|_{T_1=0} = \delta_0. \tag{85}$$

From the other hand, substituting (74) in (82) and (75) yields

$$\dot{\delta} = \Sigma + \frac{1}{2}b\left(\frac{\sqrt{\xi_2}\sqrt{\xi_3}}{\sqrt{\xi_1}}e^{\frac{1}{2}(s_1-s_2-s_3)T_1} - \frac{\sqrt{\xi_1}\sqrt{\xi_3}}{\sqrt{\xi_2}}e^{\frac{1}{2}(s_2-s_1-s_3)T_1} - \frac{\sqrt{\xi_1}\sqrt{\xi_2}}{\sqrt{\xi_3}}e^{\frac{1}{2}(s_3-s_1-s_2)T_1}\right)\cos\delta, \tag{86}$$

$$\dot{\xi}_1 = b\sqrt{\xi_1}\sqrt{\xi_2}\sqrt{\xi_3}\,e^{\frac{1}{2}(s_1-s_2-s_3)T_1}\sin\delta, \tag{87}$$

where

$$b = \sqrt{\frac{a_{23}^{\mathrm{I}}}{\Omega_1}\frac{a_{13}^{\mathrm{II}}}{\Omega_2}\frac{a_{12}^{\mathrm{III}}}{\Omega_3}}.$$

The nonlinear set of Eqs. (78), (84), (86), and (87) with the initial conditions (85) completely describe the vibrational process of the mechanical system being investigated under the condition of the internal combinational resonance and could be solved numerically.

4.2.1 Particular Case

In the particular case at $\Sigma = 0$ and $s_1 = s_2 = s_3 = s$, Eq. (78) has the form

$$2\dot{\xi}_1 + \dot{\xi}_2 + \dot{\xi}_3 = 0, \tag{88}$$

which could be integrated as

$$2\xi_1 + \xi_2 + \xi_3 = E_0, \tag{89}$$

resulting in the the following law of energy dissipation

$$E = 2\frac{\Omega_1}{a_{23}^{\mathrm{I}}} a_1^2 + \frac{\Omega_2}{a_{13}^{\mathrm{II}}} a_2^2 + \frac{\Omega_3}{a_{12}^{\mathrm{III}}} a_3^2 = E_0 e^{-sT_1}, \tag{90}$$

where E_0 is the initial energy of the system.

Introducing a new variable ξ such that

$$E_0 \dot{\xi} = b\sqrt{\xi_1 \xi_2 \xi_3} \exp(-\frac{1}{2} sT_1) \sin\delta, \tag{91}$$

Equations (75)–(77) could be reduced to

$$\dot{\xi}_1 = \dot{\xi} E_0, \tag{92}$$

$$\dot{\xi}_2 = -\dot{\xi} E_0, \tag{93}$$

$$\dot{\xi}_3 = -\dot{\xi} E_0, \tag{94}$$

the integration of which yields

$$\xi_1 = E_0(c_1 + \xi), \quad \xi_2 = E_0(c_2 - \xi), \quad \xi_3 = E_0(c_3 - \xi), \tag{95}$$

where c_i $(i = 1, 2, 3)$ are constants of integration.

Note that Eq. (88) is fulfilled automatically under the substitution of (92)–(94) in it, while the substitution of (95) in (89) results in the relationship between the constants of integration

$$2c_1 + c_2 + c_3 = 1. \tag{96}$$

Considering (95), Eqs. (84), (86), and (87) take, respectively, the form

$$G(\xi, \delta) = \sqrt{(c_1 + \xi)(c_2 - \xi)(c_3 - \xi)} \cos\delta = G_0(\xi_0, \delta_0), \tag{97}$$

$$\dot{\delta} = \frac{1}{2} b\sqrt{E_0} \frac{(c_2 - \xi)(c_3 - \xi) - (c_1 + \xi)(c_3 - \xi) - (c_1 + \xi)(c_2 - \xi)}{\sqrt{(c_1 + \xi)(c_2 - \xi)(c_3 - \xi)}} e^{-\frac{1}{2} sT_1} \cos\delta, \tag{98}$$

$$\dot{\xi} = b\sqrt{E_0}\sqrt{(c_1 + \xi)(c_2 - \xi)(c_3 - \xi)}\, e^{-\frac{1}{2} sT_1} \sin\delta. \tag{99}$$

The first integral (97) defines the stream function $G(\xi, \delta)$ such that

$$v_\xi = \dot{\xi} = -b\sqrt{E_0} \frac{\partial G}{\partial \delta} e^{-\frac{1}{2} sT_1}, \quad v_\delta = \dot{\delta} = b\sqrt{E_0} \frac{\partial G}{\partial \xi} e^{-\frac{1}{2} sT_1}, \tag{100}$$

which describes steady-state vibrations of an elastic shell attenuating with time.

Eliminating the variable δ from (97) and (99) and integrating over T_1, we have

$$\int_{\xi_0}^{\xi} \frac{d\xi}{\sqrt{(c_1+\xi)(c_2-\xi)(c_3-\xi)-G_0^2}} = \frac{2b\sqrt{E_0}}{s}\left(1-e^{-\frac{1}{2}sT_1}\right). \tag{101}$$

The integral in the left-hand side of Eq. (101) can be transformed to an incomplete elliptic integral of the first kind and can be easily calculated using special tables [30].

4.3 Internal Resonance 2:1:1

Let us consider the case (50), when $2\Omega_1 = \Omega_2 = \Omega_3$. Then eliminating secular terms in Eqs. (42)–(44), we obtain the following solvability equations:

$$2i\Omega_1 D_1 A_1(T_1) + \mu_1 \tau_1^\gamma A_1 (i\Omega_1)^\gamma + a_{13}^{\mathrm{I}} \bar{A}_1 A_3 + a_{12}^{I} \bar{A}_1 A_2 = 0, \tag{102}$$

$$2i\Omega_2 D_1 A_2(T_1) + \mu_2 \tau_2^\gamma A_2 (i\Omega_2)^\gamma + a_{11}^{\mathrm{II}} A_1^2 = 0, \tag{103}$$

$$2i\Omega_3 D_1 A_3(T_1) + \mu_3 \tau_3^\gamma A_3 (i\Omega_3)^\gamma + a_{11}^{\mathrm{III}} A_1^2 = 0. \tag{104}$$

Reference to the set of Eqs. (102)–(104) shows that, in contrast to Eqs. (57)–(59) corresponding to the case (46) when $\Omega_3 = 2\Omega_1$, while $\Omega_2 \neq \Omega_3$ and $\Omega_2 \neq 2\Omega_1$, now we obtain the set of three coupled nonlinear equations.

Representing functions A_i entering in (102)–(104) in the polar form (60) and applying the same procedure as it has been carried out above for 2:1 internal resonance, we have

$$\left(a_1^2\right)^{\cdot} + s_1 a_1^2 = \Omega_1^{-1} a_1^2 \left[a_{13}^{\mathrm{I}} a_3 \sin(2\varphi_1 - \varphi_3) + a_{12}^{\mathrm{I}} a_2 \sin(2\varphi_1 - \varphi_2)\right], \tag{105}$$

$$\left(a_2^2\right)^{\cdot} + s_2 a_2^2 = -\Omega_2^{-1} a_{11}^{\mathrm{II}} a_1^2 a_2 \sin(2\varphi_1 - \varphi_2), \tag{106}$$

$$\left(a_3^2\right)^{\cdot} + s_3 a_3^2 = -\Omega_3^{-1} a_{11}^{\mathrm{III}} a_1^2 a_3 \sin(2\varphi_1 - \varphi_3), \tag{107}$$

$$\dot{\varphi}_1 - \frac{1}{2}\sigma_1 - \frac{1}{2}\Omega_1^{-1}\left[a_{13}^{\mathrm{I}} a_3 \cos(2\varphi_1 - \varphi_3) + a_{12}^{\mathrm{I}} a_2 \cos(2\varphi_1 - \varphi_2)\right] = 0, \tag{108}$$

$$\dot{\varphi}_2 - \frac{1}{2}\sigma_2 - \frac{1}{2} a_{11}^{\mathrm{II}} \Omega_2^{-1} a_1^2 a_2^{-1} \cos(2\varphi_1 - \varphi_2) = 0, \tag{109}$$

$$\dot{\varphi}_3 - \frac{1}{2}\sigma_3 - \frac{1}{2} a_{11}^{\mathrm{III}} \Omega_3^{-1} a_1^2 a_3^{-1} \cos(2\varphi_1 - \varphi_3) = 0. \tag{110}$$

The nonlinear set of Eqs. (105)–(109) with the initial conditions (85) completely describe the vibrational process of the mechanical system being investigated under the condition of the internal resonance 2:1:1 and could be solved numerically. Introducing new functions such that $\xi_1(T_1)$, $\xi_2(T_1)$, and $\xi_3(T_1)$

$$a_1^2 = \xi_1 e^{-s_1 T_1}, \quad a_2^2 = \frac{a_{11}^{\mathrm{II}} \Omega_1}{a_{12}^{\mathrm{I}} \Omega_2} \xi_2 e^{-s_2 T_1}, \quad a_3^2 = \frac{a_{11}^{\mathrm{III}} \Omega_1}{a_{13}^{\mathrm{I}} \Omega_3} \xi_3 e^{-s_3 T_1}, \tag{111}$$

and adding Eqs. (105)–(106) with due account for (111), we obtain

$$\xi_1 e^{-s_1 T_1} + \dot{\xi}_2 e^{-s_2 T_1} + \dot{\xi}_3 e^{\ \ s_3 T_1} = 0. \tag{112}$$

Equation (112) describes the law of energy variation for this case of the internal resonance.

4.4 Internal Resonance 2:2:1

Let us consider the case (53), when $2\Omega_1 = 2\Omega_2 = \Omega_3$. Then eliminating secular terms in Eqs. (42)–(44), we obtain the following solvability equations:

$$2i\Omega_1 D_1 A_1(T_1) + \mu_1 \tau_1^{\gamma} A_1 (i\Omega_1)^{\gamma} + a_{13}^{\mathrm{I}} \bar{A}_1 A_3 + a_{23}^{\mathrm{I}} \bar{A}_2 A_3 = 0, \tag{113}$$

$$2i\Omega_2 D_1 A_2(T_1) + \mu_2 \tau_2^{\gamma} A_2 (i\Omega_2)^{\gamma} + a_{13}^{\mathrm{II}} \bar{A}_1 A_3 + a_{23}^{\mathrm{II}} \bar{A}_2 A_3 = 0, \tag{114}$$

$$2i\Omega_3 D_1 A_3(T_1) + \mu_3 \tau_3^{\gamma} A_3 (i\Omega_3)^{\gamma} + a_{11}^{\mathrm{III}} A_1^2 + a_{22}^{\mathrm{III}} A_2^2 = 0. \tag{115}$$

Representing functions A_i entering in (113)–(115) in the polar form (60) and applying the same procedure as it has been carried out above for 2:1 internal resonance, we have

$$\begin{aligned}\left(a_1^2\right)^{\cdot} + s_1 a_1^2 = {} & \Omega_1^{-1} a_{13}^{\mathrm{I}} a_1^2 a_3 \sin(2\varphi_1 - \varphi_3) \\ & + \Omega_1^{-1} a_{23}^{\mathrm{I}} a_1 a_2 a_3 \sin(\varphi_1 + \varphi_2 - \varphi_3),\end{aligned} \tag{116}$$

$$\begin{aligned}\left(a_2^2\right)^{\cdot} + s_2 a_2^2 = {} & \Omega_2^{-1} a_{23}^{\mathrm{II}} a_2^2 a_3 \sin(2\varphi_2 - \varphi_3) \\ & + \Omega_2^{-1} a_{13}^{\mathrm{II}} a_1 a_2 a_3 \sin(\varphi_1 + \varphi_2 - \varphi_3),\end{aligned} \tag{117}$$

$$\begin{aligned}\left(a_3^2\right)^{\cdot} + s_3 a_3^2 = {} & - \Omega_3^{-1} a_{11}^{\mathrm{III}} a_1^2 a_3 \sin(2\varphi_1 - \varphi_3) \\ & - \Omega_3^{-1} a_{22}^{\mathrm{III}} a_2^2 a_3 \sin(2\varphi_2 - \varphi_3),\end{aligned} \tag{118}$$

$$\begin{aligned}\dot{\varphi}_1 - \frac{1}{2}\sigma_1 - \frac{1}{2}\Omega_1^{-1} a_{13}^{\mathrm{I}} a_3 \cos(2\varphi_1 - \varphi_3) & \\ - \frac{1}{2}\Omega_1^{-1} a_{23}^{\mathrm{I}} \frac{a_2 a_3}{a_1} \cos(\varphi_1 + \varphi_2 - \varphi_3) & = 0,\end{aligned} \tag{119}$$

$$\begin{aligned}\dot{\varphi}_2 - \frac{1}{2}\sigma_2 - \frac{1}{2} a_{23}^{\mathrm{II}} \Omega_2^{-1} a_3 \cos(2\varphi_2 - \varphi_3) & \\ - \frac{1}{2}\Omega_2^{-1} a_{13}^{\mathrm{II}} \frac{a_1 a_3}{a_2} \cos(\varphi_1 + \varphi_2 - \varphi_3) & = 0,\end{aligned} \tag{120}$$

$$\dot{\varphi}_3 - \frac{1}{2}\,\sigma_3 - \frac{1}{2}\,a_{11}^{\mathrm{III}}\Omega_3^{-1}a_1^2a_3^{-1}\cos(2\varphi_1-\varphi_3) - \frac{1}{2}\,a_{22}^{\mathrm{III}}\Omega_3^{-1}a_2^2a_3^{-1}\cos(2\varphi_2-\varphi_3)=0. \quad (121)$$

The nonlinear set of Eqs. (116)–(121) with the initial conditions (85) completely describe the vibrational process of the mechanical system being investigated under the condition of the internal resonance 2:2:1 and could be solved numerically.

5 Conclusion

Free damped vibrations of a shallow non-linear thin cylindrical shell in a fractional derivative viscoelastic medium have been investigated. Nonlinear vibrations are described in the cylindrical system of coordinates by coupled three Donnell–Mushtari–Vlasov equations with respect to the three displacements.

The proposed analytical approach for investigating the damped vibrations of the nonlinear cylindrical shell subjected to the conditions of the internal resonance has been possible owing to the new procedure suggested in this paper, resulting in decoupling linear parts of equations with the further utilization of the method of multiple scales for solving nonlinear governing equations of motion.

It has been shown that the phenomenon of internal resonance could be very critical, since in the thin shallow circular cylindrical shell under consideration the internal resonance is always present. Thus, damped vibrations occur under the different conditions of the internal resonance, in so doing the type of the resonance depends on the order of smallness of the fractional derivative entering in the equations of motion of the shell. That is why all possible cases of the internal resonances of the order of ε have been examined in the present paper, namely: two-to-one, one-to-one-to-two, one-to-two-to-two, and combinational resonances of the additive and difference types, while the internal resonances of the order of ε^2 should be considered in detail in a separate paper.

For each type of the resonance, the nonlinear sets of resolving equations in terms of amplitudes and phase differences have been obtained. It has been shown that for some types of the internal resonances there exist such particular cases when it is possible to obtain two first integrals, namely: the energy integral and the stream-function, what allows one to reduce the problem to the calculation of elliptic integrals.

Comparison of the results obtained in this paper for the nonlinear shallow cylindrical shell in the cases of the internal resonance of the order of ε with those for a nonlinear plate, the motion of which is described also by three coupled nonlinear equations in terms of three displacements [25], reveals the fact that the shell equations could produce much more diversified variety of internal resonances than the plate equations.

Acknowledgments This research was made possible by the Grant from the Ministry of Education and Science of the Russian Federation (No 7.22.2014/K).

Appendix A

$$a_{1mn}^{m_1 n_1 m_2 n_2} = \int_0^1 \int_0^{2\pi} \sin \pi m_1 x \ \sin n_1 \varphi \ \cos \pi m_2 x \ \sin n_2 \varphi \ \cos \pi m x \ \sin n\varphi \ dx d\varphi,$$

$$a_{2mn}^{m_1 n_1 m_2 n_2} = \int_0^1 \int_0^{2\pi} \cos \pi m_1 x \ \cos n_1 \varphi \ \sin \pi m_2 x \ \cos n_2 \varphi \ \cos \pi m x \ \sin n\varphi \ dx d\varphi,$$

$$a_{3mn}^{m_1 n_1 m_2 n_2} = \int_0^1 \int_0^{2\pi} \sin \pi m_1 x \ \cos n_1 \varphi \ \sin \pi m_2 x \ \sin n_2 \varphi \ \sin \pi m x \ \cos n\varphi \ dx d\varphi,$$

$$a_{4mn}^{m_1 n_1 m_2 n_2} = \int_0^1 \int_0^{2\pi} \cos \pi m_1 x \ \sin n_1 \varphi \ \cos \pi m_2 x \ \cos n_2 \varphi \ \sin \pi m x \ \cos n\varphi \ dx d\varphi,$$

$$a_{5mn}^{m_1 n_1 m_2 n_2} = \int_0^1 \int_0^{2\pi} \sin \pi m_1 x \ \sin n_1 \varphi \ \sin \pi m_2 x \ \sin n_2 \varphi \ \sin \pi m x \ \sin n\varphi \ dx d\varphi,$$

$$a_{6mn}^{m_1 n_1 m_2 n_2} = \int_0^1 \int_0^{2\pi} \cos \pi m_1 x \ \cos n_1 \varphi \ \cos \pi m_2 x \ \cos n_2 \varphi \ \sin \pi m x \ \sin n\varphi \ dx d\varphi,$$

$$a_{7mn}^{m_1 n_1 m_2 n_2} = \int_0^1 \int_0^{2\pi} \cos \pi m_1 x \ \sin n_1 \varphi \ \cos \pi m_2 x \ \sin n_2 \varphi \ \sin \pi m x \ \sin n\varphi \ dx d\varphi,$$

$$a_{8mn}^{m_1 n_1 m_2 n_2} = \int_0^1 \int_0^{2\pi} \sin \pi m_1 x \ \cos n_1 \varphi \ \sin \pi m_2 x \ \cos n_2 \varphi \ \sin \pi m x \ \sin n\varphi \ dx d\varphi,$$

$$a_{9mn}^{m_1 n_1 m_2 n_2} = \int_0^1 \int_0^{2\pi} \cos \pi m_1 x \ \sin n_1 \varphi \ \cos \pi m_2 x \ \sin n_2 \varphi \ \sin \pi m x \ \sin n\varphi \ dx d\varphi,$$

$$a_{10mn}^{m_1 n_1 m_2 n_2} = \int_0^1 \int_0^{2\pi} \sin \pi m_1 x \ \cos n_1 \varphi \ \sin \pi m_2 x \ \cos n_2 \varphi \ \sin \pi m x \ \sin n\varphi \ dx d\varphi.$$

References

1. Witt, A., Gorelik, G.: Vibrations of an elastic pendulum as an example of vibrations of two parametrically coupled linear systems (in Russ.). J. Tech. Phys. **2–3**, 294–307 (1933)
2. Nayfeh, A., Balachandran, B.: Modal interactions in dynamical and structural systems. Appl. Mech. Rev. **42**, S175–S201 (1989)
3. Nayfeh, A.: Nonlinear interaction: Analytical, computational, and experimental methods. Wiley, New York (2000)
4. Amabili, M., Pellicano, F., Vakakis, A.: Nonlinear vibrations and multiple resonances of fluid-filled circular shells, Part 1: Equations of motion and numerical results. ASME J. Vibr. Acoust. **122**, 346–354 (2000)
5. Popov, A., Thompson, J., McRobie, F.: Low dimensional models of shell vibrations. Parametrically excited vibrations of cylindrical shells. J. Sound Vibr. **209**, 163–186 (1998)
6. Nayfeh, A., Raouf, R.: Non-linear oscillation of circular cylindrical shells. Int. J. Solids Struct. **23**, 1625–1638 (1987)
7. McRobie, F., Popov, A., Thompson, J.: Auto-parametric resonance in cylindrical shells using geometric averaging. J. Sound Vibr. **227**, 65–84 (1999)
8. Avramov, K.: Nonlinear forced vibrations of a cylindrical shell with two internal resonances. Int. Appl. Mech. **42**(2), 169–175 (2006)
9. Popov, A.: Auto-parametric resonance in thin cylindrical shells using the slow fluctuation method. Thin-Walled Struct. **42**, 475–495 (2004)
10. Amabili, M., Paidoussis, M.P.: Review of studies on geometrically nonlinear vibrations and dynamics of circular cylindrical shells and panels,with and without fluid-structure interaction. ASME Appl. Mech. Rev. **56**, 349–381 (2003)
11. Kubenko, V., Koval'chuk, P.: Nonlinear problems of the vibration of thin shell (review). Int. Appl. Mech. **38**(8), 703–728 (1998)
12. Lee, Y.: Review on the cylindrical shell research (in Korean). Trans. KSME **33**, 1–26 (2009)
13. Avramov, K., Mikhin, Y., Kurilov, E.: Asymptotic analysis of nonlinear dynamics of simply supported cylindrical shells. Nonlinear Dyn. **47**, 331–352 (2012)
14. Breslavsky, I., Avramov, K.: Nonlinear modes of cylindrical panels with complex boundaries R-function method. Meccanica **46**, 817–832 (2011)
15. Goncalves, P., del Prado, Z.: Low-dimensional Galerkin models for nonlinear vibration and instability analysis of cylindrical shells. Nonlinear Dyn. **41**, 129–145 (2005)
16. Amabili, M.: A comparison of shell theories for large-amplitude vibrations of circular cylindrical shells: Lagrangian approach. J. Sound Vibr. **264**, 1091–1125 (2003)
17. Amabili, M.: Internal resonances in non-linear vibrations of a laminated circular cylindrical shell. Nonlinear Dyn. **69**, 755–770 (2012)
18. Amabili, M., Reddy, J.: A new non-linear higher-order shear deformation theory for large-amplitude vibrations of laminated doubly curved shells. Int. J. Non-linear Mech. **45**, 409–418 (2010)
19. Amabili, M.: Reduced-order models for nonlinear vibrations, based on natural modes: The case of the circular cylindrical shell. Phil. Trans. Roy. Soc. A: Math. Phys. Eng. Sci. **371**(20120), 474 (2013)
20. Rossikhin, Y., Shitikova, M.: Application of fractional calculus for analysis of nonlinear damped vibrations of suspension bridges. ASCE J. Eng. Mech. **124**, 1029–1036 (1998)
21. Rossikhin, Y., Shitikova, M.: Application of fractional calculus for dynamic problems of solid mechanics: novel trends and recent results. Appl. Mech. Rev. **63**(1), 010,801-1–010,801-52 (2010)
22. Rossikhin, Y., Shitikova, M.: Analysis of nonlinear vibrations of a two-degree-of-freedom mechanical system with damping modelled by a fractional derivative. J. Eng. Math. **37**, 343–362 (2000)
23. Rossikhin, Y., Shitikova, M.: Free damped nonlinear vibrations of a viscoelastic plate under the two-to-one internal resonance. Mater. Sci. Forum **440–441**, 29–36 (2003)

24. Rossikhin, Y., Shitikova, M.: Analysis of free non-linear vibrations of a viscoelastic plate under the conditions of different internal resonances. Int. J. Non-Linear Mech. **41**, 313–325 (2006)
25. Rossikhin, Y., Shitikova, M.: A new approach for studying nonlinear dynamic response of a thin fractionally damped plate with 2:1 and 2:1:1 internal resonances. In: Altenbach H, Mikhasev G (eds) Shell and Membrane Theories in Mechanics and Biology: From Macro- to Nanoscale Structures, Advanced Structured Material, vol. 45, Springer, pp. 259–280 (2015). doi:10.1007/978-3-319-02535-3_15
26. Volmir, A.: Nonlinear dynamics of plates and shells (in Russ.). Nauka, Moscow (1972)
27. Rossikhin, Y., Shitikova, M.: Free non-linear vibrations of an elastic cylindrical shell under the conditions of the internal resonance. In: Proceedings of 18th International Congress on Sound and Vibration 2011, ICSV'2011, vol. 4, pp. 2775–2782 (2011)
28. Nayfeh, A.: Perturbation methods. Wiley, New York (1973)
29. Rossikhin, Y., Shitikova, M., Shcheglova, T.: Forced vibrations of a nonlinear oscillator with weak fractional damping. J. Mech. Mat. Struct. **4**(9), 1619–1636 (2009)
30. Abramowitz, M., Stegun, I. (eds.): Handbook of mathematical functions with formulas, graphs, and tables, applied mathematical services, vol. 55. National Bureau of Standards U.S.A, Washington, D.C (1964)

MIX
Papier aus verantwortungsvollen Quellen
Paper from responsible sources
FSC® C105338

If you have any concerns about our products,
you can contact us on
ProductSafety@springernature.com

In case Publisher is established outside the EU,
the EU authorized representative is:
Springer Nature Customer Service Center GmbH
Europaplatz 3, 69115 Heidelberg, Germany

Printed by Libri Plureos GmbH
in Hamburg, Germany